Microwave Remote Sensing for Oceanographic and Marine Weather-Forecast Models

NATO ASI Series

Advanced Science Institutes Series

A Series presenting the results of activities sponsored by the NATO Science Committee, which aims at the dissemination of advanced scientific and technological knowledge, with a view to strengthening links between scientific communities.

The Series is published by an international board of publishers in conjunction with the NATO Scientific Affairs Division

A	**Life Sciences**	Plenum Publishing Corporation
B	**Physics**	London and New York
C	**Mathematical**	Kluwer Academic Publishers
	and Physical Sciences	Dordrecht, Boston and London
D	**Behavioural and Social Sciences**	
E	**Applied Sciences**	
F	**Computer and Systems Sciences**	Springer-Verlag
G	**Ecological Sciences**	Berlin, Heidelberg, New York, London,
H	**Cell Biology**	Paris and Tokyo

Series C: Mathematical and Physical Sciences - Vol. 298

Microwave Remote Sensing for Oceanographic and Marine Weather-Forecast Models

edited by

Robin A. Vaughan

Department of Applied Physics and
Electronic & Manufacturing Engineering,
University of Dundee, Dundee, Scotland, U.K.

Kluwer Academic Publishers

Dordrecht / Boston / London

Published in cooperation with NATO Scientific Affairs Division

Proceedings of the NATO Advanced Study Institute on
Microwave Remote Sensing for Oceanographic and
Marine Weather-Forecast Models
Dundee, U.K.
August 14–September 3, 1988

ISBN-13: 978-94-010-6715-7   e-ISBN-13: 978-94-009-0509-2
DOI:10.1007/978-94-009-0509-2

Published by Kluwer Academic Publishers,
P.O. Box 17, 3300 AA Dordrecht, The Netherlands.

Kluwer Academic Publishers incorporates the publishing programmes of
D. Reidel, Martinus Nijhoff, Dr W. Junk and MTP Press.

Sold and distributed in the U.S.A. and Canada
by Kluwer Academic Publishers,
101 Philip Drive, Norwell, MA 02061, U.S.A.

In all other countries, sold and distributed
by Kluwer Academic Publishers Group,
P.O. Box 322, 3300 AH Dordrecht, The Netherlands.

TABLE OF CONTENTS

The power of microwave remote sensing for studying the oceans of the world was demonstrated conclusively by the SEASAT mission in 1978. Since then, no further satellite-flown instruments have been available to provide further data of this type. However, the proposed launch of ESA's ERS-1 satellite will lead to a new set of active microwave instruments being flown in space in 1990.

Even though similar data has been obtained from aircraft-flown instruments - SAR, scatterometers, altimeters etc. - a great deal of activity has been taking place to develop the necessary expertise in handling and analysing such data when it comes on-stream from ERS-1 and from subsequent satellites.

It was against this background that the Scientific Affairs Division of NATO again agreed to sponsor an ASI in Dundee in 1988. Its purpose was to review existing knowledge of the extraction of marine and atmospheric geophysical parameters from satellite-gathered microwave data and to enable scientists to prepare themselves and their computing systems to utilise the new data when it becomes available. The importance of the data is largely as input parameters to assist in the fitting of boundary conditions in large computer models. The course was concerned more with the non-imaging instruments, that is with passive radiometers, altimeters and scatterometers, than with the (imaging) synthetic aperture radar.

The lectures were concerned with general background and some discussion of the instruments used, supporting instruments, the nature and format of the data generated, the extraction of geophysical parameters, the nature of oceanographic and weather-forecast models and the results that can be obtained and can be expected to be obtainable in the future from such models. Seminars on related special topics were also presented by a few of the participants.

Considerable effort was put into the design of the course given at the summer school to give balanced coverage of the topic, and it is particularly disappointing that these proceedings do not contain the texts of all the lectures given and thus does not reflect the balance of the course. Publication has been delayed in order to try to cajole some of these lecturers to produce material, and our sincere apologies are due to the majority of the lecturers who produced their text in good time and in good faith. No manuscripts were received from E Mollo-Christensen (general principles of microwave scattering by the sea surface), W Alpers (modelling of physical processes), G J Komen (WAM) and S R Brooks (wind scatterometers). A Hollingsworth and O M Phillips decided that since the material they covered has now been published (see references below) they did not wish to publish them again. E Barrett could not attend the summer school in person, and his lecture was delivered by his colleague J Bailey.

The present summer school is the latest in a series on leading-edge topics in remote sensing. The Dundee summer schools, held biennially since 1980, developed as a result of a proposal to develop postgraduate training modules made at the General Assembly of the European Association of Remote Sensing Laboratories (EARSeL) in 1977, and have now become an institution in the field of remote sensing. The broad ambitions of the original initiative were summarised in the proceedings of the first summer school (Cracknell 1981) and have subsequently been reviewed at European Workshops in Education and Training held in Lyngby (Cracknell 1986) and in Helsinki (Vaughan 1989).

Three of the five courses have had NATO as their main sponsor, and all five have been generously supported by the Council of Europe, EARSeL and the European Space Agency (ESA). With the exception of 1984, they have all covered some aspect of marine applications, which is, of course, the speciality of Dundee, and have all been directed by Arthur Cracknell. The proceedings of the previous summer schools have now become valued reference books, presenting both reviews and state-of-the-art information, and it is to be hoped that the present book will take its place alongside these others.

THE DUNDEE SUMMER SCHOOLS

1980   Remote Sensing Applications in Meteorology,
        Oceanography and Hydrology

1982   Remote Sensing Applications in Marine Science
        and Technology

1984   Remote Sensing Applications in Civil Engineering

1986   Remote Sensing Applications in Meteorology and
        Climatology

1988   Microwave Remote Sensing for Oceanography and
        Marine Weather Forecast Models

From these summer schools also has been developed a range of practical exercises, many contributed by the lecturers, using digital data, computer print-out, hard copy photographs and images, image processing systems and computer simulations. These have been designed to illustrate the range of data and applications. Some of these are at present being assembled into a workpack to be distributed eventually by ESA as a "Technical Training Manual". This is in line with the feeling amongst all members of the EARSeL working group that such material should be made available to as wide an audience as possible.

The field of remote sensing seems to be particularly fraught with abbreviations and mnemonics, and in past proceedings these have been cross-referenced in the index. To try to tidy this up, and to help the reader, this time these have all been collected together into a glossary of acronyms which will be found at the back, after the index. This is not guaranteed to be completely comprehensive nor does it list abbreviations not used in this book.

The gratitude of the organisers must go not only to the sponsors of this summer school, but also to the many people who worked hard to make it a success - the organising committee, the lecturers, the technical and secretarial assistants, the domestic staff of Chalmers Hall of Residence, and, of course, the students themselves.

I should like to conclude by giving my own very sincere thanks to all those who have contributed to the production of this volume - the authors, the publishers and all those who have helped in the typing. An unfortunate result of producing a book like this from camera-ready material is the inevitable non-uniformity of presentation. It also precludes the exercise of any real editorial powers. I hope that you, the reader, will be understanding if you find cause to criticise and appreciative if you find the contents educative.

Robin Vaughan
Dundee 1989

References

A P Cracknell, 1981, "Remote Sensing Applications in Meteorology, Oceanography and Hydrology". Ellis Horwood, Chichester.

A P Cracknell, 1983, "Remote Sensing Applications in Marine Science and Technology". D Reidel, Dordrecht.

ESA, 1985, "Remote Sensing Applications in Civil Engineering" ESA SP-216, European Space Agency, Paris.

A P Cracknell, 1986, Proceedings of Workshop in Education and Training, Lyngby.

A Hollinsworth, 1986, "Objective Analysis for Numerical Weather Prediction", in "Short and Medium Range Weather Predictions": collected papers presented at the WMO/IUGG NWP Symposium, Tokyo. J Met Soc, Japan, special issue ed T Matsuno, pp11-59.

x

O M Phillips, 1988, "Radar Returns from the Sea Surface - Bragg Scattering and Breaking Waves" J Phys Oceanog, 18, 8. pp1065-1074.

R A Vaughan, 1987, "Remote Sensing Applications in Meteorology and Climatology", D Reidel, Dordrecht.

R A Vaughan, 1989, "The Dundee Summer Schools", Proceedings of Workshop in Education and Training, Helsinki.

ORGANISATION

Organising Committee

Professor A.P. Cracknell (Dundee), Director
Professor R. Frassetto (Venice), Co-Director
Mrs J.E. Callison (Dundee), Secretary
Dr R.A. Vaughan (Dundee), Treasurer
Professor W. Alpers (Bremen)
Dr P.A. Davies (Dundee)
Dr T.M. Guymer (Wormley)
Dr A.O. Tooke (Dundee) Social Programme
Mr G.R. Whyte (Dundee) Technical Arrangements

Technical Assistance

Mr I. Ballingal
Miss P. Cracknell
Mr I. Durajczyk
Mr D. Gray
Mrs M. Laurence
Miss J.N. Low

1	P.A. Volz	19	R.A. Brown	37	X. Durrieu de Madron
2	J.A. Leese	20	O.M. Phillips	38	S. Kasturi
3	A.O. Tooke	21	M.R. Ramesh Kumar	39	D. Balicki
4	A.P. Cracknell	22	G. Kocasoy	40	P.J. Minnett
5	R.A. Vaughan	23	C. Mauritzen	41	L. Hus
6	J.M. Anderson	24	P.E. Bjerke	42	D. Djavadi
7	J.E. Callison	25	J. Gosink	43	J.A. Lowrenzzetti
8	K. Kouzai	26	H.J. Brosin	44	V. Chopra
9	J.N. Low	27	C.J.D. Tavares	45	R. Kuzniar
10	Y. Yu	28	J. Prajs	46	P.C. Oliveira
11	W. N.Bowman	29	K.M. Baird	47	R.L. Phillips
12	P. Kowaliczuk	30	J.J. Martinez-Benjamin	48	M.A. Al-Taee
13	I.T. Hunter	31	T.S. Uyar	49	T.H. Guymer
14	M. Candouna	32	C. Simmer	50	A.J. Andrews
15	A.C. Bijlsma	33	W. Huang	51	P. Trela
16	W. Hsieh	34	B.J. Topliss	52	R.G. Caves
17	D. Offiler	35	F. Paulet	53	J.P. Thomas
18	A. Eriksson	36	R. Lalbeharry	54	A.C.M. Stoffien
				55	L.A. Breivik

ON THE ROLE OF SCIENCE IN PREPARING OPERATIONAL USES OF EARTH OBSERVATIONS FROM SPACE

R. Frassetto
C.N.R.-ISDGM
1364 San Polo
30125 Venice
Italy

ABSTRACT. An overview of the strategy and policies of research to face the problems of global change including climate and environmental alterations at all scales from global, to regional and local, is given with particular emphasis on the role of Earth observation from satellites as the major source of information in the coming decades.

The objective is to encourage scientists to direct their research keeping in mind the ongoing international large scale programs and experiments having for objective a better knowledge of the anthropogenic modifications and natural variabilities of the system Earth.

The role of European countries and space agencies in this subject is discussed.

1. GLOBAL CHANGE

In front of the pressing responsibility of science to understand, simulate and predict the global changes of the system Earth which may threat life and activity of mankind in the near future, it is useful for all research scientists to be aware of the overall problem and is appropriate to direct their research to a better knowledge of the many interdisciplinary problems of the system and to establish the role of remote sensing in this endeavor.

In the system Earth all mechanisms of different time and space scales are interrelated, therefore any effort to deepen the understanding and to describe a single process or feature is a contribution to the general problem, and vice versa.

Changes in weather, climate and in the environment's quality and dynamics have been recorded in different parts of the planet earth showing tendencies that are caused by anthropic

1

R. A. Vaughan (ed.), Microwave Remote Sensing for Oceanographic and
Marine Weather-Forecase Models, 1–22.
© *1990 Kluwer Academic Publishers.*

action and may become so serious to compromise human life in the near future. Since the 1979 conference in Stockholm on Global Climate, CO_2 increase in the atmosphere, recorded over several years at the Mauna Loa (Pacific) and several other stations, represents the most serious threat. Other gases are contributing to the green house effect with alterations of the thermo-dynamics and of the quality of the earth's natural environment.

From in situ measurements and analysis of human activity parameters one can try to estimate future trends but it is difficult to discriminate trends from the multiple fluctuations.

2. MODELS

The basic tools for prediction are mathematical models which can simulate, describe, and predict natural processes. Models in turn, need the proper input, and, to become effective, they must parameterize and simplify the mechanism under study, they must be calibrated, verified, and tuned on reasonably long-term tests.

Time series of global observations from space, with selected simultaneous in situ observations are indispensable for the models to become reliable. The assimilation of the data from this variety of observing platforms into the numerical models is a complex technique which is presently under development.

Although computer facilities are becoming of greater capacity and speed, in most cases they are still insufficient to deal with the complexity of computations. Simplification is therefore the only solution which requires an intuitive art.

A variety of preliminary models have tried to simulate global effects of the increase of CO_2 in the atmosphere starting from different approaches. They have shown an increase of global surface temperatures in the next sixty years assuming a doubling of CO_2 concentrations in the atmosphere during the same years.

The results range from +1.5 to 5.0 °K. Other models show increases in humidity by 40% and rainfall by 15%. Indications are that the major changes should occur in the temperate latitudes of the Northern Hemisphere, above the 40 degree latitude. These results are obviously not realistic. There is a need for a better parameterization.

One major missing information is the cloud solar radiation feedback. This and many other global information are expected from future second generation earth observation satellites provided that global coverage and continuity will be achieved.

3. OCEAN ROLE

The role of the ocean in climate and environment changes is mainly represented by the large exchange of energy with the atmosphere through air-sea fluxes at the interface, heat absorption and redistribution by means of ocean circulations at all time and space scales.

The average global heating rate of the oceans is about 173 Pegawatts (Mason, 1988). This heat is released into the atmosphere at the ocean surface by long wave radiation (78%), conduction (sensible heat) 5% and evaporation (latent heat) 17%. Nearly all of the incoming radiation is absorbed in the top 200 meters, about half of which in the first centimeter. Most long wave (IR) absorption occurs in the first centimeter.

Because of its large mass, specific heat and thermal inertia, the ocean represents a reservoir of heat and CO_2, and acts as a buffer. The ocean is the main source of moisture for the global hydrological cycle and largely controls the atmospheric circulation, the precipitation and climate. Oceans transport great quantities of heat horizontally with currents and gyres, and in slow vertical convergences over long periods of time.

The solar heating, the surface fluxes, thermohaline (density) differences and wind stress with the Coriolis force are the main forcings, constricted by bottom topography of the oceans.

Wind stress fields and interface fluxes data can be derived from satellite data. The low bit rate data of scatterometer, microwave radiometers and altimeters of next generation satellites, beginning with ERS 1, will furnish the first set of data even if not complete, for an improvement of present numerical modeling.

The ocean-atmosphere boundary layers where the exchange of energy (heat, momentum) occurs are the most representative field of investigation of the present and next decade research. Coupled air-sea interaction models are under study for global and regional scales. Currently planned studies deal with the physics of the upper ocean, the air-sea gas exchange and the interactive roles of physical, chemical and biological processes.

4. INTERNATIONAL ORGANIZATIONS AND PROGRAMS

It is only by means of large international programs of research and experiments that these ambitious objectives can be reached.

The major international organizations have discussed the Earth System problems proposing programs and actions with clear definition of objectives, missions, research activities and experiments as well as technical developments to improve observations and the use of data. Among these are ICSU, WMO, UNEP, IOC, IUGG, NATO, CCE to name the major ones who have organized committees, panels, working groups, symposia and workshops. NASA, ESA and NASDA are joining in this endeavor with their Earth Observation Programs trying to reach an agreement on an integrated Earth Observation System, in terms of cost and benefit for the users.

Intercomparison of data from different sources will be one of the difficult problems to be solved. The space agencies are trying to meet the requirements of the major global experiments which will take place in the coming decade. They represent in fact the most coordinated joint efforts to properly use the data from space.

The major experiments designed for the 1990s which involve the ocean and atmosphere are the WCRP, TOGA, WOCE and JGOFS and the last one, conceived for the decade 2000, the Global Energy and Water Cycle Experiment (GEWEX) which will incorporate a large part of the results of other experiments. Perhaps the major research program which originated from ICSUS is the International Geo-Biosphere Program. IGBP places high priority on global tropical wind measurements, precipitation, runoff and radiation fluxes (see Annex I).

A strong numerical model development program is the core of these experiments. The early planning which is underway is important also to define the new requirements for the space observing systems even beyond the first proposed polar platforms.

New instruments need to be developed taking advantage of the fast progressing technology of microwave and laser systems such as:

- Doppler lidar profilers for tropospheric wind observations
- Active rain radar in low earth orbit for precipitation
- Nadir sounding lidars for cloud and radiation fluxes
- Passive atmospheric sounders and lidars operating in differential absorption mode for boundary layer properties.

Optical, thermal and microwave sensors are all needed to observe different variables of the earth surface which will eventually converge in synergisms in future operational systems.

Operational agencies such as the meteorological services, with their EUMESAT organization, are defining requirements to deal with the changes in climate and natural environment and to deal with real time data analyses.

For many years to come the major role in the creation of an environmental operational system is that of science through scientific understanding and complex numerical model development for simulation and forecasting.

The return of this expensive activity of such large dimensions, is in the indispensable and accurate information needed for decision making of governments, managements, administrations having the responsibility of safeguarding the environment for the survival of mankind and of its welfare.

5. EARTH OBSERVATION STRATEGY OF ESA

Well aware of the fact that Europe must take a significant role and responsibility in preventing dangerous degenerations of the natural environment of the planet Earth, and demonstrating that the Earth observation from space is the major indispensable synoptic monitoring system to cover horizontal space on a routine basis, ESA is making a great effort to propose and continuously adjust a program of objectives and strategy of Earth observation for a couple of decades. Objectives and strategies suggested by the community of scientists will have to meet and accept the technical, industrial, political and economical factors of the member nations of ESA.

Satellites for Earth observations as well as their ground segments operation are indispensable but are expensive.

The responsibility of the Executive Committee of ESA, in accordance with the strategy planned by and with NASA and NASDA, relies on conceiving a practical system which can satisfy, to the greatest possible extent, the requirements of science and operation to reach the high priority objectives at an acceptable cost.

The industrial and economical interests of each member state of ESA, in turn, must converge to an optimized compromise and come to a final agreement for action.

This will mean delays in decision making. Reiterated discussions however, may lead, (it is hoped) to a suffered but cleverer solution.

Governments are facing the necessity to assume their responsibility and to decide on a substantial economical effort to realize a common system of Earth observation for use over national and regional European areas as well as over the globe. This effort is due, as a consequence of the uncontrolled degradation of the natural environment since the beginning of the industrial era and of the increasing demands of modern society.

A proposal for an Earth observation program until the year 2000 is shown in Table I.

6. DATA MANAGEMENT - ROLE OF SCIENCE

Data from space platforms and from long term observations on land, ice and ocean including ships and aircrafts, are costly. The large mass of data which can be collected from these observational systems needs accurate calibrations and validations and must reach absolute geophysical values to be intercomparable. High quality data must then be interpreted with the proper algorithms, classified and catalogued for practical access by the users.

A variety of models require data in real time for fast prediction or data in delay mode for hindcasts, scientific studies, statistical elaborations, recording of long term variabilities.

Some instruments, such as the synthetic aperture radar (SAR) of satellites have data rates of several hundred magabytes per second and they can be grouped in joint frequencies (X,L,C bands). To organize an immense amount of data and to prepare reliable products for the users represent the most cumbersome, expensive and time consuming program that must be studied and prepared well ahead of the satellite launching time.

The classification of data to be stored in archives and data banks over a reasonable length of time must also be made evaluating the character and final use of each set of data.

Large episodic (a decade?) global experiments such as TOGA, WOCE, JGOFS will organize their own data processing and archiving and will select those which should be kept over periods longer than a decade at the World Data Banks in a practical, easily accessible format.

Several agencies, international and national, will have the task to organize similar activities for each particular use.

UNEP, for example, has established a Global Environmental Monitoring System (GEMS) which falls in five categories:

- health related monitoring
- climate related monitoring
- renewable resources monitoring
- ocean monitoring
- international disaster monitoring.

TABLE I

ESA Earth Observation Long-Term Planning

Schedule of Future Missions

PROJECT	1988	1989	1990	1991	1992	1993	1994	1995	1996	1997	1998	1999	2000	COMMENTS
MOP														Approved EUMETSAT Programme
ERS-1														Approved
ERS-2														Proposed
Aristoteles														Dual Launch with ERS-2
POEOP A1														
POEOP B1														
POEOP A2														Launch post 2000
POEOP B2														Launch post 2000
MSG														
Earthnet Optional Prog.														Complementary to the Mandatory Programme
EOPP														Approved (except extension)

LEGEND: ▨ Phase B/C/D   ▲ = Launch   ☐ = Phase E   ▨ = EOPP   ▨ = EOPP Ext.   ▨ = EOPP Ext.   ▨ = Earthnet Opt. Prog.

GEMS includes GRID (Global Resources Information Data Base) and is placed in Geneva, Switzerland.

WMO also has organized the World Climate Data Program (WCDP) encouraging each country to rescue, digitize, quality control, store, retrieve and use climate data.

Elements of the climate system are: upper air, surface climate, ocean surface and subsurface, cryosphere, radiation budget, atmospheric composition, hydrosphere, land and vegetation, proxy and solar data.

We have given just a few examples of presumably coordinated programs of data management; in order to avoid duplication of expensive efforts, when these are not a choice for national classified reasons, there is a need of frequent international discussions, collaboration and interconnection.

A must for planning and operating expensive Earth observation satellites is to have a documented evidence that the return is sound and the cost justified.

The major present role of scientists should be to help the Space Agencies to decide on priorities for science and applications and to study the algorithms and techniques needed for a more accurate interpretation and use of data, and to prepare the basis for future practical applications.

7. THE MEDITERRANEAN - A SCIENTIFIC MODEL SEA

Since the early scientific investigations of the Mediterranean Sea at the beginning of the century, it was found that most mechanisms, processes and phenomena occurring in the large oceans could be studied in this marginal closed sea in convenient time and space scales and with the assistance of a moderate climate.

During the last few decades a variety of international oceanographic experiments took place particularly in the western basin of the Mediterranean and the Straits of Gibraltar and Sicily (Medoc series of cruises, NATO, U.S. and French studies of Gibraltar and Sicily straits, and German Med cruises).

Mathematical models to describe thermohaline, geostrophic and wind forced circulations, energy fluxes, waves and wind fields, storm surges, bottom and intermediate water formation, are being studied in this decade on the basis of new, more accurate measurements from satellites, ships and fixed platforms or coastal stations of the Mediterranean. The major international field measurement programs under way are those of the Eastern Mediterranean (P.O.E.M., IOC sponsored), of the Western Mediterranean (W.M.C.E., NORDA sponsored) and of the

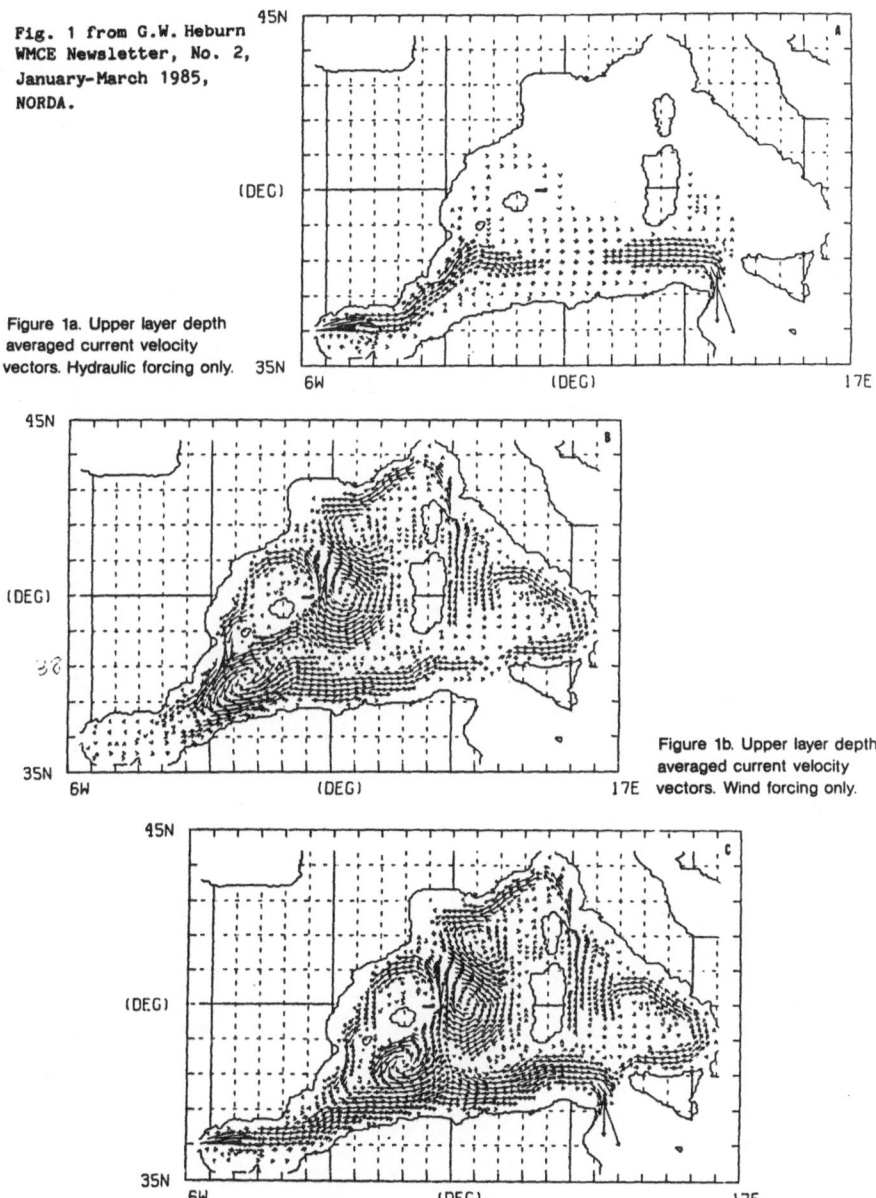

Fig. 1 from G.W. Heburn
WMCE Newsletter, No. 2,
January-March 1985,
NORDA.

Figure 1a. Upper layer depth
averaged current velocity
vectors. Hydraulic forcing only.

Figure 1b. Upper layer depth
averaged current velocity
vectors. Wind forcing only.

Figure 1c. Upper layer depth averaged current velocity vectors. Combined hydraulic and wind forcing.

10

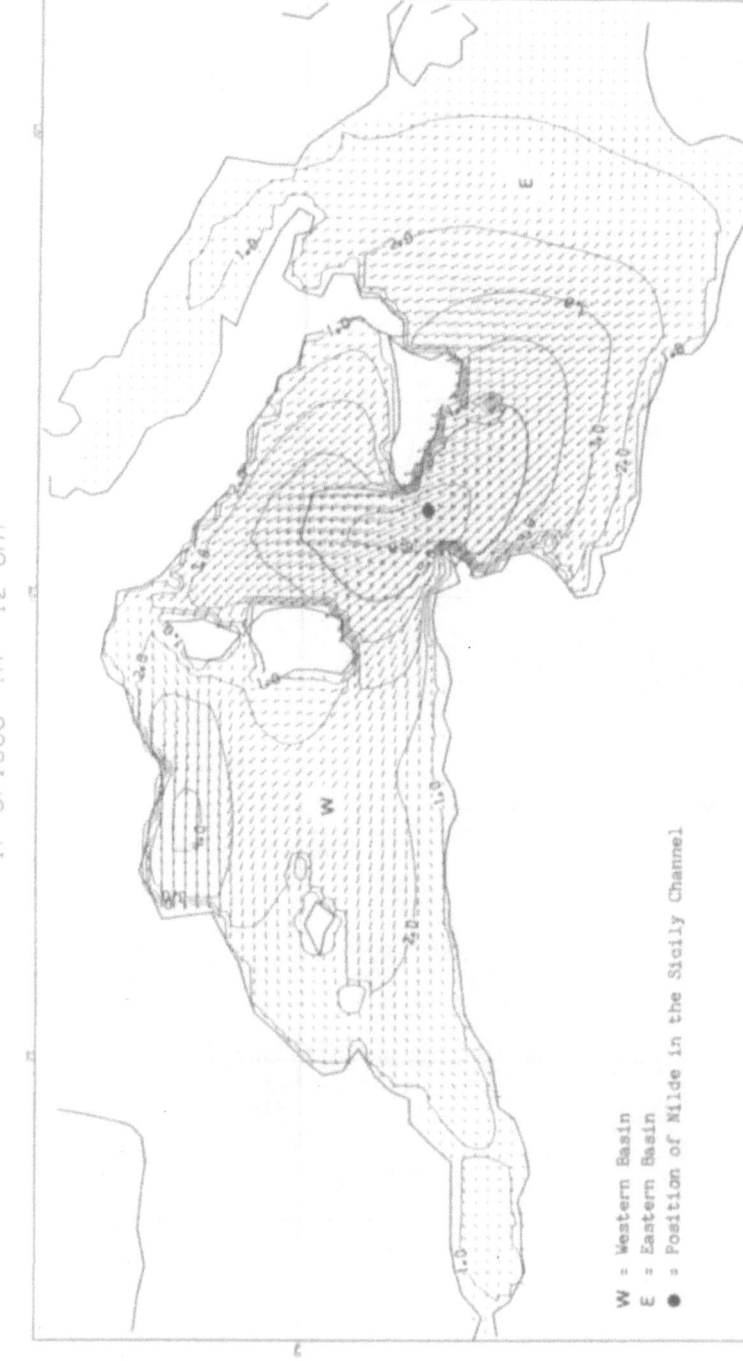

Fig. 2. The wave field during the storm of March 1, 1986 from the wave model using ECMWF data (after Bertotti, Cavaleri and Lionello, 1988).

regional scale WAM (WAve Model).

Further more, because of its very limited tidal oscillations and numerous nodal lines or amphidromic points, the Mediterranean is a convenient calibration area for satellite altimeters and because of the variability of orographically conditioned winds, it is a good test area for scatterometers and SARS.

The main forcings of the Mediterranean are: the hydraulic forcing from the Atlantic through the Gibraltar Strait and the wind stress (shown in Fig. 1, G. Heburn, 1987) esitmated from climatological traditional atlases

The significant wave height and the wind field of the Mediterranean described by WAM for a meteorological case of March 1, 1986 (Cavaleri et al., 1988) is shown in Fig. 2. The wind field is reconstructed from data of the European Centre of Medium Range Weather Forecasts (ECMWF).

Fig. 3 shows instead a wind field reconstructed on a 50 km grid from data of Seasat of 2 October 1978. This analysis (by Zecchetto, 1986) demonstrates the potential of the scatterometer to describe with good resolution and a much better representation of the wind field pattern over the Mediterranean at the time of a satellite passage. This kind of information should be fed in the surface wind models in future application activities using the new generation satellites. At the present time a great deal of research is needed to improve the wind spectral accuracy of the scatterometers and the data assimilation processes into models.

POEM models and measurements are describing the physics of circulation, thermal variability and energy fluxes of the eastern Mediterranean where evaporation exceeds rain and river runoff, and where the intermediate water of the Mediterranean (having high salinity and high temperature) is formed and carried in all basins of the Mediterranean mainly by geostrophic-thermohaline circulation and bottom geometry forcing. Fig. 4 shows the international cruises and hydrographic station area of POEM 1987.

Energy fluxes have been studied for the area of the Adriatic Sea and the Tyrrhenian using wind stress curl from Seasat data. This study is promising as a means to study climatological variations of evaporation (Fig. 5 by Zecchetto, 1987). Wind stress monthly averages (Fig. 6) and monthly total heat fluxes have been analysed (Fig. 7). These elaborations will then be used in ocean circulation models (A. Bergamasco (ISDGM-CNR), 1987) and in coupled ocean-atmosphere models under study by A. Navarra, IMGA-CNR.

From global change test models, which need a careful verification and reappraisal, an increase of the atmospheric

Fig. 3. Elaboration of Seasat scatterometer
data. Dealiased wind vectors on a
50 km grid represent a case of cyclogenesis more accurately than the
meteo-forecast map (Zecchetto, 1987).

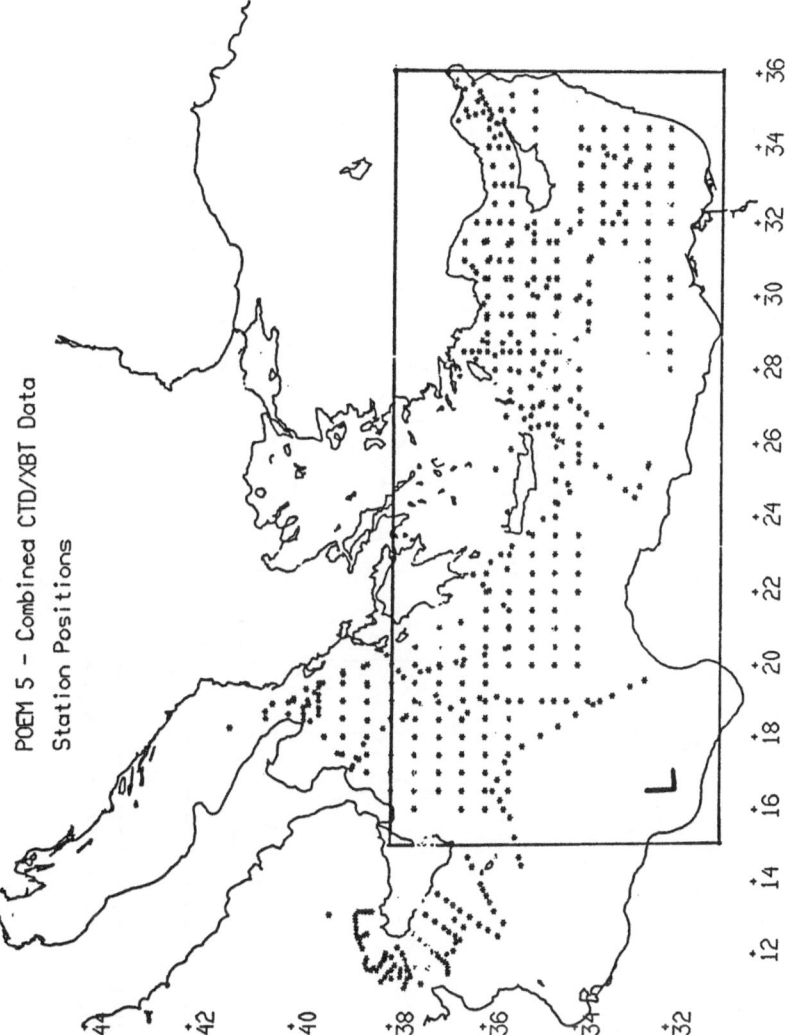

POEM 5 - Combined CTD/XBT Data
Station Positions

Fig. 4

14

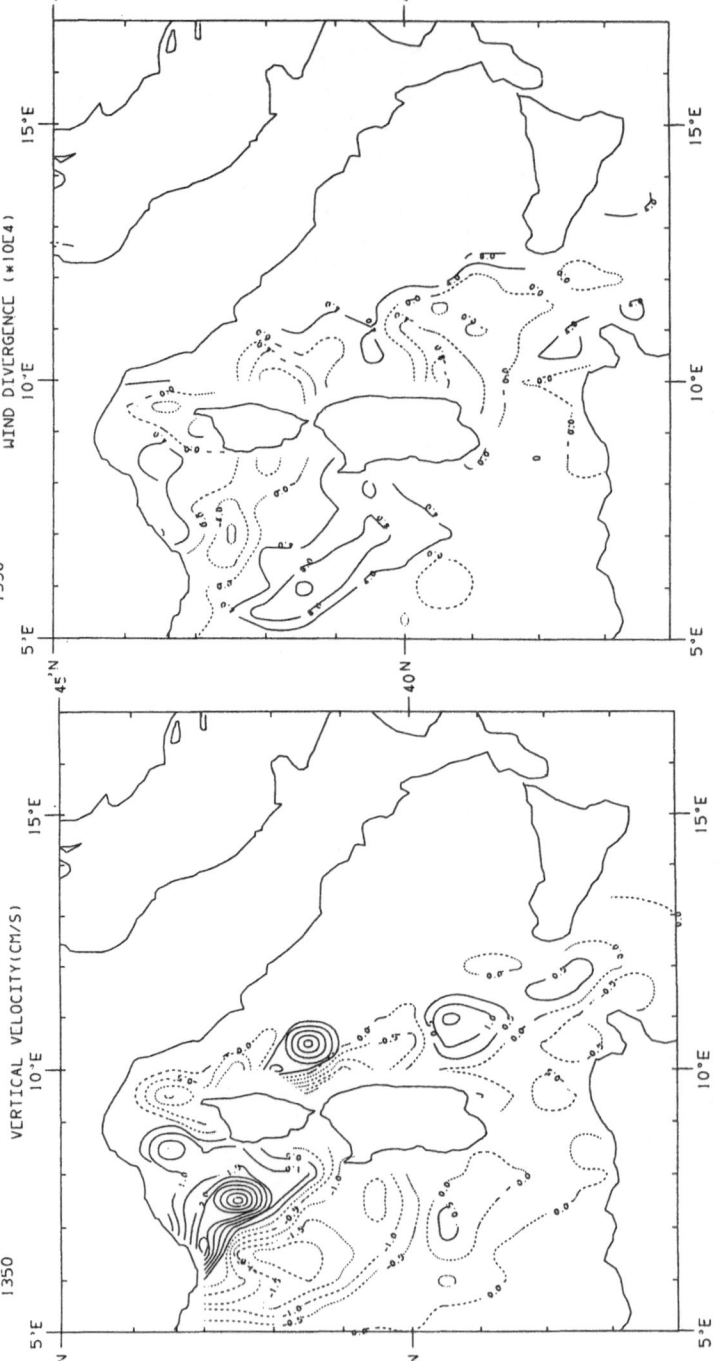

Fig. 5. Scatterometer derived products – Ekman pumping (from S. Zecchetto, 1985). Future applications of such information are expected to be used in coupled ocean-atmosphere models.

temperature over the Mediterranean Sea has been shown by W. Bach and V. Munster 1988 (Fig. 8). In about sixty years the air temperature should increase from 4 to 5 degrees K.

8. THE CASE OF VENICE'S SAFEGAURD - SUBSIDENCE AND POLLUTION

A typical example of a problem of modern society is the safegaurd of the historical city of Venice, Italy, an integer old marine city of high cultural value which suffer from local to global changes in the environment and must be preserved for posterity.

Born around 900 A.D. and developed in the Renaissance, Venice reached its splendor in the sixteenth-seventeenth century with a flourishing Mediterranean commerce with the Orient, enriched with art treasures. The city has practically not changed its texture and external beauty since then.

With the onset of the industrial era in the thirties however, the territory facing the lagoon of Venice filled up with chemical and oil refineries and a power plant. Clean water needed for large hotels and industrial firms was pumped from artesian wells exploiting different deep groundwater producing subsoil compaction and, as a consequence, an artificial subsidence of the city at a rate of centimeters per year.

Water and air pollution, the worst of which is sulfur dioxide from refineries, marine aerosols and acid rain all contributed since 1930 to change the environment of the Lagoon of Venice and to menace the structures of the city. Beautiful buildings 400 to 600 years old need now complete restoration, consolidation, waterproofing to retain their stability and leftover beauty. External decorations in stone are losing their image at a rate of a few percent per year.

Such a delicate aged structure suffers not only from local urban, industrial and agricultural, pollution problems, but from the interconnected global change in climate and environment as well.

For an unlucky coincidence the Venetian area suffers from plate tectonic effects with a natural subsidence of more than 10 cm per century. Further more the global mean sea level increases at a rate of 10-15 cm per century at the present time. The cumulative integrated effect is the rise of m.s.l. to more than 20 cm/century relative to the plane of the city.

Storm surges therefore are forcing floods into the city at ever increasing frequencies causing the penetration of brackish, polluted water into the capillaries of bricks and stones, creating erosions, weakening their resistance and changing their

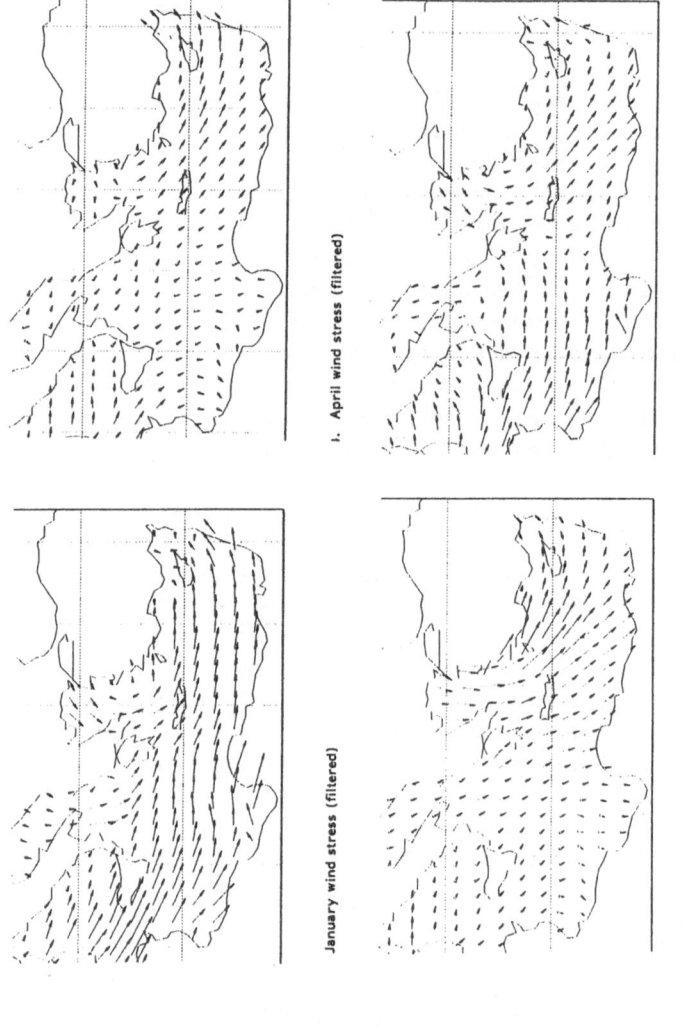

January wind stress (filtered)

i. April wind stress (filtered)

August wind stress (filtered)

November wind stress (filtered)

Fig. 6. Climatological wind stress averages. Individual stresses were averaged by month over the 20 year base period 1950-1970 to obtain climatology. Smoothed versions of the monthly wind stress estimates, appropriate for forcing numerical models, were also calculated. (From Paul W. May, NORDA, 1982)

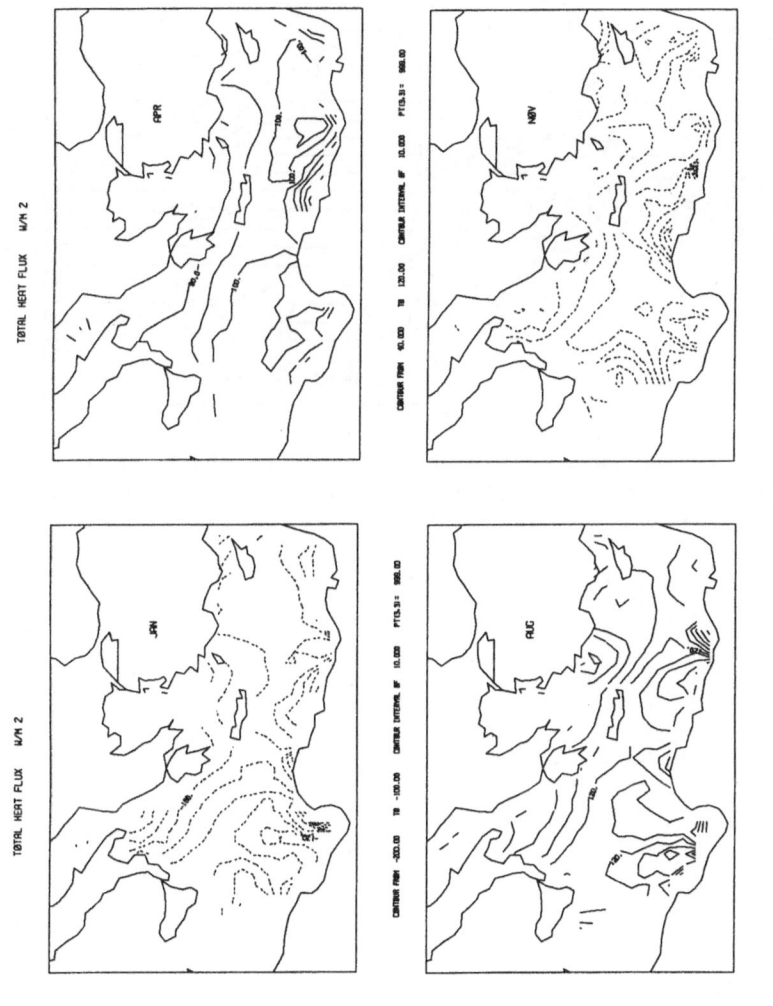

Fig. 7. Monthly averages of Mediterranean total heat fluxes in watts/m^2 are represented for 4 months. The data including, long and short wave radiation and sensible and latent heat fluxes were taken from Paul W. May, NORDA, 1982.

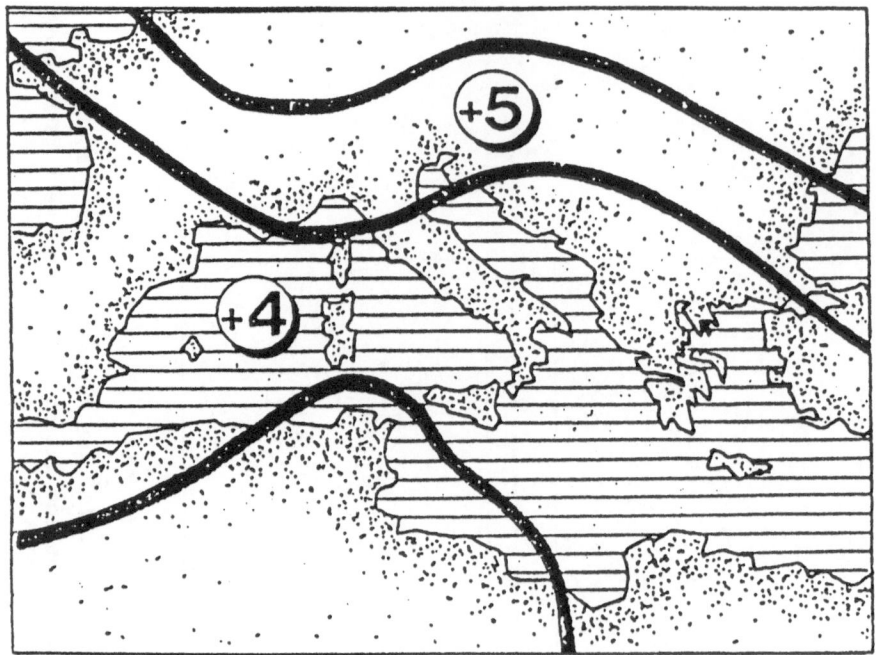

Fig. 8. Predicted increase of winter temperatures about 50 years
from now after models of W. Bach and U. Munster (personal
communication).

aesthetic character.

Models have been developed, verified, tuned over routine applications to predict the occurrence of floods to warn the local citizens and are now used by municipal forecasting services. The necessary protection from floods by means of dikes and gates instead, is still in the phase of experiments and projects.

Projects of this kind must take into account global climate and environment variations at time scales of decades to centuries to last and be effective for a reasonable length of time.

The global, regional and local character of atmosphere and oceans and their alterations in time concern therefore also the problem of Venice which becomes a complex model within the general context of the planetary systems change.

The safeguard of this historical center with its technical, economical, social, political and managerial activities relies on timely decisions.

Decision making needs accurate information from scientific research. The role of science and technology is to understand, simulate and predict the geophysical and environmental phenomena, interpret them and provide the ideas for sound preventive interventions.

To reach this objective measurements from in situ networks, from satellites and aircrafts, and mathematic model studies have been undertaken since the last two decades with successful results which are beginning to be used on operational mode.

In conclusion the main geophysical threats of the city are the floods and the natural subsidence; the anthropogenic threats are the pollution of atmosphere, water and soil.

Remote sensing monitoring from space will be an important tool for the control and prevention of undersirable changes, over long periods of time.

REFERENCES

Bergamasco, A., 1988, Preliminary Analysis of data and modelling for POEM (Physical Demography of the Eastern Mediterranean) CNR-ISDGM - Venice, Italy.
Cavoleri, 1986, CNR-ISDGM - Venice, Italy. Unpublished Internal Report.
Heburn, G.W., 1985, WMCE Newsletter No. 2. Code 331 NORDANSITC, Mississippi USA 39529.
Zecchetto, S., 1986, EARSeL WG11 - Technical Report to ESA.
Zecchetto, S., 1987, Use of Scatterometer Data in ocean circulation model for pollution monitoring, Nato Fellowship Report, CNR-ISDGM - Venice, Italy.

Activities of ICSU and other international organizations of direct relevance to the IGBP

Programme	Status	Subject	Sponsor
A. Lithosphere			
1. International Lithosphere Programme (ILP)	Ongoing (1981-) Tectonics	Solid-Earth Physics;	ICL; IUGG; IUGS; ICSU
2. Global Geoscience Transects (GGT)	Ongoing (1985-)	Geology; Geophysics	IUGG; IUGS; ICL
3. International Geological Correlation Programme	Ongoing (1973-)	Geology;Solid- Earth Physics; Geochemistry	IUGS; Unesco
B. Biosphere			
1. A Decade of the Tropics	Ongoing (1983-92)	Tropical Biology	IUBS
2. The Biogeochemical Cycles and their Interactions	Ongoing (1974-)	Biology; Meterology Oceanography; Chemistry	SCOPE
3. Bioindicators	Ongoing	Pollution of Environment; Biogeochemical Cycles	IUBS
4. Man and the Biosphere (MAB)	Ongoing	Ecology	Unesco; IUBS INTECOL; SCOPE
5. International Satellite Land Surface Climatology Project (ISLSCP)	Ongoing	Variations in Physical and Biological Land Surface Characteristics	COSPAR; IAMAP; UNEP; JSC of WMO-ICSU
C. Oceans, Atmosphere, Hydrology			
1. World Climate Research Programme (WCRP)	Ongoing	Meteorology; Oceanography	WMO; ICSU
2. Tropical Oceans and Global Atmosphere (TOGA)	Ongoing (1985-94)	Meteorology; Oceanography	WCRP; IOC; SCOR
3. World Ocean Circulation Experiment (WOCE)	1990-95	Oceanography; Meteorology; Circulation; Air-Sea Inter- action; Sea Ice	WCRP; CCCO IOC; SCOR

Programme	Status	Subject	Sponsor
4. International Hydrological Programme (IHP)	Ongoing 1975-	Hydrology	Unesco; WMO; UNEP; IAHS; IUGG
5. Global Energy and Water Cycle Experiment (GEWEX)	Proposed	Transport and Distribution of Water and Energy	WCRP
6. Joint Global Ocean Flux Study (JGOFS)	Proposed	Air-Sea Interaction	SCOR
7. International Satellite Cloud Climatology Project (ISCCP)	Ongoing (1983-)	Cloud Cover; Solar Radiation Fluxes	IAMAP; COSPAR; SCAR; JSC
8. International Global Atmospheric Chemistry Programme (IGAC)	Proposed	Atmospheric Trace Constituents; Atmospheric Photochemistry; Aerosols	IAMAP; IUGG
9. Integrated Management of Coastal Systems (COMAR)	Ongoing	Coastal/Marine Ecological Processes	Unesco; IABO; IUBS
D. Solar-Terrestrial			
1. Middle Atmosphere Programme (MAP)	Ongoing (continuation proposed)	Solar - Terrestrial Physics; Atmospheric Physics	SCOSTEP
2. Polar and Auroral Dynamics (PAD)	1985-90	Auroral Particles; Particle Physics; Physics of Aural Formation	SCOSTEP; SCAR
3. Solar-Terrestrial Energy Programme (STEP)	Proposed	Plasma Physics; Solar Physics	SCOSTEP; URSI IGAG
4. World Ionosphere-Thermosphere Study (WITS)	1987-89	Global Dynamics of Ionosphere-Thermosphere	SCOSTEP
5. Solar Interplanetary Variability (SIV)	1988-90	Transition of Sun and Inter-Planetary Medium	SCOSTEP

Programme	Status	Subject	Sponsor
E. Monitoring and Data			
1. Global Environmental Monitoring System (GEMS)	Ongoing	Meteorology; Environmental Sciences; Ecology	UNEP; WMO; IOC; ICES; FAO; SCOPE
2. World Digital Data-base for Environmental Science (WDDES)	Proposed	Hydrology; Bathymetry; Terrain; Coastline; etc.	ICA; IGU
3. World Soils and Terrain Database (SOTER)	Proposed	Soils;Terrain	ISSS; UNEP
4. Multi-Satellite Thematic Mapping Project	Ongoing	Geology; Mineral Resources	CODATA
5. World Data Centers (WDC)	Ongoing	Solar and Geophysical Information	ICSU
6. Federation of Astronomical and Geophysical Services (FAGS)	Ongoing	Astronomical and Geophysical Observations	ICSU
7. Monitoring the Sun Earth Environment (MONSEE)	Ongoing	Solar - Terrestrial Interaction Data	SCOSTEP
8. World Climate Data Programme (WCDP)	Ongoing	Atmosphere - Ocean - Cryosphere - Terrestrial Earth Science Climate Data	WMO; ICSU; UNEP

GENERAL PRINCIPLES OF RELEVANT SATELLITE SYSTEMS

J. Askne
Department of Radio and Space Science
Chalmers University of Technology
412 96 Göteborg
Sweden

ABSTRACT Earth observation satellites launched in the last 25 years have made important contributions to the understanding of our planet. Based on 10 years of experience from Seasat and Nimbus-7 the possibilities to study quantities of e.g. oceanographic interest by means of microwaves have increased considerably. In the near future an important oceanographic satellite ERS-1 will be launched and a large number of satellites is planned for the future. For all the microwave sensors are fundamental. This chapter includes a description of the characteristics of satellite systems for earth observations by means of microwaves together with a review of different satellite systems

1.   Objectives of Earth Observation from Space

The objectives of earth observation systems is to provide data in order to understand the earth as a natural system. Only by means of satellites can we obtain global coverage. The global coverage is a definite advantage over ground–based instruments although the continuity of the monitoring may be lost if not one system is replaced by another. Temporal and spatial scales of the the physical phenomena of interest may vary considerably. If we consider for example meteorological phenomena we have variations on temporal scales from seconds to thousands of years and spatial scales from meters to thousands of kilometers. In oceanography we have a similar situation. Polar satellite systems may provide information on spatial scales down to at least tens of meters and temporal scales of days up to several years, while geostationary satellites may provide temporal scales of the order of minutes. By using several satellites the temporal scale can be extended to shorter as well as longer time periods.

The sensors on–board oceanographic satellites are mainly working in the microwave part of the spectrum. A main advantage of microwave sensors is their ability to look through clouds and to be independent of sunlight. This means that one can obtain data regularly. A still more important reason to use microwaves is that some physical phenomena can only be observed by means of microwaves such as sea surface winds and ocean waves. Microwave sensors are then of fundamental importance for oceanographic applications.

25 years of satellite observations of the earth have provided much information of different fundamental phenomena and over the years the need for understanding the earth as a complete system has developed. The Earth system consisting of land, oceans, and atmosphere is linked together by a number of complicated processes. We have much to learn about each individual system but we will in the future more and

R. A. Vaughan (ed.), Microwave Remote Sensing for Oceanographic and Marine Weather-Forecase Models, 23–44.

more concentrate on the whole system. However this will only occur if scientists have easy access to data sets including several different sensors as well as in situ data over long time periods. This means that the data must be easily accessible in the form of an immense geographical information system.

2. Satellite Properties of Importance for Remote Sensing

Considering general principles of satellite systems a few words should be mentioned about important technical constraints.

The basic is of course the limitations in weight. The payload will increase considerably from of the order of 300 kg to 2 − 4 tons for the next generation of system in the 1990's. In the meantime the total weight of the satellite has increased from 1 ton to 12 tons. This change, due to the developments in the capability of the rockets, means possibilities to increase the number of sensors and their complexities considerably, but still weight is a fundamental constraint.

Power for the instruments is normally supplied by solar cells, backed up by batteries to store energy when the satellite is in the shadow of the earth. Particularly the requirements of the synthetic aperture radar sensors on the power source are considerable, with a peak power for Seasat of 1 kW, for ERS−1 of 4.8 kW, and for the polar platforms of 13.5 kW. In contrast to these figures the peak power for the Geosat radar altimeter is 20 W.

Another fundamental and critical factor for many instruments is the aspect control system for accurat pointing. This is particularly the case for high resolution instruments. For meteorological applications high microwave frequencies are used and the required pointing accuracy is of the order of .5 arcmin. Consequently the pointing systems are more and more elaborate.

Finally the data transfer rate is a very important constraint for the construction of SAR−systems, with a transfer rate of 105 Mbits/s for ERS−1, a figure increased by a factor 2 for the polar platforms.

3. Some Principles of Observations

A measurement system should be able to detect, differentiate and recognize properties on the basis of spatial properties such as form and geometry and temporal properties. A signature is also recognized by its properties of reflectivity, spectral properties and polarization properties. Therefore we need to develop our understanding of the interaction between the wave and the medium and we may meat the objectives by developing more advanced multisensor, multifrequency, multipolarization systems with higher resolution and accuracy.

Normally we wish to observe a number of observables on a global basis for a long period of time. These will be used to derive maps of the mean fields, as well as to measure temporal and spatial fluctuations. It is fundamental to make this type of measurements during a long time under as similar conditions as possible. On the other hand there are measurements that will focus on specific problems and which will require detailed in situ measurements together with the satellite measurements. Results of these measurements will be used in model developments of the processes. The goal must be to develop from simple models describing a particular phenomenon to predictive models that allow us to infer the responses of the earth to changes in the environment. Some models can only be of a statistical nature. They reproduce the statistical properties of the system so that the system interactions are described correctly. Although the physical basis of many of the interesting measurements are well understood, for example radar scattering from ocean surfaces, there may still be many problems associated with corrections for special situations like high wind speeds, enclosed seas etc. to be solved. There are of course also problems which can not be solved by observations from space only.

4. Microwave Instruments

The interest in microwave sensors is partly due to their ability to propagate through the atmosphere without being influenced very much. This is illustrated by the atmospheric transmission over wavelengths of interest for remote sensing, see Fig. 1

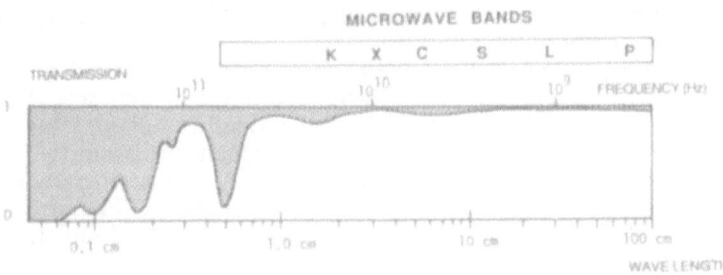

Fig. 1 Illustrates atmospheric transmission in the microwave part of the electomagnetic spectrum.

4.1 FREQUENCIES.

By tradition from World War II the different wavelength regions have been denoted as given below

Table 1: Microwave frequency bands		
Band	Frequency, GHz	Wavelength, cm
P–band	0.225– 0.39	133 – 77
L–band	0.39 – 1.55	77 – 19
S–band	1.55 – 3.90	19 – 7.7
C–band	3.90 – 6.20	7.7 – 4.8
X–band	5.75 – 10.9	5.2 – 2.8
K_u–band	10.9 – 18.0	2.8 – 1.7
K_a–band	18.0 – 36.0	1.7 – 0.8

The interaction between the electromagnetic wave and the medium varies with the wavelength. Normally the penetration into the earth is larger at longer wavelength and there may occur resonances between natural objects and an electromagnetic wave with a wavelength of the same order as the dimensions of the medium.

4.2   INSTRUMENTS

Two basic measurement principles are used, passive measurements and active. In the passive case no wave is transmitted but only natural black–body radiation is recorded, while in the active case the sensor is transmitting its own illumination. Some characteristic properties of the microwave sensors will be given below.

Microwave radiometers measures the microwave emission from the earth which is indicative of the temperature and composition of the surface and the intervening atmosphere.

Radars provide their own illumination and can be used to measure both the distance to a reflecting surface and the character of the surface. Particularly water yields a strong radar return. The strength of the radar return signal decays with distance, which influences the height of the satellite.

In synthetic aperture radar (SAR) the coherence of the radar is used, together with the motion of the satellite, to synthesize a large aperture for high spatial resolution. The resolution is in principle independent of the satellite height. This is in contrast to all the other sensors with a spatial resolution dependent on the physical dimensions of the antennas and the satellite height. The SAR like other radars is sensitive to changes in the dielectric constant and surface roughness. Future SAR–systems may include three frequencies (L–, C–, and X–band) with quad–polarization and multiple–incidence angles.

Radar altimeters measure the distance to the average ocean surface by means of a pulse directed straight down and by measurement of the time delay of the pulse. With knowledge of the geoid the geostrophic circulation can be determined. The surface roughness can be measured from the strength of the radar return and significant wave height by the shape of the return pulse. Elevation and roughness of land ice and sea ice can also be determined.

Microwave scatterometers measure microwave energy reflected back from small wavelets caused by the wind. Wind stress is obtained from the amplitude of the return signal, and wind direction is determined by combining measurements from several antennas.

4.3   SPATIAL SAMPLING

The resolution of a (diffraction limited) sensor is a fundamental property determined by the size of the receiving aperture (antenna), D, and the wavelength, λ. The beam–angle θ is approximately given by λ/D. If we compare observations at some different wavelengths we will note the difference, see Table 2 below, where the ground resolution, Δ, is given by $\Delta \simeq (\lambda/D)h$. The height, h, of the satellite assumed to be 800 km.

Table 2: The resolution of diffraction limited sensors

Wavelength λ	Diameter D	Resolution Δ
0.5 μm	1 cm	40 m
10 μm	1 cm	800 m
3 cm	1 m	24 km

(The values are not entirely relevant for the optical case)

The synthetic aperture radar applies computer processing to use the return echo from the area Δ to synthesize an antenna as large as Δ. This means that the resolution is approximately D independent of the satellite height. In practise the resolution of a SAR is of the order of 25 – 35 m.

A fundamental property of a coherent radar like the SAR and the altimeter is the speckel i.e. the interference between many coherent radar returns from different parts of the resolution cell. This speckle is a serious problem and in order to reduce the effect a number of radar returns are averaged. In the SAR–case we speak about multiple looks. A resolution of 30 m is typically a 4–look image.

Sensors may be divided in two categories, scanning and non–scanning. The microwave radiometer is normally a scanning sensor as illustrated in Figure 2 below.

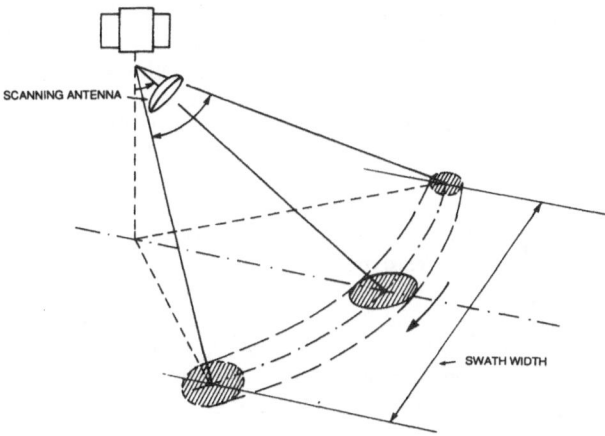

Fig 2. Illustrating a scanning sensor. In the future several antenna beams will be used simultaneously in a pushbroom configuration.

5. Orbits, Orbit Accuracy

The orbits chosen for a satellite depend on many factors including properties of the microwave sensors. For remote sensing it is often desirable to have the illumination geometry as constant as possible or generally the influence of the sun as constant as possible. This is accomplished with a sun–synchronous orbit. The altitude of the satellite is another fundamental parameter affecting the satellite repeat pattern and sensor properties like output power, resolution, and swath. Orbits can be adjusted to produce precisely repeating ground traces in different intervals. In order to obtain the possible synergism effect between different sensors they may have to be flown on different satellites as some are side–looking and some are nadir looking or one has to accept a certain time lapse between the measurements.

It is very important to know the orbit with high accuracy. This is the case for example with the altimeter which determines the height between the satellite and the earth surface. But it is also important for the SAR. In order to determine the exact geographical location of the pixels in the SAR–image it is important to know all the properties of the satellite in its orbit. Different methods are used to determine the orbit. A lidar is used together with a reflector onboard the satellite, or radio waves can be used to measure the distance between control points and the satellite. The Global Positioning System is another possibility.

5.1   THE KEPLER PROBLEM FOR SPHERICALLY SYMMETRIC BODIES

The two–body problem was solved by Kepler. In general the Kepler problem can be split in two parts; the motion of the centre of the mass and the relative motion. The relative distance between the bodies is governed by

$$\ddot{r} = - \nabla U(r) \tag{1}$$

where U(r) is the gravitational potantial which in the case of the motion of two spherically symmetric bodies is given by: $U(r) = - GM_e/r$, where G is the gravitational constant and M_e is the total mass of the earth and the satellite, i.e. approximately the mass of the earth. If the earth can not be assumed spherical the gravitational potential can be described in terms of trigonometric functions and associated Legendre polynomials. Depending on the starting velocity we can obtain elliptical, parabolic, or hyperbolic orbits. The elliptical orbit is illustrated in Fig 3 and described by

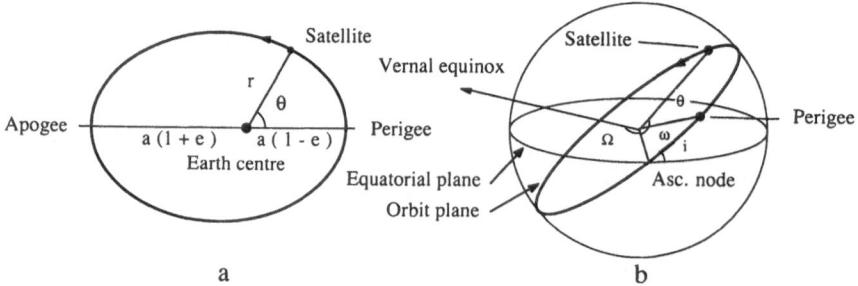

a   b

Fig 3 Illustrating the elliptic orbit and the orbital plane

$$r = \frac{a(1 - e^2)}{1 + e \cos\theta} \tag{2}$$

where
a = the semi–major axis of the ellipse
e = the eccentricity of the ellipse
θ = the angular position of the satellite in its orbit

The point of closest approach is known as the perigee and the point of maximum distance as the apogee. The period T for the satellite to travel round its orbit is

$$T = 2\pi \cdot (a^3/GM_e)^{1/2} = 2\pi a/v_0 \tag{3}$$

where v_0 is the mean horizontal velocity. If we have a circular orbit, e=0, then v_0 is the speed of the satellite.

Exercise: Determine v_0 by using $a = R+h$ and $h = 800$ km ($GM_e = 3.98603 \cdot 10^{14}$ m^3/s^2, and R = semi–major axis (equatorial radius) of the earth = 6378.14 km).

5.2   ORBIT PLANE OF A SATELLITE

The elliptical orbit has to be defined relative to fixed directions in space and the ellipse orientation within the orbital plane has to be defined. For this purpose three parameters are introduced, cf Fig. 3.
i, the inclination of the orbital plane to the earth equatorial plane
Ω, the right ascension of the ascending node (the crossing between the ascending orbit and the equatorial plane)
ω, the angular distance of perigee around the orbit, measured from the ascending node.

5.3   PERTUBATIONS DUE TO OBLATENESS OF EARTH

The above expressions are based on the assumption of a spherical earth. However this is not the case in practice and pertubations therefore occur in the elliptical orbit. The main correction is due to the oblateness and in this case the gravitational potential is given by, see (Kaula, 1966)

$$U(r) = - \frac{GM_e}{r} \left[1 - J_2 P_{20}(\sin\varphi)(\frac{R}{r})^2 \right] \tag{4}$$

where $J = 0.00108263$, P_{20} = Legendre polynomial, and φ = latitude. (We have included only the first term of an series)
The most interesting effect due to the oblateness is that the orbital plane starts to rotate and that

$$\Omega = - \frac{3}{2} \cdot \frac{v_0 J_2/a}{(1-e^2)^2} \cdot (\frac{R}{a})^2 \cdot \cos i \tag{5}$$

For a given orbit height it is possible, by a suitable selection of the inclination i, to compensate for the earth rotation of 360 deg about the sun in one year. This results in a sun–synchronous orbit in which the satellite crosses the equator at the same local solar time on each pass throughout the year. For a sun–synchronous orbit i is usually around 100 deg, i.e. the orbit is almost polar. The orbit is characterized by the equator crossing time, which is fixed by the launch conditions or by later course adjustments. For the reference orbit of ERS–1 the mean local solar time at descending node will be 10.30 within 1 minute and the polar platforms will probably have a crossing time between 13:00 and 14:00 for an afternoon platform and 9:00 and 9:30 for a morning platform.

Due to the oblateness of the earth, there is also a small correction to the orbital period, the time between to consecutive ascending nodes

$$\frac{2 \cdot \pi}{T} = \frac{v_0}{a} \{1 + \frac{3}{4} \cdot \frac{J_2}{(1-e^2)^2} \cdot (\frac{R}{a})^2 \cdot [(3 \cos^2 i - 1)\sqrt{1 - e^2} + 5\cos^2 i - 1]\} \tag{6}$$

By an adequate choice of the orbital parameters it is possible to force the satellite to repeat the same pattern after a given number of days. For a sun—synchronous orbit an integer number of the orbital time should be equal to an integer number of days.

It should also be mentioned, that there are other corrections to the satellite orbit e.g.
— atmospheric drag
— radiation pressure from the sun
— third—body pertubations (e.g. from the moon)
However for low—altitude earth orbits between 500 km and 1000 km such effects are of less order than effects due to the oblateness.

5.4   SENSOR TRACES

The nadir and off—nadir traces can be derived by providing the geometrical transformation between the satellite coordinates, look direction of sensor and the latitude and longitude on the earth. These relations are somewhat complex and we refer to Duck and King (1983).

5.5   ERS–1 ORBIT PARAMETERS

The parameters of the ERS–1 reference orbit are given below in Table 3. Some properties associated with other possible orbits with different repeat cycles are given in Table 4. The SAR coverage in the reference orbit is illustrated in Fig 4
Of some particular interest is the 35 day repeat orbit for which the the ground track is displaced by 80 km per orbit which means that the SAR—swath will cover the entire earth. However it should be pointed out that the SAR can only be used for less than 10 minutes per orbit and only when a ground station is within view.

Table 3: Orbital parameters of the ERS–1 reference orbit

semi—major axis	7153.135 km
mean inclination	98.5227°
mean eccentricity	0.001165
mean argument of perigee	90°
mean period	6027.95 s
repeat cycle	3 days
mean local solar time at desc. node	$10^{30} \pm 1$ min

Table 4: Parameters of the suggested orbits for ERS–1

Cycle	Orbits per day	a(km)	$i(^0)$	T(min)	Δ_s(km)	
3	$14\frac{1}{3}$	7153.135	98.5527	100.465	932	Ref. orbit
7	$14\frac{2}{7}$	7169.056	98.5896	100.800	401	Highest
11	$14\frac{4}{11}$	7143.050	98.4805	100.253	254	Lowest
17	$14\frac{6}{17}$	7146.605	98.4954	100.328	164	
20	$14\frac{7}{20}$	7149.584	98.4994	100.348	140	
29	$14\frac{10}{29}$	7149.305	98.5067	100.385	96	
35	$14\frac{12}{35}$	7149.962	98.5094	100.398	80	

Note: Δ_s denotes the distance at the equator between two adjacent tracks.
With a 35 day repeat cycle we can obtain world SAR coverage.

6. Data and Information Systems

More and more complex space systems with a great number of sensors and observation possibilities put a demand on the administration of the satellite. Although most data will be taken in a routine manner there will be situations where choices of view angle, spectral channels, scan pattern, resolution patterns have to be changed for specific test situations. The effective use of a satellite will put very high demands on the planning.

With all the sensors planned for future satellite the data flow from space will give rise to serious problems. For future systems it has been estimated that the average data flow from space will be 120 Mbs which means that the data and information system must cope with roughly 1 terabyte of data each day and store this with a host of derived products in a data base which has to be kept for up to 15 years.

Looking back we can conclude that Seasat, launched 10 years and unfortunately with a lifetime of only 3 months, produced data of interest until this day. In spite of its limited time of operation only those SAR images of immediate interest have been analyzed. Although processing of all SAR–scenes never has been the intention this illustrates in some way the serious problems we will obtain in the future if not precautions are taken in time. Some of the problems are of course based on the fact that standard methods to deduce the interesting data from ocean or ice imaging has not yet been developed for more than a limited number of products. The next few years until the launch of ERS–1 in 1990 researchers have a great responsibility to develop methods necessary to show the full usefulness of the coming system.

For European users Earthnet handles remote sensing data. Presently the following products can be obtained through Earthnet: Seasat, Nimbus–7 coastal zone colour scanner, Heat Capacity Mapping Mission, Metric camera, Meteosat, NOAA AVHRR/TOVS, MOS–1

7. Some Relevant Non—European Satellite Systems

The development of radar technology goes back to World War 2 but it was not until the mid—sixties that a NASA—supported study drew the attention to the applications in oceanography. In this study nearly all present—day applications were identified.

Microwave observations of the oceans have taken place at least since 1968 when the Russian satellite Cosmos 343 was launched with nadir—looking microwave radiometers. There has been a large number of Russian satellites since then, however the results are not very well known in the Western World. As an example Cosmos 1500 from 1983 can be mentioned.

The first U.S. space experiments gathering ocean—surface data were conducted from Skylab in 1974 when a microwave radiometer, scatterometer, and radar altimeter were used. Since then microwave instruments were flown on GEOS—3, Nimbus 5 and 6 before the well—known launch of Seasat in June 1978.

In § 8 and § 9 we will concentrate on the European efforts by describing ERS—1 and the discussions around the European Polar Platform. In § 7 below we will summarize properties of some other satellite systems for oceanographic applications and equipped with microwave sensors.

7.1 SEASAT

Seasat, launched on 28 June 1978 and operating only 105 days due to a power failure, was a remarkable satellite. It was the first satellite with an imaging SAR system used as a scientific sensor, and the first satellite with a scatterometer (after tests on Skylab). Seasat was also equipped with a radar altimeter and the Scanning Multichannel Microwave Radiometer, SMMR, also launched on Nimbus—7,cf below. Seasat was a satellite designed for oceanographic applications. It was a satellite ahead of its time and it seems like we will have to wait until the polar platforms in order to obtain a sensor package as complete as that on Seasat. Until ERS—1 there will be no satellite with a SAR.

Orbit: Most oceanographic satellites have a sun—synchronous orbit but not geodetic satellites and not Seasat. During the first two months Seasat had a 17 day repeat orbit with inclination 108 deg. The semi—major axis was 7163 km and the eccentricity 0.0010. The nodal period was 100.75 min. The last month the semi—major axis was changed to 7169 km and the eccentricity to 0.0008. Seasat then had a 3 day repeat orbit and a nodal period of 100.75 min.

Instrument parameters: Seasat had a L—band SAR (1.28 GHz) with 25 m resolution and a swath width of 100 km. It used HH polarization and incidence angles between 18.8^0 and 25.9^0.
The scatterometer at 14.6 GHz used four antennas, two at each side, VV and HH polarization, and a resolution of 50 km over a swath of 500 km.
The radar altimeter working at 13.5 GHz had a pulse limited footprint of 1.7 km and a range precision of 10 cm.
For SMMR parameters see Nimbus—7.

Geophysical parameters: Evaluation of the obtained results has produced the following table (Lame and Born, 1982):

Table 5 SEASAT Evaluation Summary

Sensor	Observable	Demonstrated Accuracy (1σ)	Demonstrated Range of Observable
Altimeter	altitude	8 cm (precision)	$H_{1/3} < 5$ m
	significant wave height ($H_{1/3}$)	10% or 0.5 m	0 to 10 m
	wind speed	2 m/s	0 to 10 m/s
Scatterometer	wind speed	1.3 m/s	4 to 26 m/s
	wind direction	16°	0° to 360°
Scanning multichannel microwave	sea surface temperature	1.0°C	10° to 30°C
radiometer	wind speed	2 m/s	0 to 25 m/s
	atmospheric water vapor	10% or 0.2 g/cm²	0 to 6 g/cm²
Synthetic aperture radar	wave length	12%	wave length ≥ 100 m
	wave direction	15°	0° to 360°

Note: Demonstrated accuracy, as used here, means the the SEASAT data has been found to compare with conventional or in situ measurements within the stated value.

7.2   NIMBUS–7

Nimbus–7 was launched 24 October 1978. The SMMR has been functioning until 1988 when the belt driving the antenna misfunctioned. The objectives were to conduct experiments in the pollution, oceanographic and meteorological disciplines and to refine atmospheric measurement capabilities demonstrated on previous Nimbus satellites. Nimbus–5 in 1972 and Nimbus–6 in 1975 preceded Nimbus–7, all equipped with scanning microwave radiometer. The Scanning Multichannel Microwave Radiometer, SMMR, was designed to determine sea surface temperature, wind speed, atmospheric water vapour and atmospheric liquid water parameters, sea–ice type and concentration.

Orbit: Nimbus–7 had a circular sun–synchronous orbit with an altitude of 955 km and an orbital period of 104.16 min. The inclination was 99.2 deg.

Instrument parameters: The SMMR instrument measured the thermal microwave radiation from earth at five dual linearly polarized frequencies, 6.6, 10.69, 18, 21, and 37 GHz with a sensitivity varying between 0.9 and 1.5 K. The SMMR had a 79 cm, mechanically scanned offset parabolic reflector and the beamwidths at the different frequencies varied from 4.2 to 0.8 deg.

7.3   COSMOS 1500

As an example of Russian microwave remote sensing satellites Cosmos 1500 launched in September 1983 can be mentioned. Such satellites have been used since some time for example for ice mapping and ship routing in arctic waters. Since this satellite two or three similar satellites have been launched, the last in 1988.

Orbit: The altitude is 650 km and the inclination 82.0°.

Instrument parameters: There is a scanning 37 GHz radiometer with a swath width of 650 km and a foot print of 17 km together with a non scanning microwae spectrometer at 3.5 GHz (footprint 85 km), 22 GHz (footprint 20 km), and 37 GHz (footprint 17 km). A side looking radar at 10 GHz with vertical resolution has a swath width of 475 km and a resolution of 2.1 km along track and 0.9 km cross track. There is also additional instruments in the visible and a multichannel scanning radiometer in the infrared with a swath width of 1930 km and a resolution of 1.7 × 1.0 km.

7.4 GEOSAT

Geosat is an US Navy satellite launched on March 12, 1985 and with a program scheduled to end in April 1989. The program has two objectives: to exploit the environmental potential of the altimeter during the first 18 months while the satellite was in a geodesy orbit, and then to maximize the quality and quantity of data that could be used for operational physical oceanography during the rest of the mission. The primary classified mission was to provide a gravity database over the global ocean which can be used for improved map compensation for geoid height, vertical deflections, and in–flight gravity, detection of possible bathymetric hazards to sub-merged navigation etc. The Geosat exact repeat mission has an orbit optimized for the operational measurements of fronts and eddies. The exactly repeating orbit allows long–term averaging to give an accurate local mean surface along the tracks. This data is freely available through NOAA National Environmental Satellite Data and Infor-mation Services in Washington, D.C.

Orbit: Geosat was launched into an 800 km orbit, 108⁰ inclination orbit with a 3 day near–repeat ground track. On October 1 1986 the orbit was changed into a 17–day exact repeat orbit (1 km accuracy) optimized for oceanographic data. This orbit is essentially the same as that used for Seasat.

Instrument parameters: The frequency of the altimeter is 13.5 GHz and a 1–m para-bolic antenna is used. The beamwidth is 2 degrees to compensate for the attitude control system using a gravity–gradient system with an accuracy of 1 degree. A 2 kW amplifier was used on Seasat but had a limited lifetime. Instead Geosat uses a 20 W amplifier and the uncompressed pulse width was increased from 3.2 to 102.4 us and the noise figure of the receiver was improved.
The altimeter provides three basic measurements specified as
− An altimeter height precision of 3.5 cm for a 2 m significant wave height (SWH)
− SWH: 10% of the SWH or 0.5 m, whichever is greater.
− Wind speed: 1.8 m/s over the range 1 to 18 m/s

An altimeter is designed primarily for ocean measurements and can loose track of the height when the satellite is crossing over ocean–land boundaries. Geosat has a more responsive tracking circuit than Seasat and is then producing much more data over sea–ice, land–ice as well as land. As a measure of precision the difference in ele-vation where passes cross over the same sub–satellite point can be examined. Zwally et al has found the standard deviation for cross overs on the Greenland ice sheet to be 1.61 m, while the corresponding value for Seasat was 1.06 m probably depending on a better knowledge of the radial error of the Seasat orbit.

7.5 SSM/I

The SSM/I is the Special Sensor Microwave Imager is a joint program of the U.S. Navy and Air Force consisting of a passive seven channel, four frequency microwave radiometer. It is flown on the Defense Meteorological Satellite Program operational spacecraft as an all–weather oceanographic and meteorological sensor.

Orbit: The orbit used is a 833 km circular sun–synchronous orbit with a 101 minute period and 98.7 deg inclination

Instrument parameters: SSM/I uses a conical scanning with a 45 deg nadir angle of scan and 53.1 deg incidence angle. The total swathwidth is 1394 km. In comparison to SMMR flown on Seasat and Nimbus–7 the SSM/I will have one less frequency (6.6 and 10.6 GHz in SMMR are replaced by 85.5 GHz in SSM/I). The SSM/I will have a more frequent global coverage by about a factor of four than the SMMR (about once very 6 days)

Geophysical parameters: The goal is to determine the following parameters with a geometric resolution of 25 km:
- ocean wind speed   2 m/s over 3 to 25 m/s
- sea ice cover   12% over 0 to 100%
- sea ice age   1st year vs multi–year
- sea ice edge   12.5 km
- precipitation   5 mm/hr over 0 to 25 mm/hr
- cloud water 0.1 kg/m^2 over 0 to 6 kg/m^2
- liquid water 2.0 kg/m^2 over 0 to 6 kg/m^2
- soil moisture   dry–moist, wet–saturated

SSM/I is an operational tool based on the experience with SMMR, but for research purpose the future AMSR proposed for HMMR will have a geometric resolution almost a factor 10 better.

7.6   MOS–1

MOS–1 stands for Japan's First Marine Observation Satellite. The satellite was launched on 19 February 1987. It is the first in a 13–year series planned to be operational in the 1990s. MOS–1 has a lifetime of two years and will be followed by MOS–1b in 1989
By an agreement with the Japanese Space Agency, NASDA, raw data or system corrected data will be distributed by Earthnet. There is no tape recorder onboard MOS–1 but Maspalomas, Fucino and Kiruna have been selected as groundstations covering Europa.
The objectives of the satellite are to determin parameters like sea surface colour, vegetation, land use, to determine suspended sediment, stratospheric water vapour, earth and sea surface temperatures and to determine water vapour content, liquid water content, ice, snow etc.

Orbit: MOS–1 has a sun–synchronous orbit with an altitude of 908.7 km and eccentricity 0.004 deg. The inclination is 99.1 deg and the period 103.2 min. The descending node time is 10:00 – 11:00 (final decision missing). It has a 17–day exact repeat orbit. The satellite will be controlled so its orbital ground track remains within 20 km of its nominal location.

Instrument parameters: MOS–1 carries instruments operating in the visible, infrared and microwave part of the spectrum. The Multispectral Electronic Self–Scanning Radiometer, MESSR, scans a swath of 100 km in four spectral bands in the visible and near infrared and provides images with a resolution of 50 m. The visible and thermal–infrared radiometer, VTIR, images a swath of 1500 km in one visible and three thermal–infrared bands, with a resolution of 900 m to 2700 m. The Microwave Scanning Radiometer, MSR, is a conical, passive microwave imager which scans a swath of 317 km at two frequencies, 23.8 GHz and 31.4 GHz with spatial resolutions of 32 and 23 km respectively.

7.7 N–ROSS

N–ROSS stands for the Navy Remote Ocean Sensing System and is an U.S. Navy satellite planned to be launched in June 1989. However the plans for the satellite will probably be cancelled, and the information is included here mainly as an illustration af an interesting satellite system. The primary mission of the satellite were to provide all–weather global oceanic data to support numerous fleet activities. The data was planned to be available within 3 hours after acquisition at the processing facility.

The products supposed to be obtained from N–ROSS are
– sea surface wind
– sea surface temperature (night/day – all weather)
– sea ice (cover, thickness, age)
– ocean waves (height, direction)
– ocean surface topography (fronts and eddies)

Orbit: A sun–synchronous orbit of about 833 km altitude with an inclination of 98.7 deg and a 5:30 descending local equatorial crossing time. The nodal period was planned to 101.3 min and the repeat cycle to one day.

Instrument parameters: The N–ROSS sensors were planned to consist of
– a scatterometer, essentially similar to the Seasat design but with improved discrimination between wind–direction aliases and an increased swathwidth from 500 to 600 km.
– an altimeter, essentially the same as on Geosat
– the Special Sensor Microwave Imager, SSM/I, flown on the DMSP–satellite
– a new low frequency radiometer, LFMR, which is a dual–frequency and polarization radiometer designed for measurement of sea surface temperature with an accuracy of 0.5 K at a surface resolution of 10 km although 1.0 K and a surface resolution of 25 km are considered acceptable.

7.8 TOPEX AND POSEIDON

The TOPEX and POSEIDON mission is a NASA–CNES satellite to be launched in 1991. The objectives are related to ocean circulation and the satellite is equipped with a single frequency radar altimeter and a 2–frequency radar altimeter resp, a microwave radiometer and a precise positioning system.

Orbit: The orbit is non sun–synchronous in order to allow the investigation of some of the tidal components. The reference altitude is 1335.5 km with an inclination of 65⁰ and a repeat cycle of 10 days.

Instrument parameters: The frequencies used for the altimeter are 13.6 and 5.3 GHz. The pulse limited footprint is 2.2 km.

7.9   JERS-1

JERS-1 is the first Japanese Earth Remote Sensing satellite and it will be launched in 1992. It is equipped with an L-band synthetic aperture radar and visible and near IR radiometer.

Orbit: The orbit is sun-synchronous with an altitude of 568 km.

Instrument parameters: The resolution of the synthetic aperture radar is planned to 20 m and the swath width to 75 km. Polarisation is HH and angle of incidence 44 deg.

7.10   RADARSAT

Radarsat is a Canadian satellite expected to be launched in 1994. It has an expected lifetime of 5 years. Radarsat has a rather advanced and complete SAR as payload although only at one frequency, C-band, and one polarization, HH. The objectives are to produce better ice and weather forecasts to reduce navigational risks and make resource exploration safer and more effective. The satellite will also be quasi operational by providing surveillance of Canada's high Arctic, scanning the region once per day, merging the data with other spatial data for the same area and providing the information to users (such as ships) within a few hours of the satellite's passage. Ocean applications include improved weather and sea state forecasts, detection of ships and oil spill. Land applications include identification of geological features important in mineral and petroleum exploration, soil moisture and forest changes.

Orbit: Radarsat is to be launched by an U.S. rocket to a circular, sun-synchronous orbit with an altitude of 792 km and an inclination of 98.6 deg. The period will be 100.7 min and the ascending node will cross the equator at 18:00 local time. The repeat cycle will be 16 days with as sub-cycle of 3 days.

Instrument properties: The flexibility in the design of the SAR is provided by two features. The beamforming is obtained by an adjustable phase of the elements across the antenna width. The ground range resolution is varied by using two pulsewidths of about 17.3 and 11.6 MHz with resolutions in the range from about 20 to 30 m, and by an additional bandwidth of 30 MHz to provide the finer resolution required for certain applications.

Radarsat provides 5 different SAR modes, and illustrated in Fig 4, below.

Table 6: SAR modes of Radarsat

Mode	Swath width (km)	Resolution (m)	Looks
Basic	100+	28×25	4
Scansar	500	100×100	8
Wideswath	150	28×40	4
Fine res.	40+	10×10	1
Experimental[1]	100	28	4

[1] Incidence angle smaller than 20 or larger than 50 deg.

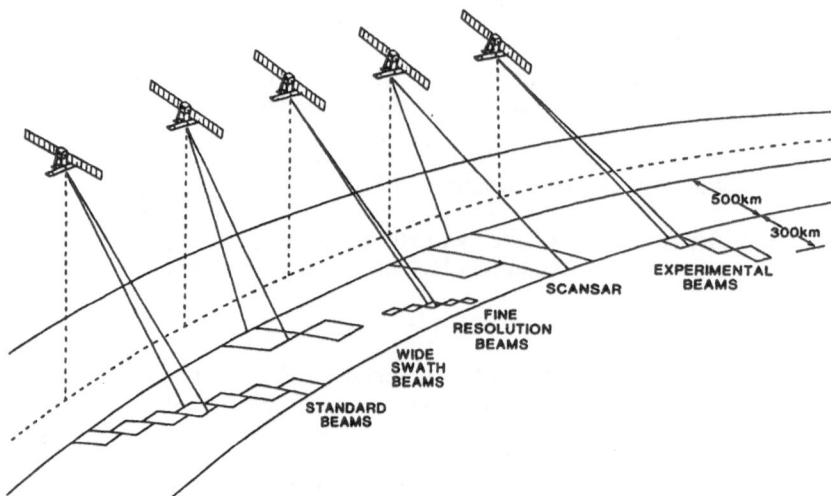

Fig 4 Illustrating different SAR–modes of Radarsat.

With 14 orbits a day, and the ability to record 20 min. of data in light and 8 min in eclipse, the RADARSAT could produce about 1600 100x100 km square scenes per day. The need for efficient handling, product generation, dissemination and archiving of the large amounts of that is staggering.

8.   European Satellite Systems – The ERS–1 Satellite

In this section and the next the European plan for development of satellite systems for primarily oceanographic applications will be summerized. By giving some more details in connection to the above described non–European satellites the trend in the design developments will be illustrated.

8.1   ERS–1 MISSION OBJECTIVES.

ERS–1 stands for the first European Remote Sensing Satellite. It is a satellite under the auspices of the European Space Agency, ESA and a program approved by 13 countries. The development started in January 1985 and the launch is planned to April 1990. The expected lifetime of ERS–1 is 2 to 3 years and there are plans for an ERS–2 satellite, due to the need for continuity of the availability of data over extended periods. The financial envelop for ERS–1 is approximately 750 MAU at 1986 economic prices (1 AU is presently 0.7 pound) and with the extra costs made at national levels for instruments, processing and achieving facilities, and ground stations the ERS–1 programs are estimated well above 850 MAU (1986 price level). The cost associated with ERS–2 is 355 MAU.   The ERS–1 mission objectives are of both a scientific and application nature and it is aimed primarily to investigations in the field of oceanography, including global ocean and coastal zone processes. The ERS–1 objectives are declared to be:
– Establish, develop and exploit the coastal, ocean and ice applications of remote sensing data.

– Increase the scientific understanding of coastal and global ocean processes, together with the monitoring of polar regions.
– To develop on an experimental basis the use of SAR data over land.
  Significant advances in fields like physical oceanography, glaciology and climatology are anticipated. The scientific areas of interest include:
– Oceanic circulation at local, regional and global scales
– Mesoscale features such as eddy fields
– Wind and wave interaction
– Sea surface topography and marine geoid
– Air–sea interaction/heat exchange/energy budget
– Data assimilation into numerical models (atmosphere, ocean, ice)
– Ice sheet mapping and dynamics
– Climate modelling and variability
– Coastal bathymetry
– Interaction mechanisms between microwave radiation and various targets (ocean, ice, vegetation, bare soils ..)
  However not only scientific aspects are of interest but applications as well, an aspect being more and more important in the justification of the large costs associated with satellites. Applications based on the expected forecasting improvements of ocean and ice geophysical parameters include offshore petroleum activities (oil exploration and production activities), ship routing, and fishing activities. The synthetic aperture radar offers possibilities like maritime traffic surveillance in coastal areas, detection of oil pollution and prediction of trajectories as well as monitoring of land renewable and non–renewable resources.
  However before the scientific and application oriented missions there is an important phase of geophysical data validation including calibration/validation of the individual sensors as well as the synergism between different sensors. Consequently the first three months of the satellite life will be devoted for validation/calibration over specific sites.

8.2   FEATURES OF THE ERS–1 SATELLITE

8.2.1   *Payload* Priority has been given to a comprehensive set of radar instruments designed to observe the surface wind and wave structure over the oceans and to provide high resolution images of the earth's surface. The core instruments have been complemented with an infra–red radiometer for sea–surface temperature monitoring and with a microwave sensor for atmospheric corrections. For accurate tracking information passive laser retro–reflectors are used as well as a special microwave unit. For more details of the instruments, see below.

8.2.2   *Platform* ERS–1 will make use of the same platform as developed for the French SPOT program and adopted to specific ERS–1 requirements by including for example
– yaw steering capability, allowing compensation for earth rotation to ease wind scatterometer data processing
– geodetic pointing of the yaw axis (i.e. local normal pointing) instead of geocentric pointing to reduce errors for radar altimeter measurements
– roll tilt capability to allow on an experimental and time–limited basis only, operation of the full imaging SAR mode at larger incidence angle (35 degrees at mid–swath in contrast to the normal 23 degrees).

8.2.3.   *Orbit configuration* ERS–1 will be launched to a sun synchronous orbit by the Ariane rocket from Kourou in French Guiana. During the 3 first months of its life it will be placed in a 3–day repeat cycle orbit with a mean altitude of 780 km. This reference orbit is planned to pass over a measurement tower outside Venice and is sometimes called the Venice–orbit. ERS–1 has the potential of changing its orbit up to 6 times during its life time (limited by the hydrazine aboard). By changing the semi–major axis by 26 km the nodal period will change from 14 4/11 orbits per day for the lowest orbit to 14 2/7 orbits per day for the highest orbit. The repeat cycle will then change between 3 and 35 days. All orbit repeat cycles (except 7 days) have a 3–day subcycle. Any repeat cycle change will take around 8 days until a stable well–known orbit is known. This has of course consequences for the measurement planning. (ESA has provided a PC–based visualization program of the properties of the ERS–1 orbit illustrating swath widths for different repeat cycles.)

8.2.4.   *Attitude control system* ERS–1 has a three–axis stabilized earth pointed spacecraft with the yaw axis pointed toward the local vertical. The maximum errors per axis are 0.21 deg for the yaw axis and 0.11 deg for the pitch and roll axis.

8.2.5.   *Orbit control facilities* For a number of important scientific investigations such as mean and time–variability of sea surface topography it will be necessary to reconstitute the ERS–1 orbit or the radial component of the orbit with an accuracy of decimeters. The orbit control is primarily maintained by means of precise tracking facilities of two kinds. The satellite is equipped with a laser retrofeflector consisting of so–called corner cubes mounted on a half sphere. A number of laser ranging stations are planned to be used to determine the distance to the satellite. The highest accuracy can be obtained by means of laser ranging, but microwave methods can be used independently of clouds. The traditional method is Doppler ranging. However for high orbit accuracy corrections have to be made for tropospheric water vapour and ionospheric electron density. The PRARE system, Precise Range And range Rate Equipment, includes two frequencies, X– and S–band in order to correct for the electron density. Ground meteorological quantities or even a ground–based microwave radiometer can be used to correct for water vapour. The predicted ranging accuracy of PRARE is 5–10 cm.

8.2.6.   *Orbit accuracy* The orbit restitution data will be provided with an accuracy of 15 to 50 m depending on direction. The possibility of providing orbit restitution at an accuracy of a few decimeters or better may be provided.

8.3   ERS–1 MICROWAVE SENSORS

The core instruments consist of a C–band active microwave instrumentation acting as a SAR or a scatterometer and a radar altimeter. These instruments are supported by an along–track scanning radiometer and a microwave radiometer as well as the precise range and range rate equipment and the passive laser corner retro–reflector.

8.3.1.   *AMI* A C–band Active Microwave Instrumentation (AMI) at 5.3 GHz and combining the functions of a Synthetic Aperture Radar, SAR, a wave mode and a wind mode. The aim is to measure wind fields and wave spectrum over the oceans and taking all–weather high resolution images over polar caps, coastal zones and land areas.

8.3.2. *Radar altimeter* A Radar Altimeter, RA, operating at 13.7 GHz with the aim to determine altitude, significant wave height, ocean surface wind speed, and various ice parameters.

8.3.3. *Along–track scanning radiometer and microwave radiometer* The core instruments are complemented by an Along–Track Scanning Radiometer and Microwave Sounder, ATSR/M. The ATSR is a scanning three channel infrared imaging sensor (3.7, 11 and 12 μm) which allows the same spot on the Earth's surface to be viewed with a spatial resolution of 1 x 1 km at two different angles (0 and 55 degrees). This permits a more accurate correction for atmospheric temperature measurements than previously as long as the atmospheric along track variations can be considered negligible. The predicted accuracy is 0.5 K over a 50 x 50 km square with 80% cloud cover. The Microwave Sounder is a nadir looking two–channel radiometer operating at 23.8 and 36.5 GHz. It is designed to measure the vertical column water vapour content and cloud liquid of the atmosphere within the 22 km footprint of the radar altimeter. In conjuction with the ATSR measurements the tropospheric range corrections for the altimeter can be accurately determined with a predicted precision of 2 cm.

8.4   DATA DISSEMINATION (GROUND SEGMENT)

8.4.1. *Ground Stations* There are four primary ground stations, Kiruna, Fucino, Gatineau, and Maspalomas. ESA has signed an agreement with NASA for the access to SAR data from the Fairbanks station. A number of other stations are presently under negotiation to receive ERS–1 data e.g. in Australia, Brazil, India, and Japan and a number of stations in other countries are as well anticipated as well as a station for coverage of the Antarctica. The ERS–1 communication link has two X–band channels, one channel for high bit rate (105 Mbps) used for AMI in imaging mode, and one for low bit rate, including all other instruments data at a rate of 1093 Kbps or at a rate of 15 Mbps at the output of the onboard tape recorder which has a capacity of 6.5 Gbits. The Kiruna ground station has a S–band link for telemetry and telecommand.

8.4.2. *EARTHNET Fast Delivery Product Stations* The different ground stations as well as cataloging of available data and data products will be coordinated by the ESA/Earthnet program office in Frascati. The goal is that the fast delivery, FD, products are produced and distributed to the nominated centers in a country within 3 hours. However distribution channels with a slower turn around such as the use of a suitable ECS/SMS channel and of technology and equipment developed in the framework of the Apollo project are as well investigated. The nominated center in each country in its turn delivers the data to the end users. The FD products are based on generally accepted algorithms for the main products of ERS–1. International agreement on formats on CCT and on photographic paper have been achieved as well as software for accessing the formatted data.
– The SAR image FD product consists of images over 100 x 100 km with 20 m (range) x 16 m (azimuth) interpixel distance with a spatial resolution better than 33 m for all points in the image in both range and azimuth. 16 bits per pixel or 8 bits out of 16.

– The AMI wave mode FD product is obtained by using the SAR system every 200 or 300 km for an area of 6 km by 5 km. After an image is derived the power spectrum is determined by a fast Fourier transform. Up to 150 spectra are generated per orbit.

– The wind scatterometer FD product is obtained by converting the measurement to sigma naught measurements with 50 x 50 km resolution excluding cells with a large percentage of land coverage. The sigma naught values are then converted to wind vector (speed and direction) estimates based on results from various measurement campaigns. Up to six different solutions exist in the data inversion and assumptions of a continuous wind field or some extra meteorological information is used to derive the wind field without ambiguity.

– The radar altimeter FD product is based on 3 quantities related to the time delay and shape of the return echo from which the ground processing derives geophysical quantities based on the Brown model (see e.g. G.S. Brown et al). From the radar return one obtains an estimate of the wind speed based on the average power of the return, an estimate of the significant wave height from the estimation of the standard deviation of the leading edge slope and an estimate of the altitude from the time delay. An atmospheric look–up table updated once a month provides propagation delay and attenuation correction for this standard product while the ATSR/M can be used for further corrections. Special processing and achiving facilities, PAFs, in France, U.K., Germany and Italy will be responsible for:

– archiving/retrieval of raw data and data products
– support in the geophysical validation of mission and products
– generation of precision and thematic products.

The different PAF:s have taken a responsibility for different products:

France: RA (ocean), AMI (wind and wave)
U.K.: SAR, RA (ice/land), ATSR/M, AMI (wave)
Germany: SAR, Precision orbit determination, RA (geophysical products)
Italy: SAR, LBR over Mediterranean.

Proposed products from the PAFs are available.

The pricing policy for ERS–1 data is presently under discussion and it can be anticipated that there will be different policies for participating and non–participating states.

8.5   Planned experiments

During the lifetime of a satellite like ERS–1 there will be many contrasting requirements. The 13 participating states have given their national requirements and a large number of investigators (151) answering on the Announcement of Opportunity issued late 1986 have been accepted. Presently the constraints are under investigation. Out of the 151 accepted responses we find 34 concerning oceanography, 19 about air–sea interaction, 13 about polar ice, 5 about marine monitoring, 4 about coastal phenomena.

9.   Polar Platforms

9.1 BACKGROUND

Together with the Space Station a number of Polar Platforms are envisaged, two planned by NASA, two by ESA and one by NASDA. The polar platforms were to be serviced in orbit but budget constraints seems to mean that at least for ESA in–orbit servicing is disregarded and more conventional expendable satellites capable of two tonnes of instruments and launched every three years may be chosen. All the platforms will carry operational instruments involved alternatively in meteo–ocean–climate or exposed land mass missions. According to plans the first ESA platform should be launched in 1997, but delays are probable. The first mission is centered around meteo–ocean–climate and would cover the requirements of the operational and research communities in such areas as operational weather, marine and ice forecasting, physical oceanography, glaciology, atmospheric research and climatology. The nominal altitude for the near–circular sun–synchronous orbit is 824 with a design range between 700 and 850 km. Local time for descending node will be 8:30 to 10:30 and for ascending node 13:00 to 14:30. A number of sensors have been discussed and the choice between different alternatives will be made in the near future. Here some of the microwave sensors for oceanographic applications will be mentioned as they are illustrating the principal development in the field.

9.2 PLANNED MICROWAVE INSTRUMENTS FOR OCEANOGRAPHIC APPLICATIONS

During 1988 basic decisions will be taken concerning what instruments should be developed for the European first polar platform. The candidates under discussion are shortly described below.

9.2.1.   *ALT–2* An altimeter derived from the ERS–1 concept with applications in studies of ocean, ice and atmospheric as well as for Solid Earth.
This advanced terrain tracking radar altimeter will operate in an adaptive pulse limited mode in a continuous and autonomous way over all type of surfaces. A two frequency (13.8 GHz plus a band to be decided) radar altimeter including a passive radiometer mode at 23.8 and 36.5 GHz is currently under study.

9.1.2.   *SCATT–2* A C–band scatterometer derived from the ERS–1 model for oceanographic and meteorological studies. Compared to the ERS–1 design a number of improvements are being considered such as the provision of a double–swath instrument to improve the coverage and frequency of observation of the oceanic zones, a better capability for the removal of wind direction ambiguity, a larger wind speed range capability and possibly a higher spatial resolution.

9.1.3   *SAR–C* A synthetic aperture radar evolved from the ERS–1 instrument operating in C–band for land, ice and ocean studies. Compared to the ERS–1 instrument it shall provide several operating modes depending on observational requirements. It shall provide signatures as function of polarisation and incidence angle and provide an increased capability for global surveillance.

9.1.4   *MIMR* A Multi–band Imaging Microwave Radiometer intended for land, atmospheric, oceanographic and glaciological applications including ice characteristics like extent, type and concentration and ocean surface parameters like wind speed, sea surface temperature, salinity, and oil spill. Planned frequencies are 1.4, 6.8, 10.65, 18.7, 22.355, 36.5, and 90 GHz. Vertical and horizontal polarization are planned to be included and the spatial resolution varies from 10 to 2 km. The swathwidth is 1400 km. In order to reach the goals rather large antennas are required (of the order of 20 m for the lowest frequency).

44

10. Conclusion

The technological developments in connection to remote sensing satellites are very rapid and a new generation of satellites is presently planned while we are analyzing data from the first generation of microwave remote sensing satellites and waiting for the second generation. The need to deepen our understanding of the basic phenomena involved in microwave remote sensing, the need to develop the interpretation technique, and the need to prepare the society for the future possibilities is immense.

11. Acknowledgement

The material for this presentation has been drawn freely from publications and reports (some of them mentioned below) distributed by ESA and NASA as well as other organizations. Particularly the material distributed in connection to the Announcement of Oppertunity for ERS–1 and the EOS have been used. The author is grateful to a number of individuals supplying information and comments about the satellites systems and to L. Ulander for comments on the manuscript and contributions to the section on orbits.

12. References

BROWN, G.S., STANLEY H.R., and ROY N.A.: The wind speed measurement capability of spaceborne radar altimeters, IEEE J.Oceanic Eng. 6, 1981

DUCK K.I. and KING J.C., Orbital Mechanics for Remote Sensing, in Manual of Remote Sensing 2nd edition, Ed. R.N. Colwell. American Society of Photogrammetry, 1983

ESA: Announcement of opportunity for ERS–1, Technical Annex: ERS–1 System Description, May 20, 1986

Johns Hopkins APL, Technical Digest, April–June 1987, Vol 8, No.2: The Navy GEOSAT Mission

KAULA, W, Theory of Satellite Geodesy, Waltham, Mass. Blaisdell, 1966

LAME, D.B. and BORN G.H., SEASAT Measurement system evaluation: Achievements and limitations, J.G.R., 87 (C5), 3175 – 3178, 1982

NASA: EOS Science Steering Committee Report, Volume II, From Pattern to Process: The Strategy of the Earth Observing System.

SATELLITE RADAR ALTIMETERS

Chris. G. Rapley
University College London
Mullard Space Science Laboratory
Holmbury St Mary
Dorking, Surrey, UK

ABSTRACT. This paper provides a brief overview of the operating principles and technology of space borne radar altimeters designed to measure ocean surface topography for ocean dynamics studies.

1.   INTRODUCTION.

The primary purpose of a satellite radar altimeter is to measure the vertical distance between itself and the Earth's surface. Height profiles are derived from the time sequence of measurements as the satellite progresses around its orbit. The data can be used to address a surprisingly wide range of applications over oceans, ice and land. However, the main applications to date have been studies of the marine geoid and ocean dynamics, with goals of ~3cm and 15cm for the height precision and accuracy respectively. Since the instrument is highly quantitative, the full exploitation of the data requires a detailed understanding of its technical operation, the interaction of the microwave pulse with the ocean surface, and the propagation of the microwave signals through the atmosphere.

2.   ALTIMETRY BASICS

The principle of altimetry is illustrated in figure 1. A brief pulse of microwave radiation is transmitted towards the ground and the two-way time delay t of the echo is recorded. The height h of the altimeter above the surface is derived from :

$$h = 0.5ct \qquad (1)$$

where c is the propagation velocity.

For altimeter satellites h is chosen to (a) limit the effects of *atmospheric drag* and the higher order components of the Earth's *gravity field*, both of which affect the accuracy of the orbit ephemeris derived from the satellite tracking data ($h \geq 500$km), and (b) to satisfy limitations on *transmitter power* and the required *signal-to-noise ratio* of the radar echo. The altitude generally lies in the range $500 \leq h \leq 1500$km.

R. A. Vaughan (ed.), Microwave Remote Sensing for Oceanographic and Marine Weather-Forecase Models, 45–63.
© *1990 Kluwer Academic Publishers.*

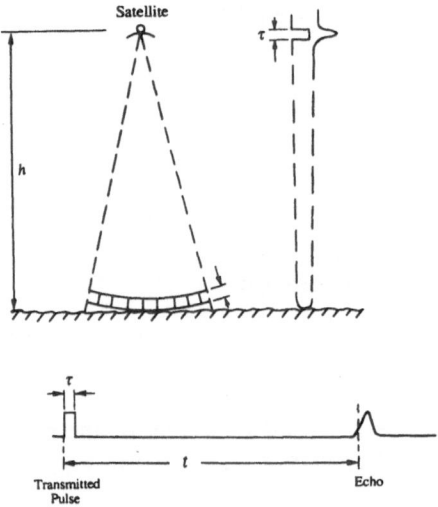

Figure 1 : The principle of Altitude measurement

Taking the ERS-1 value for h (777km) and using c_0 = 2.9979250 x 10^8 ms⁻¹ for the propagation velocity of electromagnetic waves in vacuuo, we find that the two-way time delay t is 5.18ms.

To measure ocean dynamics signals, we need to measure h to a precision of a few centimeters (say 3cm, which is equivalent to 1 part in ~ 3 x 10^7). Thus the uncertainty in the timing Δt must be ≤ 0.2ns.

3.   RANGE RESOLUTION REQUIREMENT

A short-duration pulse τ is required in order to (a) permit the range (height) to be measured precisely, and (b) to resolve and measure surface roughness (*ie* ocean waves).

The difference in delay time Δt for echoes from point targets separated in range by Δh is given by :

$$\Delta t = 2 \, \Delta h \, / \, c \qquad\qquad (2)$$

In order to resolve the separate echoes, we require :

$$\tau \leq 2 \, \Delta h \, / \, c \qquad\qquad (3)$$

Generally, the smallest wave amplitude of interest is ~ 0.5m, hence we require :

$$\tau \leq 3 \times 10^{-9} \, s$$

This is a factor ten or so greater than the timing precision requirement. However, knowing the instrument response to the surface, this can be fitted to the echo waveform data to obtain a more precise range measurement estimate. By averaging the estimates from as many as 1000 individual echoes, the ~3cm range precision can be achieved.

4. OPERATING FREQUENCY SELECTION

The effective transmitted pulse duration τ and bandwidth B of a radar are related by :

$$\tau = 1 / B \qquad (4)$$

To achieve a 3ns effective pulse duration, a bandwidth of ~300MHz is necessary. The operating frequency of an altimeter must be greater than a few hundred MHz for this to be possible. In practice, the maximum bandwidth available at a given centre frequency is limited by international agreement. The same international agreements restrict the uses of the bands within the microwave spectrum, further constraining the options for space altimetry.

Other factors influencing the selection are as follows :

- *Atmospheric absorption* : Above ~15GHz the absorption of the atmosphere begins to introduce a significant two-way loss, which will be variable, depending primarily on the water vapour content of the troposphere.

- *Technology* : Devices capable of generating and amplifying microwave signals become technologically more demanding and costly as the operating frequency is increased. This is particularly true if high transmit powers are required.

- *Antenna size* : The relationship between beam-limited footprint size L, antenna diameter D, and operating frequency f is given by :

$$L = (hc) / (fD) \qquad (5)$$

  A small footprint size can be achieved by using a high operating frequency or a large diameter antenna. Accommodation and mass considerations on-board the spacecraft favour the use of as small a diameter antenna and hence as high an operating frequency as possible.

- *Ionospheric refraction* : The propagation of microwaves through the ionospheric plasma is dispersive, the associated time delay decreasing with f^2. The magnitude of the correction required for the ionospheric delay thus reduces with increasing operating frequency.

- *On-board conflicts* : The transmit / receive frequencies (and harmonics) of other instruments, microwave ranging or Doppler systems, and the spacecraft command and data system must be avoided to prevent radio interference problems.

Since it is not generally possible to satisfy simultaneously all the criteria, a compromise must be reached. For the satellite altimeters flown or approved for future missions to date, the operating frequency selected has been in the microwave Ku band at ~13GHz (*eg* for the ERS-1 altimeter $f = 13.8 \pm 0.150$ GHz)

5. OPERATING MODE

An altimeter may be designed to operate in one two distinct modes. Conceptually the simpler is the *beam-limited* mode in which the antenna response defines the measurement footprint on the Earth's surface (figure 2).

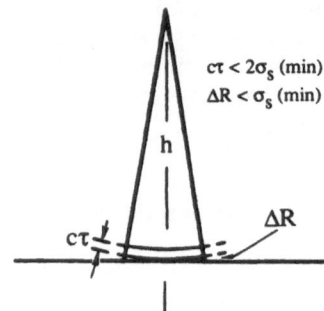

Figure 2 : Beam-limited operation

In order to resolve the minimum surface roughness of interest (rms deviation σ_s[min]), we require a transmitted pulse duration $\tau \le 2\,\sigma_s$[min] / c (from equation 3). For beam-limited operation, we also require that the slant range at the edge of the beam-limited footprint (BLF) should exceed the range at nadir by less than σ_s[min] in order to ensure that under all circumstances the entire footprint is illuminated by the transmitted pulse simultaneously. It is easy to show that, to a good approximation, ΔR, the excess slant range at the edge of the footprint, is given by :

$$\Delta R = (h\,\theta^2) / 2 \qquad\qquad (6)$$

where θ is half the antenna two-way beamwidth. Thus, to satisfy the beam-limited condition :

$$\theta \le (\,2\,\sigma_s[\min] / h\,)^{1/2} \qquad\qquad (7)$$

At a given frequency, the antenna beamwidth (2θ) is inversely proportional to the antenna diameter. To first order :

$$\theta = c / 2fD \qquad\qquad (8)$$

Thus :

$$D \geq (c / 2f) \ (h / 2 \ \sigma_s[min] \ )^{1/2} \qquad\qquad (9)$$

For h = 777km, σ_s[min] = 0.5m, and f=13.8 GHz (*ie* the ERS-1 values) we find that D ≥ 9.6m! Such a large antenna would be very expensive, would be difficult to accommodate, and would need to be deployed in orbit. It is thus not the preferred solution.

A further problem with beam-limited operation is the need to know the pointing direction of the antenna boresight very accurately. This is necessary in order to correct the measured range for off-pointing from the nadir direction. For an error in the height estimation ≤ Δh, it can be shown that the pointing direction must be known to an accuracy Δθ given by :

$$\Delta\theta = (\ 2\Delta h / h \ )^{1/2} \qquad\qquad (10)$$

For Δh = 3cm, and h = 777km, we find Δθ ≤ 57 arcsec. The practical difficulties in setting up and maintaining the alignment of an antenna to a reference frame within the spacecraft to this level of accuracy are enormous. Furthermore, a sophisticated and costly star-tracker Attitude Control System (ACS) would be required.

In contrast, the alternative *pulse-limited* mode of operation uses a small antenna, and has a much more relaxed requirement on antenna and spacecraft pointing. In this case, a broad antenna beam is used (figure 3).

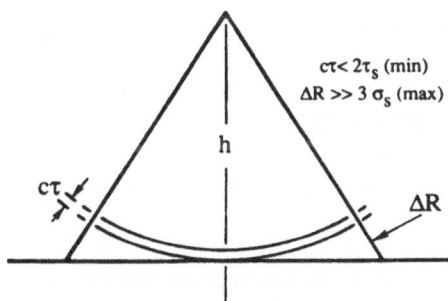

Figure 3 : Pulse-limited operation

The pulse of microwave energy propagates away from the antenna as part an expanding spherical shell. As it intersects the ground, the region illuminated "spreads out" across the beam-limited footprint, giving rise to an echo with a rapid initial rise and a long decay. The range (height) measurement is made by timing to the leading edge of the echo waveform and corresponds to a small area (the pulse limited footprint or PLF) within the BLF. As we shall see later, the diameter of the PLF is determined by the pulse duration τ (hence the name), and by the surface roughness. Note that a short value of τ is required in order to resolve surface roughness as before.

The requirement for pulse-limited operation is that the excess slant range at the edge of the beam-limited footprint ΔR should exceed the extreme (say 3 standard deviations) of the *largest* surface roughness likely to be encountered. Thus :

$$\theta \geq (6 \, \sigma_s[max] \, / \, h)^{1/2} \tag{11}$$

$$D \leq (c \, / \, 2f) \, (\, h \, / \, 6\sigma_s[max])^{1/2} \tag{12}$$

For ocean waves, $\sigma_s[max] \sim 5m$. Hence $\theta \geq 0.7deg$, and $D \leq 0.9m$.

The range measurement is made (naturally) to the nearest point on the surface, relaxing the pointing requirements. In practice, these are determined by their effect in distorting the later part of the echo profile.

The ERS-1 pulse-limited altimeter has an antenna diameter of 1.2m and a pointing stability of 0.16deg rms.

6. TRANSMIT POWER REQUIREMENT

In order to operate, the altimeter must receive a detectable echo signal relative to receiver noise. This is determined by the Signal-to-Noise Ratio (SNR) which can be estimated using the *Radar Equation* and the receiver performance characteristics.

The power per unit area received at the centre of the antenna footprint at the surface is given by :

$$F_i(R) = P_t \, G \, / \, 4\pi \, R^2 \tag{13}$$

where P_t is the transmitter power, G is the gain of the antenna and R is the range (height).

The microwave reflectivity of the surface is characterised by the backscatter coefficient σ^0, which describes the fraction of the incident radiation scattered into a notional 4π solid angle. The power backscattered from the footprint is thus :

$$F_b(R) = F_i(R) \, \sigma^0 \, A_f \tag{14}$$

where A_f is the footprint area.

The echo power per unit area received by the altimeter is :

$$F_r(R) = F_b(R) \, / \, 4\pi \, R^2$$

$$= (P_t \, G \, / \, 4\pi \, R^2) \, (\sigma^0 \, A_f) \, (1 \, / \, 4\pi \, R^2) \qquad (15)$$

The power collected by the antenna is :

$$P_r = F_r(R) \, A_e \qquad (16)$$

where A_e is the antenna effective collecting area.

Antenna theory gives :

$$G = 4\pi \, A_e \, f^2 \, / \, c^2 \qquad (17)$$

Combining equations 15, 16, and 17 and assuming no atmospheric loss :

$$P_r / P_t = \, (f^2 A_e^2 \, \sigma^0 \, A_f) \, / \, (4\pi \, c^2 \, R^4) \qquad (18)$$

For pulse-limited operation (see later) :

$$A_f = \pi h c \tau \qquad (19)$$

Thus :

$$P_r / P_t = \, (f^2 A_e^2 \, \tau \, \sigma^0) \, / \, (4ch^3) \qquad (20)$$

For ERS-1 $A_e = 1.13m^2$, $t = 3 \times 10^{-9}s$, and $P_r / P_t = 1.3 \times 10^{-15} \sigma^0$. Alternatively, we may express the power ratio as $-149 + \sigma^0$ (dB).

Over the ocean at K_u band $6 \leq \sigma^0 \leq 20dB$ with a most likely value of 11dB. Thus a typical value for the ratio of received power to transmitted power is -138dB, with a worst case (ocean) of -143dB. This is an enormous ratio, requiring that the altimeter be capable both of transmitting very high power signals and receiving extremely faint echoes.

The receiver thermal noise P_n is given by :

$$P_n = kTB = kT \, / \, \tau \qquad (21)$$

where k is Boltzmann's constant ($=1.38 \times 10^{-23}$ WK^{-1}Hz^{-1}), and T is the effective noise temperature. Adopting T = 300K (it will usually be greater) we have :

$$P_n = 1.38 \times 10^{-12} \text{ W} \qquad (22)$$

In order for the altimeter to be able to track the surface echo as the range (height) to the surface varies around the orbit, a SNR $\geq$ 5dB is necessary. We therefore require :

$$P_r \geq 4.4 \times 10^{-12} \text{W}$$

and, for the worst case over the ocean :

$$\underline{P_t \geq 870\text{W}}$$

The limit on solid state amplifiers at K_u band is $\sim$ 20W. Travelling Wave Tube Amplifiers (TWTAs) are capable of 2kW but suffer from reduced lifetime if operated at their maximum power. Thus the requirement can barely be satisfied by current technologies.

In practice, an even greater output power is required in order to provide a margin against instrument internal losses, offpointing effects, losses in rain cells and moist clouds, and to permit operation over terrain (σ^0 [min] $\sim$ -20dB!). Also, the ability to use solid state power amplifiers would be very attractive in order to achieve extended (5-10y) operational lifetimes. Thus a means of reducing the demand on transmitter power is needed.

7. PULSE COMPRESSION

The transmit power problem is overcome by using a *pulse compression technique*, which allows a long-duration, low power but high energy pulse to be transmitted in such a way that after "compression" the effect is the same as having transmitted a short-duration, high power pulse with the same energy. Several possibilities exist, but that used in past and current instruments is *linear frequency modulation (linear FM)* sometimes referred to as *"chirp"*. In this technique, a linear frequency shift (equivalent to a parabolic phase variation) is applied to the transmitted pulse such that the signal applied to the antenna is given by :

$$V(t) = V_0 \cos(\omega_0 t + at^2) \qquad (23)$$

In practice, it is difficult to generate such a signal directly at 13GHz. Either a dispersive device (such as a Surface Acoustic Wave delay line or SAW) or a digital pulse synthesis technique are used to generate an appropriate signal at a lower frequency, after which the signal is "up converted" by mixing with a higher frequency tone. Mixing results in the product of the two signals such that :

$$V_{mix}(t) = \cos(\omega_1 t) \cos(\omega_2 t)$$

$$= 0.5\cos(\omega_1+\omega_2)t + 0.5\cos(\omega_1-\omega_2)t \qquad (24)$$

The output is passed through a filter to remove the unwanted modulation. For example, the ERS-1 altimeter SAW operates at 450MHz with a bandwidth of 150MHz. Its output is mixed with a 7.35 GHz tone and is filtered to produce a 6.9GHz linear FM signal with a 150MHz bandwidth. This is then frequency multiplied by x2 to produce the "chirp" signal applied to the antenna.

The compression ratio CR of such a signal is the ratio of the true transmit duration τ' to the equivalent transmit duration τ. ie :

$$CR = \tau' / \tau = \tau' B \qquad (25)$$

For the ERS-1 altimeter $\tau' = 20\mu s$ and $B = 300MHz$. Thus, $CR = 6000$ (or 38dB). The TWTA output power is 20W. Thus, there is a respectable margin of some 21dB to allow for losses, a higher receiver effective noise temperature and for operation over low reflectivity surfaces.

Having generated a "chirp" transmitted signal, the next problem is to compress the received echo. It is difficult to construct a dispersive filter operating directly at 13.8GHz, so a technique known as *full deramping* is generally used.

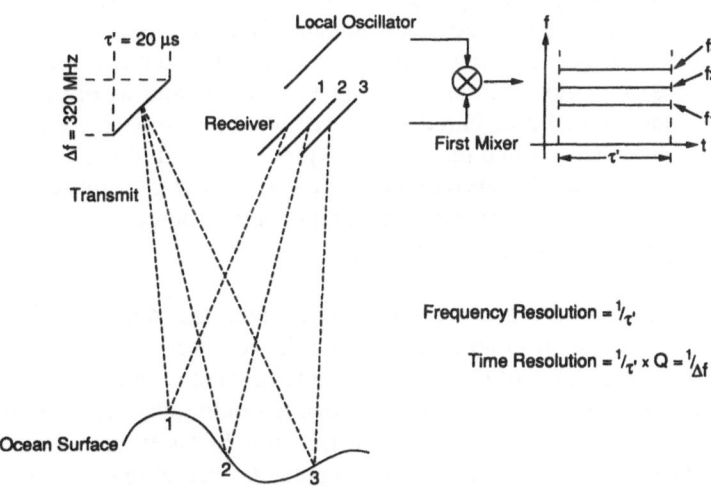

Figure 4 : The principle of Full Deramping

In radio receivers the incoming signal is usually mixed with a *local oscillator* tone to "down convert" to a lower frequency which can be processed more easily (the opposite of the "up-conversion" discussed above). The trick with full deramp processing is to mix the echo with a replica of the transmitted chirp signal rather than with a fixed frequency tone (figure 4). Signals arriving from different ranges do so with different time delays relative to the replica chirp. Each component of the echo will exhibit a fixed frequency offset Δf relative to the replica chirp as a result of their having started their frequency sweep either earlier or later. After mixing and filtering, we are left with a near-simultaneous blend of fixed tones, each corresponding to the echo signal from different ranges. The relationship between range difference ΔR and frequency difference Δf is given by :

$$\Delta f = 2Q \, \Delta R \, / \, c \qquad\qquad (26)$$

where Q is the frequency sweep rate of the chirp.

The range resolution of the altimeter is determined by the frequency resolution amongst the deramped tones. This is equal to $1 \, / \, \tau'$, giving a range resolution of $c \, / \, 2Q\tau' = c \, / \, 2B$. The spectrum analysis of the deramped signal is usually achieved by a Fourier transform method.

For the ERS-1 altimeter, B = 330MHz, and $\Delta R = 0.45$m.

A key feature of the deramp technique is that, unlike a normal radio receiver in which the local oscillator tone can be left on continuously, the replica chirp must be triggered at the instant of arrival of the echo signal from the ground. This moment must be predicted from the sequence of previous echoes, since, due to the eccentricity of the satellite orbit, the ellipsoidal shape of the Earth, and the presence of surface topography, the echo delay time changes continuously around the orbit. The prediction is fulfilled by a *range tracker* which essentially fits a simple curve to the range variation of previous echoes. Over the ocean, where the range (height) varies smoothly, the prediction can be sufficiently accurate to be used as the basic source of height data.

Note that the same on-board processor used to execute the range tracking function is generally used to control an input attenuator on the receiver to maintain the incoming signal within the linear range of the receiver and to prevent saturation. *(Automatic Gain Control or AGC)*. It may also be used to sum together individual echoes to reduce noise, and to estimate other parameters such as surface roughness from an analysis of echo pulse shape. Thus a typical block diagram of a pulse-limited altimeter using full deramp processing is as shown in figure 5.

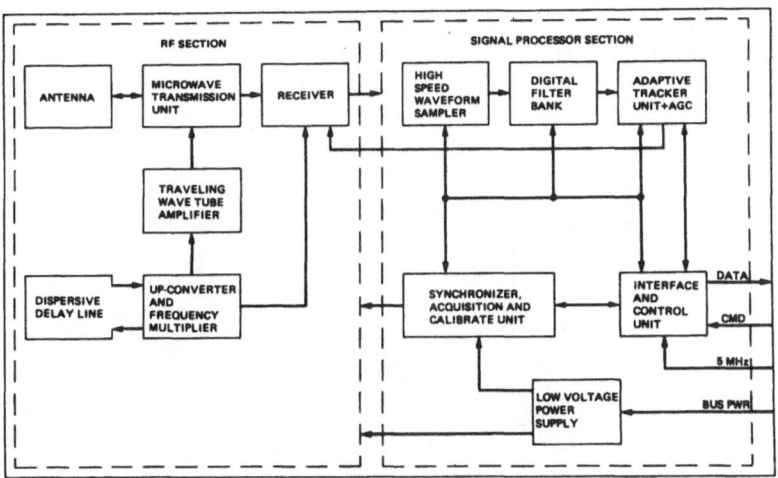

Figure 5 : A typical satellite altimeter block diagram

8. PULSE-LIMITED FOOTPRINT AND THE ECHO WAVEFORM SHAPE

Figure 6 shows schematically four stages in the development of a return echo for a pulse-limited altimeter. The expanding spherical shell of microwave energy intersects the ground firstly to illuminate an expanding disk and then, once the rear of the shell has reached the nearest point, an annulus. This results in an echo waveform which exhibits a rapid rise followed by a plateau of near constant amplitude. In fact, the plateau decays slowly due to the effect of the antenna angular response. Range (height) measurements are effectively made to the maximum radius disk or pulse limited footprint (PLF). Note that individual echoes suffer from "speckle" noise due to the coherent nature of the transmitted radiation, and that many echoes (generally 100 - 1000) are summed together to obtain a good approximation to the intrinsic profile.

Referring to figure 7, we see that at the altimeter echo power is received from the near point N from time $t = 2t_0$ to time $t = 2t_0 + \tau$ (where $t_0 = h / c$. Note that we use the effective pulse duration τ to estimate the pulse-limited footprint size). At any instant $t = (2t_0 + \Delta t)$ power is just beginning to be received from the point P' such that :

$$R'(\Delta t) = 0.5 \ (2t_0 + \Delta t) \ c$$

$$= h + 0.5 \ \Delta t \ c \tag{27}$$

Since :
$$l' = \{(R')^2 - h^2\}^{1/2} \tag{28}$$

We have :
$$A_f(\Delta t) = \pi \ h \ \Delta t \ c \tag{29}$$

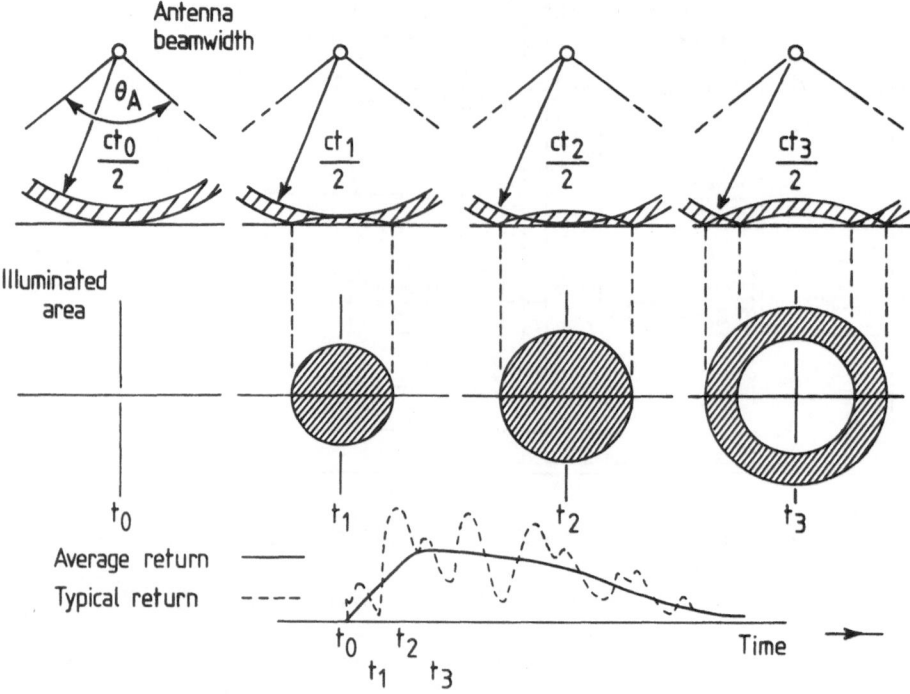

Figure 6 : Four stages in the development of the illuminated surface patch for the case of a pulse-limited altimeter. Also shown are the average return pulse waveform and a typical individual return

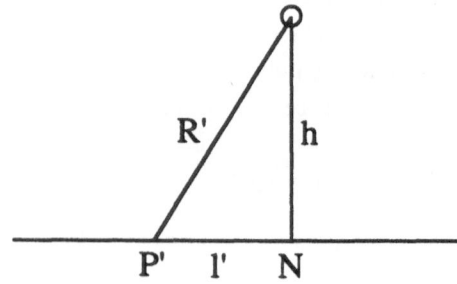

Figure 7 : The geometry for deriving the pulse-limited footprint size

where $A_f(\Delta t)$ is the area of flat surface perceived by the altimeter to be illuminated at time $(2t_0 + \Delta t)$. This area grows linearly with time, according to equation 29, until $\Delta t = \tau$, after which the disk changes to an annulus, with constant area (figure 8).

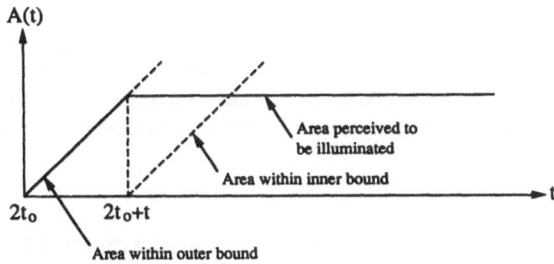

Figure 8 : Idealised echo waveform for a pulse-limited altimeter

Thus the area of the pulse-limited footprint for a flat surface without significant roughness is :

$$A_f = \pi h c \tau \qquad (30)$$

and the PLF diameter is :

$$D_f = 2(hc\tau)^{1/2} \qquad (31)$$

For the ERS-1 altimeter, $D_f = 1.7$km. (*nb* the antenna diameter required to obtain a beam-limited footprint of this size would be 10m).

The presence of surface roughness (such as ocean waves) broadens the rise time of the echo and increases the area within which surface facets appear to be illuminated simultaneously, since the crests of the waves are illuminated early whilst the troughs are illuminated late. Thus, over waves, the PLF diameter is given by :

$$D_f = 2(hc\tau'')^{1/2} \qquad (32)$$

where :
$$\tau'' = \{\tau^2 + (\ln 2 \; H_{1/3} / c)^2\}^{1/2} \qquad (33)$$

$H_{1/3}$ is the crest-to-trough height of the 1/3 largest waves. It is related to the rms waveheight σ by :

$$H_{1/3} = 4\,\sigma \qquad (34)$$

58

For the ERS-1 altimeter :

$$1.7 \leq D_f \leq 7 \text{ (km)}$$

for :
$$0 \leq H_{1/3} \leq 20 \text{ (m)}$$

(A word of caution : The surface area actually illuminated simultaneously is a factor two greater than that perceived to be illuminated simultaneously by the altimeter. This results from the increasing delay for signals returning to the altimeter from further off-axis.)

9. PARAMETER EXTRACTION

We have seen that the range resolution of a typical altimeter is ~0.5m whilst the requirement on range (height) precision is ~3cm. Furthermore, we would like to estimate the surface roughness (wave height) and radar backscatter coefficient. All of this is achieved by fitting a model of the echo waveform to the data (figure 9). Over the ocean, the radar backscatter coefficient is (inversely) related to surface wind speed. Typical examples of ocean echoes and the parameters derived from Seasat altimeter data are shown in figure 10. Wave height and wind speed validation data from the US Geosat altimeter are shown in figures 11a and 11b respectively.

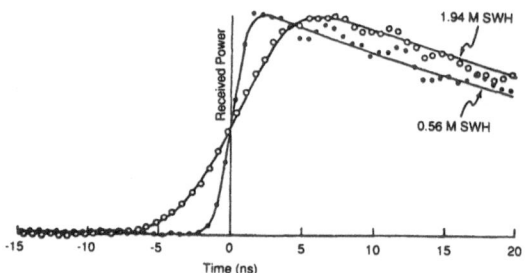

Figure 9 : Fits of an ocean waveform model to averaged echo data

10. HEIGHT CORRECTIONS

So far, we have seen how measurements of the range (height) between the altimeter and the ocean surface can be made to a precision of ~3cm. However, the scientific requirement is to know the height of the ocean surface relative to a geodetic reference surface (usually the *reference ellipsoid*, the geometric figure which provides a first approximation to the Earth's shape) to an accuracy of ~15cm. Figure 12 illustrates the problem. We shall deal with each correction in turn.

10.1 Spacecraft Centre of Gravity location

Firstly, we must know the location of the satellite's centre of gravity (cg) relative to the ellipsoid as it progresses around its orbit. Ground-based *Satellite Laser Rangers* (SLRs), *microwave rangers*, or *transponders* can provide the cg location to 3-5cm over

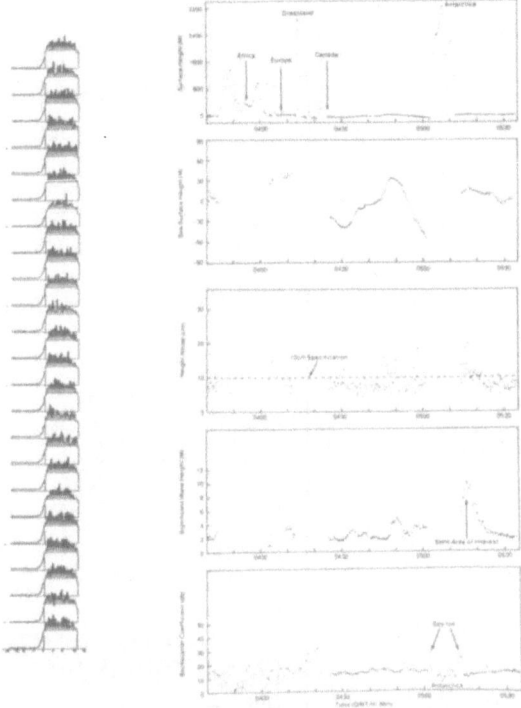

Figure 10 : Typical Seasat ocean waveforms and parameter profiles.

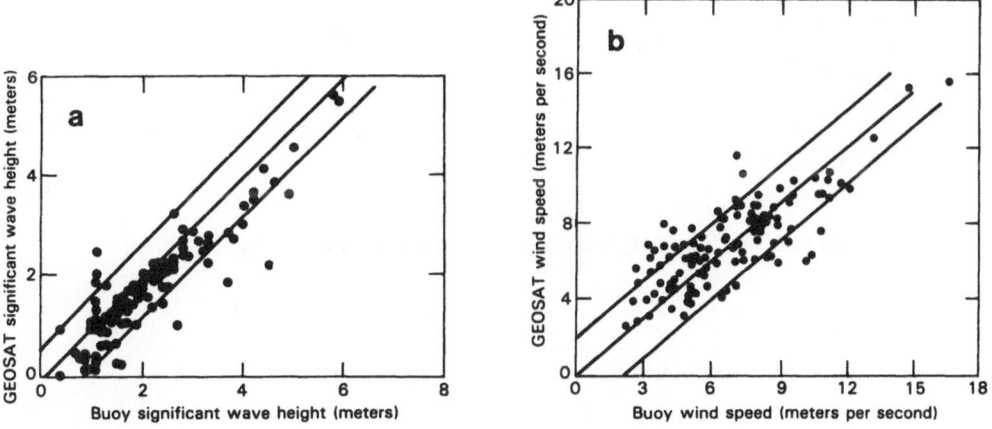

Figure 11 : Geosat wave heights (a) and wind speeds (b) compared with in situ data

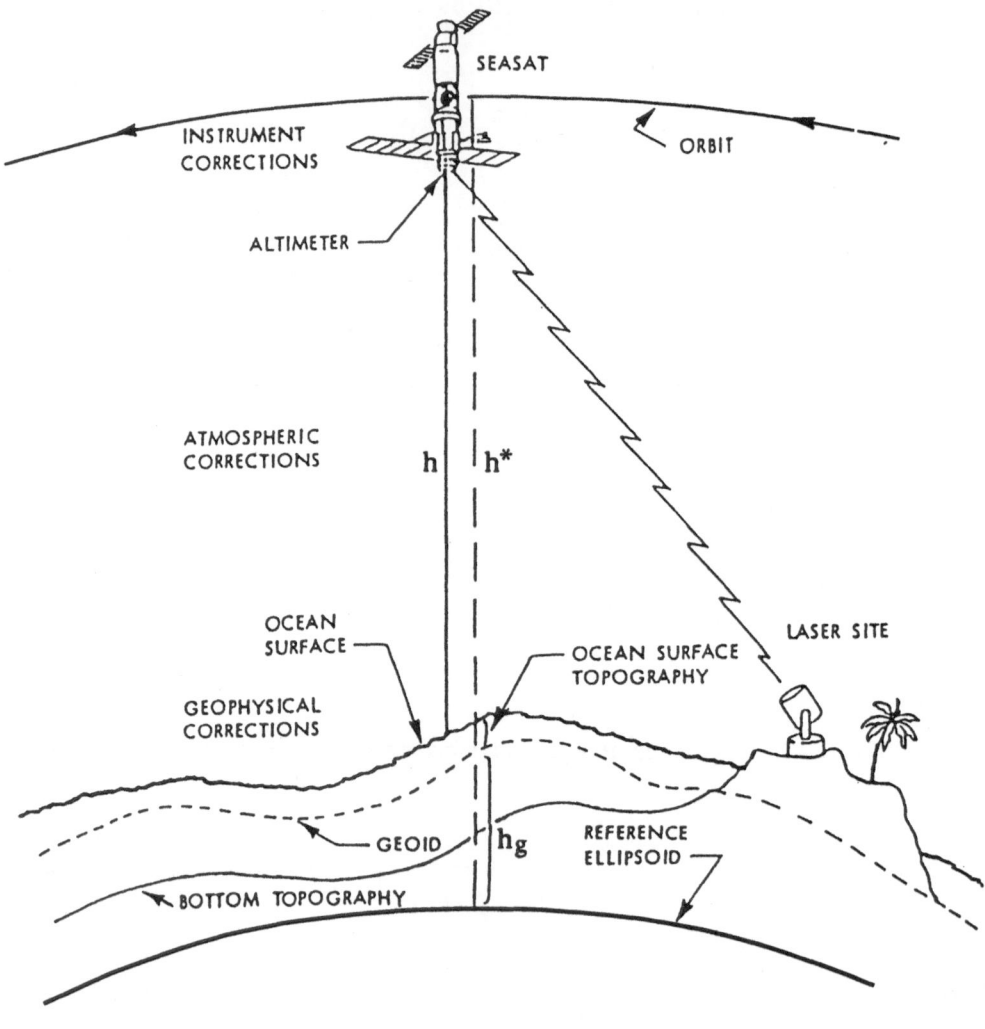

Figure 12 : The derivation of ocean surface topography

~2000km arcs. A model is then used to propagate the satellite between the measurements. The model takes into account spatial variations in the Earth's gravity field, radiation pressure (both solar radiation pressure and Earthshine), and atmospheric drag effects. It can also be used to solve for the tracking station coordinates. The very best accuracies for the orbit ephemerides currently available are ~50cm (more usually 1-2m), limited by uncertainties in our knowledge of the high spatial frequency variations in the gravity field. However, the accuracies are continually being improved as more and more satellite tracking data are incorporated into the gravity field descriptions, and the US-French TOPEX-POSEIDON mission, planned to fly in 1992, anticipates achieving the ≤15cm accuracy goal.

10.2 Instrument corrections

These include (a) the distance between the spacecraft cg and the altimeter's antenna reference plane (this can be time-dependent as manoeuvring fuel is used up or as mechanisms are operated), (b) internal timing delays within the altimeter, (c) clock frequency errors (resulting in a bias on the heights estimated), and (c) datation errors (resulting in wrongly assigned along-track locations for the height measurements).

Such errors can be dealt with piecemeal. However, a desirable precaution is to carry out simultaneous laser ranging and altimeter ranging, as shown in figure 12, so that the estimated bias in the altimeter height measurement can be verified.

10.3 Propagation corrections

The propagation velocity of electromagnetic waves depends on the *refractive index* of the medium through which they travel. Thus, a correction Δh must be applied to altimeter height estimates in order to account for the slight reduction in velocity within both the ionosphere and the neutral atmosphere.

For the ionosphere :

$$\Delta h_i = 40.3 \ N_t / f^2 \tag{35}$$

where Δh_i is in metres, and N_t is the *Total Electron Content (TEC)* per m^2 along the propagation path. N_t varies by day and night, with latitude and longitude, with season, and with solar activity, such that $10^{16} \le N_t \le 10^{19}$, giving:

$$0.002 \le \Delta h_i \le 0.2 \ (m)$$

Ionospheric models may be used to derive Δh_i to within a factor two. Alternatively, a dual frequency altimeter can take advantage of the dispersive nature of the delay to measure N_t directly.

For the neutral atmosphere, we have (to a first approximation) :

$$\Delta h_a = 2.27 \times 10^{-5} P_s + 1.723 W / T_a \qquad (36)$$

where P_s is the surface atmospheric pressure (Pa), W is the precipitable water content (kgm^{-2}), and T_a is the average temperature of the lower atmosphere (K). The two terms are often referred to as the "dry" and "wet" components of the correction. Typically, the "dry" term has a magnitude of 2m and can be estimated to within ± 0.7cm using a meteorological model, and the "wet" component has a magnitude of 6-30cm and can be measured to an accuracy of 1-2cm using data from a passive microwave sounder.

Improved corrections can be achieved if knowledge of the vertical distribution of atmospheric temperature and water vapour is available.

10.4 Surface bias

The on-board tracking equations from which the altimeter height data are generally derived assume that the ocean surface has a symmetrical height distribution of scattering facets. In practice, the height probability density function is slightly skewed (*ie* the wave crests are peaky whilst the troughs are smooth). Furthermore, the troughs are more effective at backscattering microwave energy to the altimeter. These two effects result in a bias between the true mean sea level and that derived. The net effect appears to be an overestimation of the altimeter height (underestimation of sea level) by ~0.05$H_{1/3}$. However, the effect is not yet fully understood and remains a research issue.

10.5 Tides and Atmospheric effects

Tides in the open ocean have amplitudes in the 1-2m range and may either be corrected for using a model or they may be derived from the altimeter data. Atmospheric pressure patterns cause changes in the ocean surface height (~3cm per mb) and can be corrected for using a meteorological model.

10.6 The Geoid

The *geoid* is the gravitational equipotential surface corresponding to mean sea level for a stationary, undisturbed ocean. Ocean dynamics cause the actual ocean surface to deviate from the geoid by up to a metre, and it is these signals that satellite radar altimeters are designed to measure. By placing the altimeter in a repeat orbit, the short term variations can be averaged out using data from many passes, allowing such variations to be identified relative to the mean. However, currents which are steady over the observing duration can only be derived by comparing the data with an independent geoid. These are available, but currently contain uncertainties of 1-2m over the ocean. Plans to measure the Earth's gravity field sufficiently accurately to reduce the geoid uncertainties to the decimetre level have been formulated by ESA and are under consideration. In the meantime, a variety of mathematical techniques in which the geoid and ocean dynamics signals are derived jointly from the altimeter data are under investigation.

11 CONCLUDING REMARKS

The satellite radar altimeter is a highly quantitative remote sensing instrument, capable of remarkable measurement accuracies and precisions. It has contributed greatly to our knowledge of the marine geoid, providing new information about ocean basin bathymetry and the structure of the ocean lithosphere and underlying mantle. Its ability to measure ocean dynamic signals is of great scientific interest for oceanographers and climatologists alike. The measurement of ocean currents has direct practical uses in the fields of ship routing and in the monitoring of sediment transport and pollution. Given current levels of performance, probably the most important requirement for the future is the acquisition of a continuous, self-consistent data set over many years, preferably many decades, in order to allow the detection of any long-term trends associated with climate change. Thus we look forward to ERS-1, the possibility of ERS-2, TOPEX-POSEIDON, and the series of international Polar Platforms, all of which will carry pulse-limited, ocean altimeters.

MEASURING OCEAN WAVES WITH ALTIMETERS AND SYNTHETIC APERTURE RADARS

TREVOR H. GUYMER
Institute of Oceanographic Sciences Deacon Laboratory
Brook Road
Wormley
Godalming
Surrey GU8 5UB, U.K.

ABSTRACT. A brief description of surface gravity waves, with an emphasis on those characteristics and parameters of relevance to satellite remote sensing, is followed by discussion of two techniques which have yielded important data so far - radar altimetry and synthetic aperture radar. For each sensor the basis of wave parameter estimation is presented and comparisons with conventional data are discussed. Some examples of applications are also included. Altimeters can measure significant wave height to an accuracy similar to that achievable with buoys and provide, for the first time, quantitative measurements of the spatial scales of wave fields and global climatologies. It is possible that other wave parameters, such as those representing non-linearities in waves, may also be retrievable. The use of synthetic aperture radar for wave studies is more problematic. Although very interesting high resolution imagery has been obtained, often showing wave-like features, a number of problems have to be overcome before directional wave spectra can be reliably obtained. The launch of the European Remote Sensing Satellite, ERS-1, in the early 1990s offers the opportunity to exploit these wave-measuring sensors for research and operational purposes.

1. Introduction

Before considering spaceborne sensors and their results some background information is provided on the general characteristics of ocean waves, parameters that have been found useful in describing them, and the problems with traditional measurement techniques.

1.1 GENERAL CHARACTERISTICS OF WAVES

The energy spectrum of sea surface oscillations covers a wide range of time scales. Our interest is in those periods from 0.1-30 seconds, referred to as gravity waves which are driven by the wind but have gravity as the main restoring force. The velocity of ocean waves depends on wavelength, λ_s, the phase speed being given by

$$c^2 = (g\lambda_s/2\pi)\ \tanh(2\pi h/\lambda_s) \qquad (1)$$

where h is the water depth and g is the acceleration due to gravity. In deep water (i.e. $h \geq \lambda_s/2$) $\tanh(2\pi h/\lambda_s) \sim 1$ so that

65

R. A. Vaughan (ed.), Microwave Remote Sensing for Oceanographic and Marine Weather-Forecase Models, 65–97.
© 1990 Kluwer Academic Publishers.

$$c^2_{deep} = g\lambda_s/2\pi \qquad\qquad (2)$$

From this we see that long waves travel faster, in other words they are dispersive. Wavelength is related to period (T) by

$$\lambda_s = gT^2/2\pi \qquad\qquad (3)$$

so the range of interest is from ~ 2 cm to 1300 m.

Usually a spread of periods (wavelengths) is observed which is represented as a wave spectrum (Fig. 1), where energy S(f) is plotted against frequency, f (=1/T). The action of the wind on the sea surface generates an input to the high frequency end; some of this is transferred to lower frequencies by wave-wave interactions, some is dissipated in events such as wave-breaking.

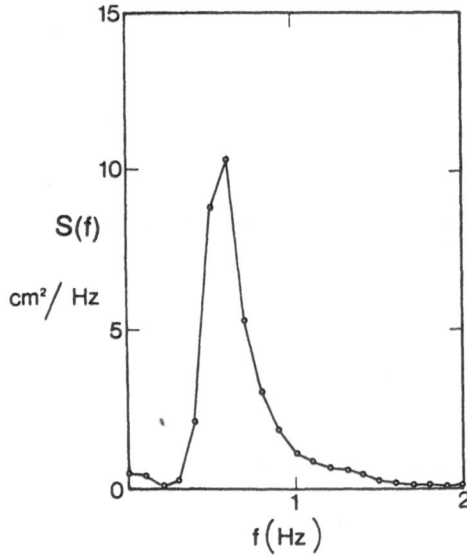

Fig. 1 Example of a one-dimensional wave spectrum.

In shallow water, where tanh () in Eq. (1) departs from unity, the propagation is constrained by the water depth; as the water shallows so the phase speed slows and refraction takes place. The most obvious example of this is when waves approaching a sloping beach at an angle are turned so that the wave crests become parallel to the shore and the wavelength is decreased.

Of more fundamental significance than the phase velocity is the group velocity, which is the speed at which a group of waves having the same frequency will travel. It is the velocity at which energy is propagated and in deep water is half the phase speed. This means that when waves are generated by a storm the energy associated with the longer wavelengths radiates away more quickly than the shorter ones. Moreover, the shorter waves suffer greater attenuation through dissipation. As a result at locations remote from the storm the first waves to arrive are the long waves which are characterised by long, nearly-parallel crests. These are swell waves and are not generated by the local wind. It is quite common for more than one swell to be observed with differences in direction both between themselves and with that of the local wind. Swell travels along great circle routes and has been tracked over thousands of kilometres. Thus a measured wave spectrum will typically have both swell and local wind-sea components. For a general discussion of ocean waves the reader is referred to Phillips (1977).

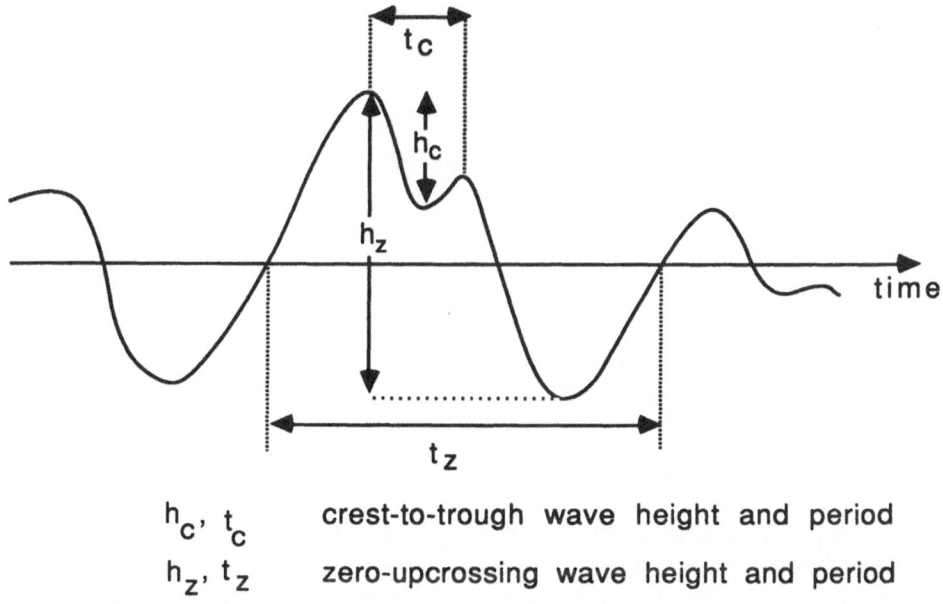

h_c, t_c crest-to-trough wave height and period

h_z, t_z zero-upcrossing wave height and period

Fig. 2 Wave periods and heights in common use.

1.2 WAVE PARAMETERS

Given a time series of wave measurements at a point useful parameters may be obtained from the statistics of surface elevation, particularly waveheight and wave period. For a simple sinusoidal wave this is easy but for a random sea the situation is more complicated and various definitions are possible. Two in common use are the crest-to-trough definition and the zero up-crossing definition (see Fig. 2) but the choice depends on the application. The mean zero up-crossing period (T_z) is the inverse of the number of up-crossings per second whereas the mean crest-to-trough period (T_c) is calculated as the average time interval between adjacent crests and troughs over the chosen period.

For waveheight a statistical measure over the period is required. A commonly used parameter is the significant waveheight, H_s, defined as:

$$H_s = 4\sqrt{m_o}$$

where m_o is the variance of sea surface elevation. (The reason for choosing this is that visual observations of waves from ships were thought to represent the mean of the highest one-third waves. By assuming a Rayleigh distribution of elevation it can be shown that

$$H_{1/3} = 4.01\sqrt{m_o} \quad)$$

m_o is also the zeroeth moment of the spectrum, where moments are defined as

$$m_n = \int \omega^n S(\omega)d\omega$$

ω being the angular frequency, $2\pi f$. The wave periods T_c and T_z can also be expressed in terms of the spectral moments

$$T_c = 2\pi\sqrt{(m_2/m_4)} \qquad (4)$$
$$T_z = 2\pi\sqrt{(m_o/m_2)} \qquad (5)$$

So far the frequency spectrum has been considered without regard to direction, θ. To fully describe the wave field it is necessary to consider the directional wave spectrum given by

$$E(\omega,\theta) = S(\omega)G(\theta)$$

where the assumption has been made that the directional spreading of the waves (G) is independent of frequency. A frequently used expression for the angular distribution is $\cos^{2r}(\theta/2)$ where r is an integer. Often directional spectra are shown as polar diagrams with frequency increasing radially outwards and with contours representing energy density.

1.3 CONVENTIONAL WAVE-MEASURING DEVICES

Although visual estimates of waves have been made for a long time from ships it is only in the last 40 years that instruments capable of deep water measurements have been used. In the early 1950s the Ship-Borne Wave Recorder was developed giving a time series of surface elevation. Improvements to this and the design of buoys such as the Waverider (in which vertical accelerations are measured) have enabled routine digital data to be recorded at many sites from which H_s, wave period and the (omni-directional) spectrum can be calculated. It is also possible to estimate the directional spectrum from instruments such as the pitch-roll buoy but the angular resolution of these is often rather poor. All of these techniques suffer the same fundamental limitations, i.e. they are point measurements (mostly in coastal regions) and sensors are vulnerable to damage from storms or passing ships.

Satellites offer the possibility of more uniform global coverage with valuable information on the spatial variations of the wave field, regardless of the weather conditions. Two types of spaceborne instruments which have yielded wave data since the 1970s are the radar altimeter and the synthetic aperture radar, both of which will be on ERS-1. However, radar measurements had previously been made from aircraft (Deacon et al. 1949; Barnett & Wilkerson 1967).

2. Radar altimeter measurements

2.1 PRINCIPLES

The idea behind the altimeter is simple. On satellites antenna size limitations restrict altimeters to pulse-limited operation in which the echo is constrained by the duration of the pulse rather than by the beam width. A short pulse of radar energy is transmitted from the spacecraft (or aeroplane) so that it is reflected from the earth's surface at normal incidence and received at the instrument (usually by the same antenna). The return power is sampled as a function of time by a number of narrow 'bins' or 'gates' which are normally about 3ns wide. (In practice it is not possible to transmit or receive pulses with such a short duration so RF techniques, such as chirp generation, must be employed; however the principle remains the same and the detailed mechanisms need not concern us here.) By accurately timing the delay between transmission and reception of the pulse the height of the spacecraft above the surface can be measured which, when combined with a knowledge of the orbit, can yield important information on the topography of the surface. The slope and amplitude of the return pulse contain information on the roughness of the surface which, over the oceans, gives the potential for extracting wave and wind parameters.

Over the oceans the echo as a function of time has a characteristic sharp rise followed by a plateau region (Fig. 3). Also shown is the way the area of ocean interacting with the pulse evolves with time. There is a flat portion before the echo is received where the only power measured is from background thermal noise. As the pulse intersects the surface the amount of power reflected increases because a larger circle on the surface is illuminated by the pulse. Once the rear surface of the pulse shell has reached the surface the reflected power reaches a maximum and in theory should remain constant as an annulus of constant area is now being illuminated; however because of antenna effects there is a slight decrease in returned power. The pulses shown in Fig. 3 are theoretical results assuming that there is no noise present; in practice this is not the case. In addition to the thermal noise mentioned above, which is small, there is also what can be described as 'fading' noise on the return signal itself. The noise on a single pulse is so large that before any analysis can be done a number of pulses must be averaged. For the present generation of altimeters pulses are emitted at a rate of one thousand per second and the echoes are normally averaged to either ten (Seasat, Geosat) or twenty (ERS-1) per second. These average waveforms are then transmitted to the ground. Before processing these averaged waveforms are usually averaged again to give one per second. For further information on radar altimetry see Robinson (1985).

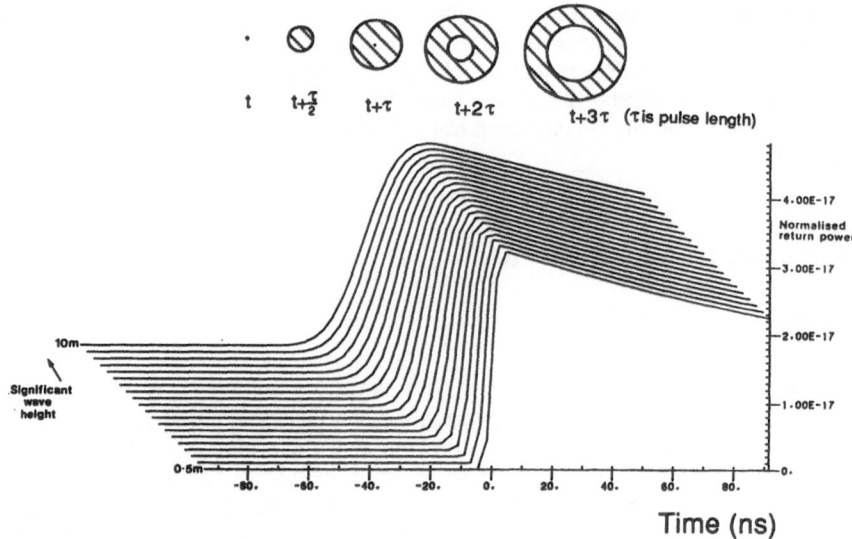

Fig. 3   Theoretical, noise-free altimeter returns for various values of
         significant waveheight, H_s. Also shown is the evolution of the
         illuminated area of sea surface.

2.2 EXTRACTION OF WAVEHEIGHT

Wave parameters are estimated from the radar altimeter data by fitting a model of the return
from the sea surface to the measured return and adjusting the parameters to obtain the
"best" fit (in some sense). It is well known that the altimeter works by specular reflection
of the radar pulse from the sea surface and that the return P_r can be expressed
mathematically as the convolution of three terms (Brown, 1977); thus

$$P_r = P_{FS} \otimes P_{PT} \otimes P_{spec}$$

where P_{FS} is the flat surface response,
         P_{PT} is the point target response,
and P_{spec} is the statistical distribution of specular reflectors on the surface of the
                    waves.

P_{PT} is assumed to be Gaussian and usually the form of P_{FS} specified by Brown (1977) is
used. However, these assumptions do not appear to be critical. P_{spec} is dependent on the
statistics of the waves on the sea surface and therefore the model of the return is dependent
on the assumptions that are made about these statistics.

If we assume that the statistics of the waves are Gaussian then the statistics of the specular reflectors are also Gaussian, with the same variance as the sea surface elevation (Barrick & Lipa, 1985; Srokosz, 1986b). The probability density function of specular points then depends only on m_0, and hence, on H_s. This means that the return may be expressed as

$$P_r = P_r (\sigma^\circ, H_s, t_0)$$

where σ° is the backscattered power,
and t_0 is the travel time of the pulse, which gives the height of the satellite above the sea surface.

Thus we can see that the primary wave parameter influencing the radar return is H_s and this is the fundamental wave parameter that can be derived from the return. The main effect of increased waveheight is to flatten out the leading edge of the return pulse (see Fig. 3) which corresponds to the pulse interacting earlier with the crests and later with the troughs than in the calm sea case.

Generally radar altimeters process the return signal to estimate σ°, H_s and t_0 in real-time using on-board algorithms known as trackers. The tracker consists of up to three loops; to keep the pulse centred in the range window, to adjust the gain control and to measure the slope of the pulse. Trackers to date have been based on the assumption that the statistics of the waves at the sea surface are Gaussian (for the sake of simplicity). Seasat, Geosat and Topex use (or will use) a weighted split-gate algorithm to analyse the return. ERS-1 and Poseidon use Sub-optimal Maximum Likelihood Estimation (SMLE) which gives minimum variance, unbiased estimates of the parameters and should allow for more accurate on-board estimation of the parameters. A short review of proposed algorithms up to and including Seasat is given in Guymer et al. (1985). Unfortunately many of these algorithms are poorly documented in the literature. Griffiths et al (1987) show how the SMLE on ERS-1 will be used to derive the geophysical parameters σ°, h and H_s from the outputs of the loops.

2.3 VALIDATION

The first space-borne altimeter was flown on Skylab in 1973. This instrument was very inaccurate and although height could be measured to a precision of about one metre it was not possible to produce useful estimates of significant wave height. Following this a more accurate instrument was flown on GEOS-3. Although not as accurate as the present generation of altimeters it was possible to derive significant wave height and compare these results with *in situ* data. The results showed agreement to ±0.5 m for $4 < H_s < 8$ m but with slightly worse results (± 0.75 m) for calmer seas. Unfortunately the satellite did not have any on-board recording and thus data are limited to those obtained within line of sight of a receiving station. A number of mobile receiving stations were available and were moved to different locations during the mission. The data are therefore 'patchy' in both space and time and they are too few to allow much analysis.

The next altimeter to fly was on Seasat. This has been, to date, the most important mission both in terms of algorithm development and exploitation of the data. Seasat lasted only one hundred days. From 1979 until 1985, when Geosat was launched, there was no available altimeter (GEOS-3 was turned off after three years, at about the same time as Seasat's demise). This meant that all available effort has been concentrated on the hundred days of Seasat data and almost all the results presented below have been derived from this dataset. The most comprehensive comparison with *in situ* data was that by Fedor & Brown (1982) who used 51 overpasses of buoys to show that altimeter waveheights were larger by 0.07 ± 0.29 m for $0.5 < H_s < 5.5$ m (Fig. 4). When the statistics were calculated separately for each platform some showed much poorer performance than others, the best results being obtained using a Waverider. In a more limited comparison Webb (1981) found agreement of ±0.12 m with a pitch-roll buoy deployed from a ship.

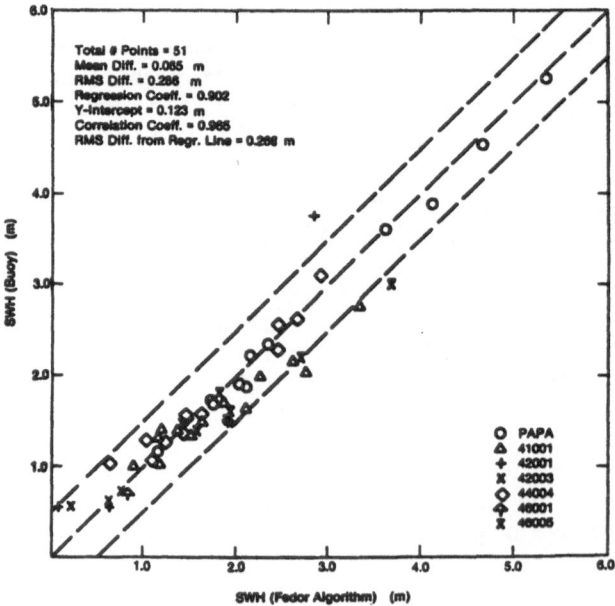

Fig. 4 Comparison of Seasat altimeter waveheight with 51 buoy measurements (Fedor & Brown, 1982).

At present an altimeter, very similar to that flown on Seasat, is flying on Geosat. Geosat is a U.S. military satellite whose only sensor is the altimeter and data from the first eighteen months of its mission Geosat data were classified. For the remainder of the mission, starting in November 1986, Geosat has been in a seventeen day repeat orbit and the data are freely available. A comparison with wave buoys (Dobson et al., 1987) showed that it could measure to ±0.5 m up to 6 m waveheight. It has been concluded that altimeters can measure H_s to better than 0.5 m or 10% (whichever is the greater) and this is a similar accuracy to that obtainable from devices.

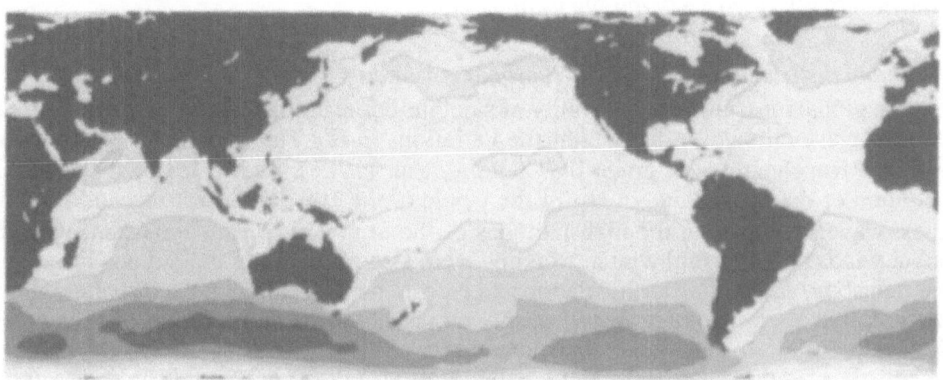

Fig. 5   Global distribution of mean significant waveheight measured by
Seasat during July - September, 1978 (Chelton et al., 1981).
Contours are at 0.8 m intervals, darkest shade corresponding to
H_S > 5.3 m, lightest to H_S <1.3 m.

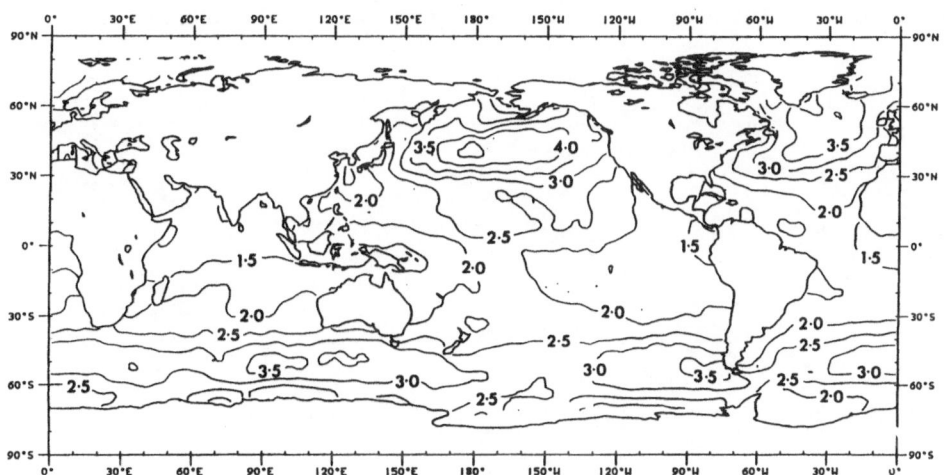

Fig. 6   Global distribution of mean significant waveheight (in metres)
measured by Geosat during January - March, 1987 (Challenor et
al., 1989).

2.4 APPLICATIONS OF ALTIMETER WAVEHEIGHTS

2.4.1 *Global distribution of H_S and its seasonal variation*

Given the global measurement capability of satellite-borne altimeters it is possible to study the global wave climate; in particular, the variations in H_S. The first such description of significant waveheight was given by Chelton et al. (1981), who calculated the global distribution of the mean value of H_S for the whole of the 100 day Seasat mission (Fig. 5). Highest waves occurred in the high latitudes of the Southern Ocean, consistent with the stronger winds of the austral winter. Maxima were also observed in the Indian Ocean, due to the monsoon, and in the North Atlantic and North Pacific. Although the distribution was broadly similar to climatologies based mainly on visual estimates the values of waveheight in the Southern Ocean were higher than anticipated.

Recently Carter et al. (1988) and Challenor et al. (1988) have analysed one year of Geosat data to look at global monthly and annual means of H_S in the period November 1986 to November 1987. For the July-September period results were very similar to Seasat's. Fig. 6 shows the distribution for January - March when wave conditions in the Northern and Southern Hemispheres are similar. This suggests that there may be less seasonal variation in the latter - a conclusion borne out by Carter et al.'s analysis in which a model of annual and semi-annual variations was fitted to altimeter data in 2° boxes for the whole of the ocean. The significance of the annual component is plotted in Fig. 7.

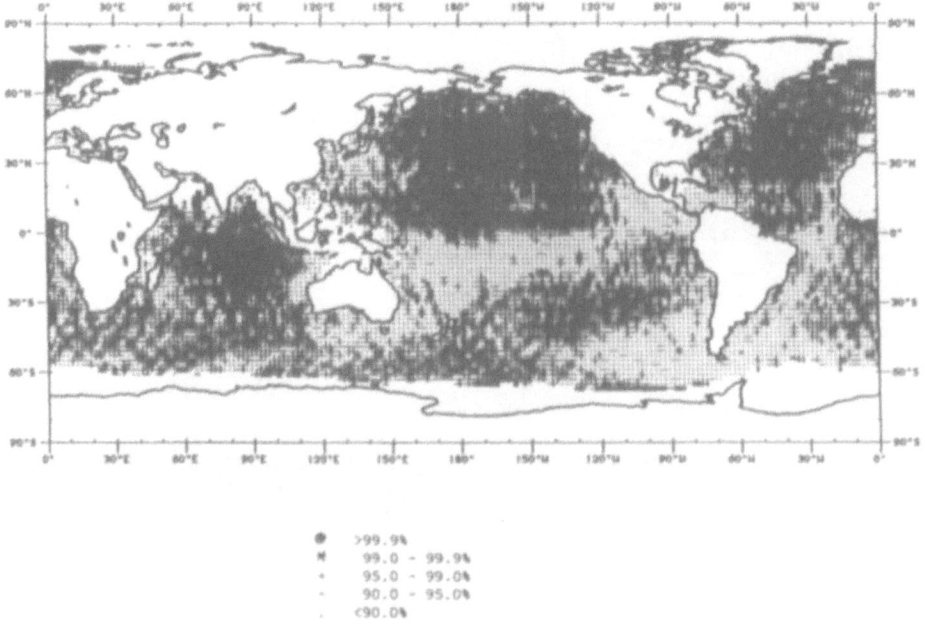

●	>99.9%
◄	99.0 - 99.9%
·	95.0 - 99.0%
·	90.0 - 95.0%
.	<90.0%

Fig. 7   Significance of annual component in Geosat waveheights
(Carter et al., 1989).

These studies show the potential for mapping the global wave climate using satellite altimeter data especially in data sparse regions such as the Southern Ocean. In the future these studies could be extended to look at the variability of H_s, on monthly, annual and interannual time-scales, so that its variance around the mean values computed to-date may be examined. This will become feasible as more altimeter data is acquired over the next few years, including ERS-1.

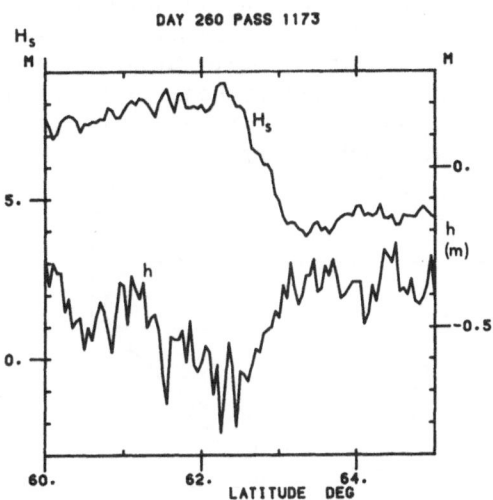

Fig. 8   An example of significant along-track variations in H_s. Also shown is a corresponding change in inferred sea level (h) due to the sea-state bias effect (Guymer & Srokosz, 1986).

2.4.2 The spatial variability of H_s

Another use of the altimeter measurements is to examine the spatial variability of H_s. This cannot be done from conventional *in situ* measurements of waves, which are generally time series measured at a single point. A few studies (Challenor, 1983; Challenor et al., 1986; Monaldo, 1988) of the spatial variability of H_s have been carried out. These show a large-scale (of the order of 1000 km) variation in H_s, associated with wave generation by storms over the oceans, as well as shorter-scale variability. One of the more interesting observations made with Seasat data is that H_s can change very rapidly in the open ocean over very short distances. Queffeulou (1983) and Guymer & Srokosz (1986) document examples where H_s changes by several metres over 50-150 km (Fig. 8). One reason for looking at the spatial variability of H_s is to enable the data acquired along the satellite tracks to be interpolated between the tracks. This may be of importance for practical applications, such as ship routeing, and also for the assimilation of the data into wave models. As most wave models have a grid spacing of the order of 200 km, or greater, these relatively rapid changes in H_s would not be resolved. These results also have implications for validation of altimeter waveheights. Data from an *in situ* measuring device on the other side of such a H_s "jump" from the satellite measurement, would not be useful for validation purposes (usually *in situ* data taken within about 100 km of the satellite track are considered good for validation purposes).

The limitation on the use of altimeter data for this type of study is the size of the footprint (typically about 7 km) which means that only variations on scales greater than that of the footprint can be examined. Nevertheless, this still enables interesting studies to be performed. Carter (pers. comm.) has examined the along-track variation of Geosat waveheights near SW England (Fig. 9). Apart from the obvious problem of land contamination and the differences arising from the fact that different tracks were sampled on different days the main feature to note is the reduction in H_s to the east of the Scillies. This is probably due to sheltering by the islands of the Atlantic swell.

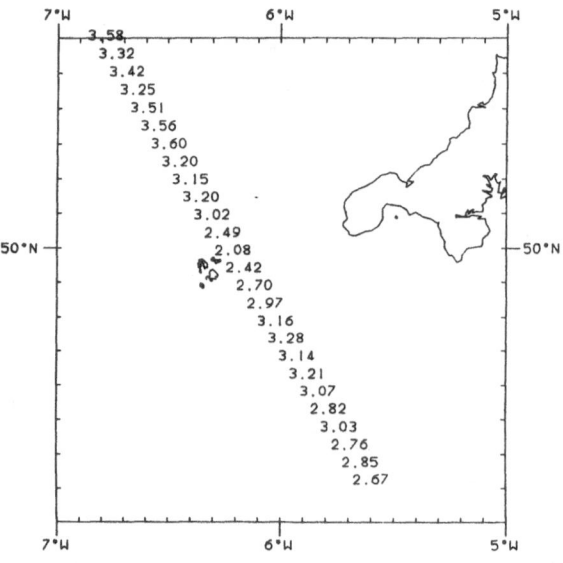

Fig. 9 Reduction of H_s in the lee of the Scilly Isles (Carter, pers. comm.).

2.4.3. *Wave climate statistics*

For some purposes there is considerable interest in deriving wave statistics from long series of records. Examples of this are the 50-year return H_s, that is the value likely to be exceeded only once in 50 years, and the chances of obtaining wave conditions calmer than a certain value for a specified period of time. These are of value in the design and deployment of coastal and off-shore structures, particularly when it is borne in mind that a 1 m increase in the estimate of 50-year H_s implies an increase in cost of £1M for building an off-shore platform.

If it is required to establish the wave climatology of a particular area of sea the present procedure is as follows. A wave measuring device is deployed in the region of interest. This is usually a buoy, such as a Waverider, but if a suitable vessel, such as a light ship or weather ship, is available a ship borne wave recorder can be used. Measurements are then taken every three hours or so (the sampling interval is sometimes reduced to one and a half hours). The deployment is maintained for at least a year, preferably longer. The data set obtained has good temporal sampling, between eight and sixteen values per day, not all of which are statistically independent, but is only relevant to a single point in space. To estimate the 50-year return H_s an attempt is made to fit particular types of probability distributions to the data and an extrapolation is carried out. Clearly the longer the time series the greater the confidence in the estimates. However, there is an additional problem. Inferences are usually required for locations other than the point at which the measurements are made so an assumption of spatial homogeneity must be made. This assumption could be tested by deploying more buoys in the region simultaneously which would be expensive and with greater chance of losing a buoy.

How might radar altimeter data be used? In contrast to a series of measurements taken at a single point the altimeter data are distributed uniformly along the sub-satellite track at a spacing of approximately seven kilometres (assuming that the data have been averaged over one second) and so the spatial sampling is therefore very good. The temporal sampling, on the other hand, is much worse than is normally achieved with buoys. At best the repeat period is once every three days so even if the selected site is on a 'cross-over' point of ascending and descending passes the maximum sampling rate is two observations every three days (the ascending and descending tracks are only separated by about eight hours so this is not equivalent to one observation every one and a half days). Indeed due to pressure from other disciplines, e.g. the measurement of ocean currents, geophysics etc., the orbit repeat period is likely to be much longer than three days, possibly seventeen or even thirty-five days. Therefore, although the altimeter produces a vast amount of wave data, (in a single day the altimeter collects the same amount of data as a Waverider recording every three hours would in twenty years of operation!) at a single point the sampling can be as poor as once every month. However, the sampling over the region as a whole can be very good. In order to utilise these data we need to be able to convert data in the space domain into their equivalents in the time domain. The necessary research has yet to be done, mainly because an altimeter has yet to be flown in a short (~ 3-day) repeat for a complete annual cycle. It is to be hoped that ERS-1 will provide the data needed to attack this pressing problem.

A further use of altimeter data in this area will be to monitor trends in H_s at various locations. The techniques referred to above assume stationarity of wave statistics over the period in question. Recently, evidence has been produced (Carter & Draper, 1988) of a possible increase in H_s off SW England during the past 25 years which could have serious implications for off-shore design criteria. Analysis of data from GEOS-3 up to ERS-1 and beyond may help substantiate this result and show the spatial extent of the increase.

2.4.4 *Joint distributions of wind and waveheight*
In designing offshore structures engineers are interested in the loads imposed by both waves and winds on the structures. One area of interest is the probability of obtaining a given waveheight and wind speed simultaneously. There is little suitable *in situ* data available to be able to provide an answer to this question. Since the power of the altimeter return can be used to derive an estimate of wind speed (Chelton & McCabe, 1985) it is

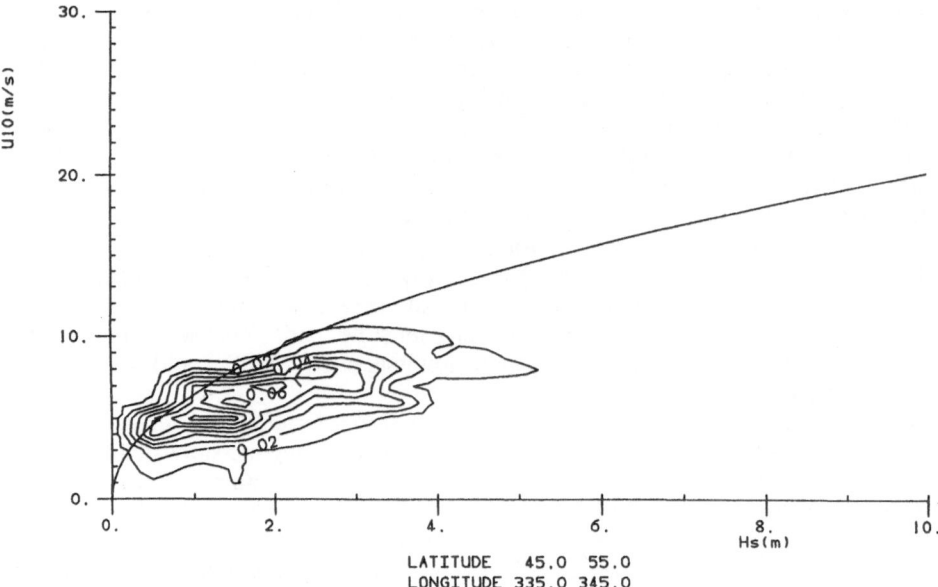

Fig. 10 Joint probability density function of Seasat wind speed and waveheight for an area in the North Atlantic (Challenor & Srokosz, 1989).

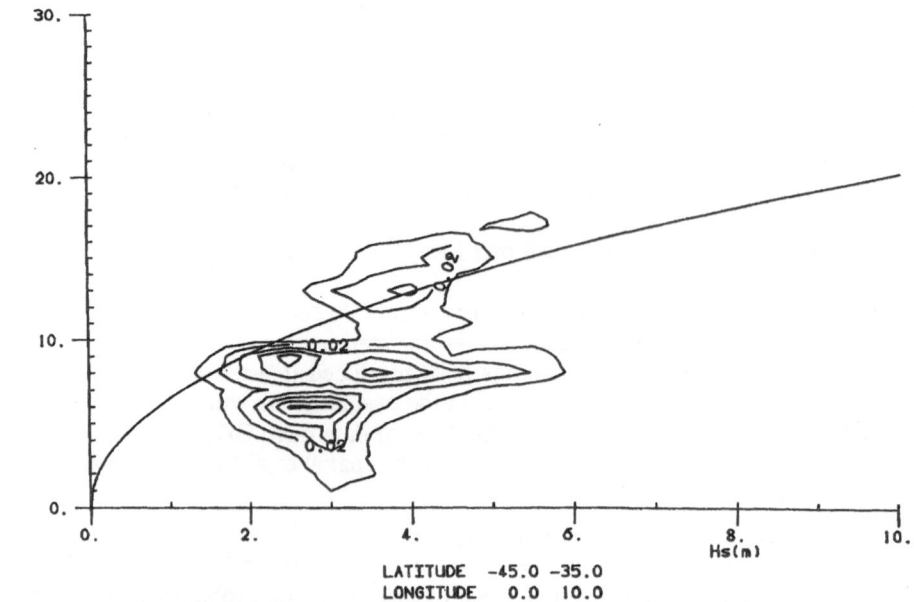

Fig. 11 Joint probability density function of Seasat wind speed and waveheight for an area in the South Atlantic (Challenor & Srokosz, 1989).

possible to obtain coincident measurements of H_s and wind speed U_{10} (here we reference the wind speed to a height 10 m above the sea surface).

Challenor & Srokosz (1988) carried out a preliminary study of the joint distribution of H_s and U_{10} using Seasat data covering the whole period of the mission. Five 10° squares in different regions of the world's ocean were chosen for study and from all the available data a joint probability density functions (pdf) for H_s and U_{10} were calculated. Figs. 10 and 11 are examples for the North and South Atlantic. Plots such as these can be used to obtain the most probable waveheights associated with a given wind speed.

Superimposed on Figs. 10 and 11 is the Pierson-Moskowitz (PM) relationship between H_s and U_{10} for a so-called "fully developed sea" (Pierson & Moskowitz, 1964).

$$H_{\text{fully developed}} = 0.0247\, U_{10}^2$$

which has been derived from PM's original formulation but with winds referred to a height of 10 m instead of 19.5 m. This curve gives a limit on the height of the waves that can be generated by the local wind. As most of the data have waveheights higher than this limit, for any given wind speed, this shows the predominance of swell (non-locally generated waves) in the measurements. In order to compute the heights of these swell waves we simply take the difference between the squares of the fully developed sea and the measured H_s, i.e.

$$H_{\text{swell}}^2 = H_s^2 - H_{\text{fully developed}}^2$$

This value gives the minimum amount of swell present since it is likely that the local sea will not be fully developed. Using $H_{\text{fully developed}}$ from before and substituting this into the above equation gives

$$H_{\text{swell}}^2 = H_s^2 - 6.20 \times 10^{-4}\, U_{10}^4$$

Mognard (1984) proposes the use of this equation to predict minimum swell. Unfortunately because it depends upon U_{10}^4 errors in U_{10} can be critical for the estimate of swell. This is illustrated in Guymer et al. (1985) where it is shown that by using winds from the passive microwave radiometer on Seasat as opposed to those estimated from the altimeter the estimate of minimum swell can be reduced from over 4 m to 0. Parsons (1979) use a different measure of swell, the ratio between H_s and $H_{\text{fully developed}}$. He does not estimate the minimum swell but only produces a flag in situations where he believes the wave field is dominated by swell.

2.5 POSSIBLE EXTRACTION OF OTHER WAVE PARAMETERS

In addition to significant wave height and wind speed there are also a number of other geophysical parameters that can be calculated from the altimeter echo. The descriptions given here will be quite brief and for further information the reader is referred to Guymer et al. (1985). All but one of these additional wave parameters are related to the estimation of mean square slope of the sea surface. This can be estimated from σ^0 through the simple equation

$$s^2 = |R_0|^2 / \sigma^0$$

where $|R_0|^2$ is the Fresnel reflection coefficient of the sea surface and σ^0 is measured in linear units not dB.

The mean square slope can be related to the frequency spectrum of the sea surface by using the dispersion relation for surface gravity waves in deep water (Eq. 2) and by making assumptions about the angular distribution of the energy it is possible to show that s^2 is proportional to m_4, which we remember is the fourth spectral moment. We can therefore estimate m_4 from the radar backscatter provided that we are prepared to assume a value for the empirical coefficient which appears in the directional distribution. From experiment it is known that this has a value of approximately 10. Given that the altimeter data can therefore yield estimates of two moments of the spectrum, m_0 and m_4, it is possible, by analogy with Eqs. (5) and (6) to conceive of a new wave period, denoted as T_A.

$$T_A = 2\pi \left(\frac{m_0}{m_4} \right)^{\frac{1}{4}}$$

This period is not one of the usual wave periods T_z or T_c (the zero-upcrossing period and the crest period) but rather their geometric mean. Fig. 12 is a plot produced by Challenor & Srokosz (1988) which compares T_A from the Seasat altimeter with nearly-coincident T_z calculated from Waverider data and provides some encouragement that the altimeter responds to wave period. Coincident estimates of H_s and a wave period would be useful for wave power studies and also in refining the partitioning of waves into wind-sea and swell components.

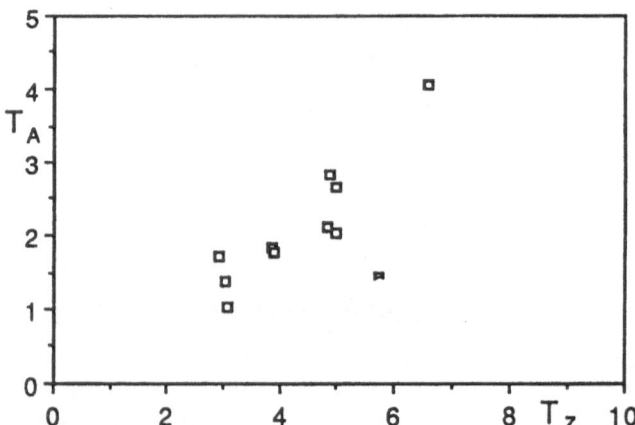

Fig. 12   Comparison of the altimeter wave period (T_A) derived from Seasat data with mean zero up-crossing period (T_z) obtained from a Waverider.

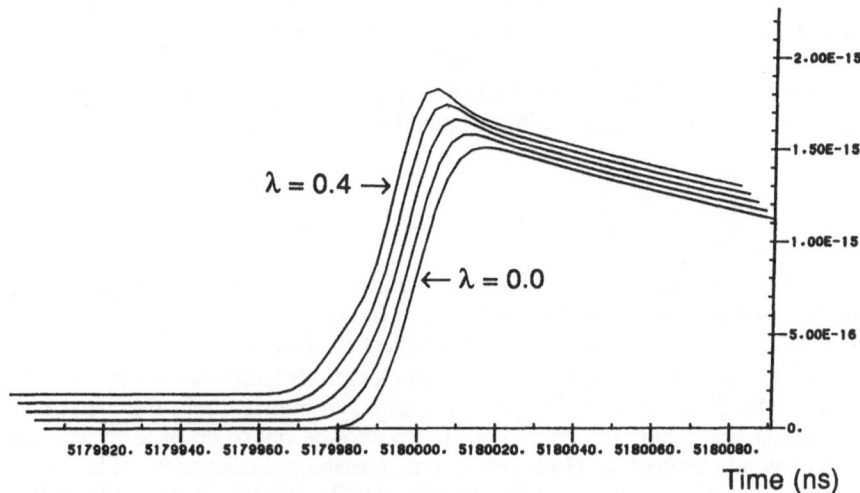

Fig. 13   The effect of sea surface skewness (λ) on altimeter pulse shape. For clarity pulses are shifted to the left with increasing λ.

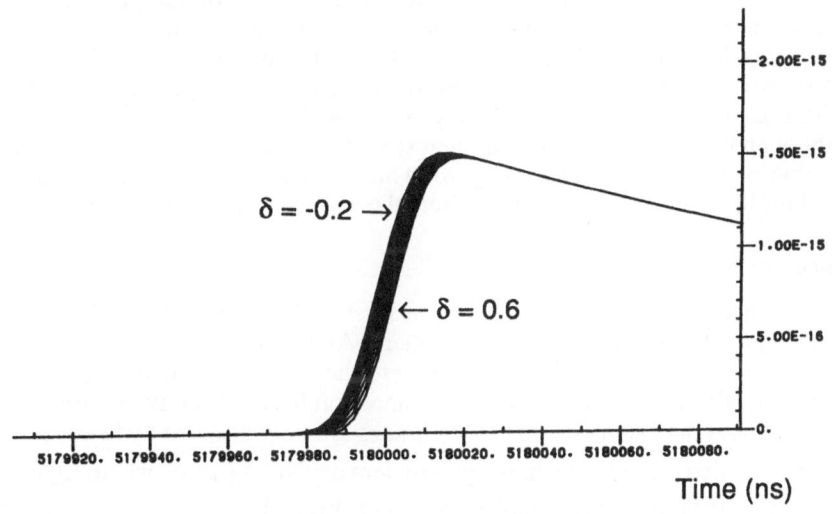

Fig. 14   The effect of the cross-skewness parameter (δ) on altimeter returns. The pulses are not off-set.

The other parameter which we can calculate using m_4 is the percentage of waves breaking. The theory behind this estimate is given in Srokosz (1986a). If we define a wave as breaking when the downward acceleration at the crest is greater than αg, where α is a constant and g is the acceleration due to gravity (Srokosz (1986a) suggests putting $\alpha = 0.4$) then the percentage of waves breaking is given by

$$\exp\left\{ -\frac{\alpha^2 g^2}{2m_4} \right\}$$

No attempt has been made to data to calculate this parameter from existing altimeter data. If it could be obtained on a global basis it would be relevant, for example, to studies of the dissipation of waves and to transfers across the air-sea interface.

So far it has been assumed that sea surface elevations follow a Gaussian distribution. If this were so the distribution of heights would be symmetric about the mean level. In the case of a linear sea (where the motion of the sea surface can be considered as a sum of sine curves) this would be correct. However, in practice, the sea surface is not linear. The crests are 'sharper' than a sine curve and the troughs are 'flatter'; this is particularly the case for steep waves. The distribution of the heights of specular points for a slightly non-linear sea has been derived by Srokosz (1986b). Once this distribution has been obtained it can be substituted into the three-fold convolution described in 2.2 and hence a 'Brown' model can be derived as shown by Srokosz (1986b). The important difference with the linear case is that two new parameters, λ and δ, are included. The effect on the pulse shape of varying these parameters is shown in Figs. 13 and 14. λ is the coefficient of skewness for the sea surface elevation, the physical meaning of δ is rather more complex. This δ is a form of 'cross-skewness' coefficient but has no readily available physical interpretation. These non-linear parameters can be estimated by maximum likelihood.

Although these parameters are of interest in their own right their main application is the estimation of sea state bias (see Fig. 8). This is an important effect which needs to be allowed for in using an altimeter to measure the surface slopes arising from ocean currents.

2.6 PROBLEMS

Although microwave sensors are regarded as having all-weather capability they are affected to some degree by the intervening atmosphere. Attenuation of echo strength is caused by cloud droplets and by rain and is more severe the higher the radar frequency. However, theoretical work by Barrick & Lipa (1985) and a preliminary study by Srokosz & Guymer (1988) of Seasat data suggests that H_s estimates are only affected by heavy rain (> 15 mm hr[-1]). The problem is more acute for parameters depending on $\sigma°$ where light to moderate rain (~ 5 mm hr[-1]) may cause significant errors. Fig. 15 is an example of the variation of H_s and $\sigma°$ through a rain area, as identified from the Seasat SMMR.

Caution is also required when using H_s values at high latitudes where sea-ice might be present. Even small amounts can be sufficient to distort the pulse shape (because of highly specular returns) leading to erroneous estimates.

Another problem is due to mispointing of the antenna. This is a particular problem with Geosat which has no attitude control system. Deviations from nadir pointing of several tenths of a degree are quite common and this appears to have a greater effect on σ^0 estimates than on H_s. Occasionally the mispointing becomes large enough for the altimeter to lose lock; ~ 5 - 10 % of the Geosat data have been lost in this way.

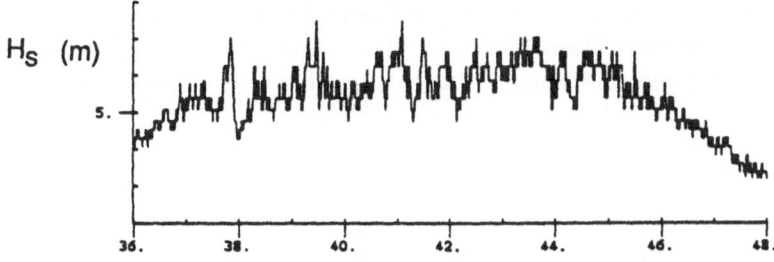

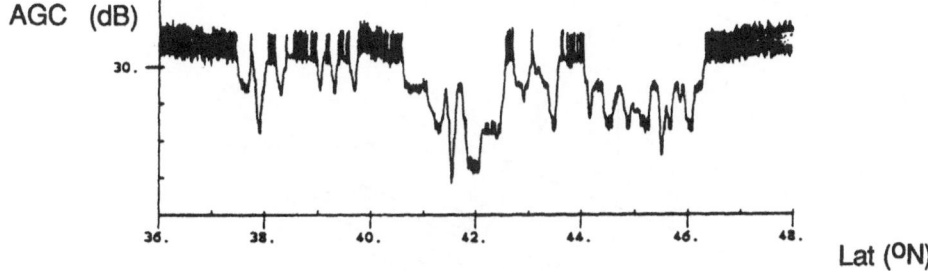

Lat (°N)

Fig. 15 Variation of H_s (upper) and Automatic Gain Control (related to σ^0) (lower) as the altimeter track crossed a rain area.

3. Synthetic aperture radar measurements

3.1 PRINCIPLES

Synthetic aperture radars (SAR) have been flown on three civilian space missions - Seasat (1978) and two Shuttles (1981 and 1984). A SAR image is basically a high resolution map of the radar reflectance (backscatter) properties of the ocean surface modified by surface motion effects. Since the dielectric properties of the ocean surface are relatively uniform, variations in reflectance are due primarily to variations in surface roughness.

Synthetic aperture radars use the varying Doppler shift of the reflected signal introduced by motion of the satellite to separate the return signals from different points on the surface, when these points are closely spaced in the along-track, or azimuth, direction (Fig. 16). Separating targets in the across-track, or range direction, is by means of their range difference in the swath being mapped which must therefore be off to one side of the satellite. Range measurements can be very precise, to a few metres, using techniques similar to those in altimetry, though the number of measurements to be made is here much greater. Use of the Doppler shift in the along-track direction gives much greater resolution than is implied by the usual diffraction limit which is related to the physical size of the radar antenna or 'aperture'. In order to achieve a resolution of a few metres a real aperture would need to be 10 km or so in length, which is clearly impracticable. By summing the coherent radar returns over 1 to 2 seconds an aperture of the desired length can be 'synthesised' with an antenna only 10 m long (remembering that the spacecraft has a speed of ~ 7 km s^{-1}.

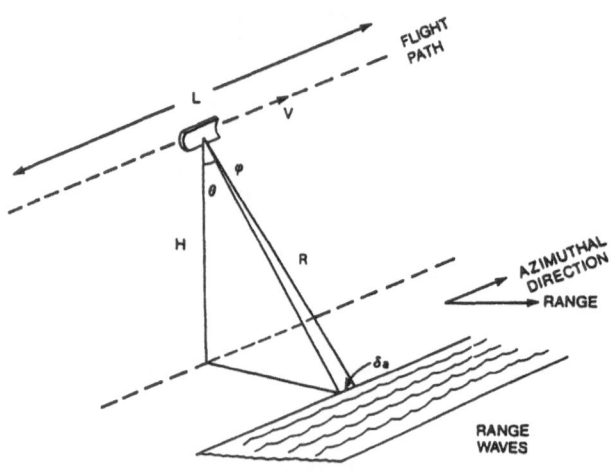

Fig. 16   Geometry of Synthetic Aperture Radar operation showing
range-travelling and azimuthal-travelling waves.

The synthetic aperture method brings the along-track resolution down to a few metres, comparable to the achievable range resolution. Both these resolutions are independent of range (sensitivity permitting) and allow extremely precise radar mapping of the Earth's surface from space. Radar images suffer from the speckle effects peculiar to coherent imaging systems, which cause strong variations in the signal observed at any pixel. In practice, resolution can be degraded to suppress this, though this will cause the loss of some small scale features. For Seasat the along-track resolution limit was increased from 6 to 25 metres to allow smoothing to "4 look" imagery. This simplified the processing, and gave a match to the 7 metre range resolution, implying 20 metre ground resolution when projected from an incidence angle of 20°.

Satellite SAR's can, in principle, provide images of the sea surface along continuous swaths of the order of 100 km width at a resolution of about 20 m. At present a limitation is imposed by the high power requirement of the instrument, which in the case of Seasat, for example, restricted the operating time to 10 minutes per orbit. A similar restriction will be imposed on the operation of the ERS-1 SAR. The high bit rate (about 100 Mbits per second) of the instrument precludes storage of the data on board a satellite, so that images can be obtained only within line of sight of a receiving station. This limited the coverage of the Seasat SAR and will similarly affect future free-flying satellites including ERS-1 when being operated in full imaging mode. For further details on the principles behind SAR see Robinson (1985).

The intensity of the SAR return signal at any pixel is a measure of the radar reflectivity or cross section, which over the ocean is strongly affected by the wind speed. SAR pulses are directed obliquely at the surface and for a mirror-like calm sea all the energy is scattered away from the radar beam. Consequently these pixels will appear dark. Rough seas appear bright as some backscatter takes place along the beam. However, one of the main interests in SAR for oceanography lies in the patterns of image intensity on scales of hundreds of metres to hundreds of kilometres and how these relate to oceanic phenomena. Before considering how waves are imaged it is useful to consider the effects of a moving target on SAR data.

3.2 SAR IMAGING OF MOVING FEATURES

A SAR processor employs the change of phase with time of a point target, that is, its Doppler frequency, to locate its azimuthal position in the image plane. If the target is moving, however, it will provide its own Doppler offset and as a result, the processor will misposition the target on the image plane.

It can be shown (e.g. Robinson, 1985) that the spatial mispositioning in the azimuthal direction is given by

$$\Delta x = (R/V) \, u_r \tag{6}$$

where R and V are the altitude and velocity of the satellite and u_r is the component of the target velocity resolved along the direction of the radar beam.

For Seasat (and similarly for ERS-1) the R/V ratio ~ 130 s so that a target (say, a ship) moving at a radial speed of 5 ms^{-1} (10 knots) would be displaced in the image plane by 650 m. This effect is clearly observable in many Seasat SAR images as a displacement of a moving ship from its (relatively) stationary wake. Obviously any motion of the sea surface itself, including the orbital motions associated with waves, could lead to an inferred spatial shift.

3.3 IMAGING OF WAVES

It is generally (but not universally) accepted that SAR pulses are Bragg scattered by ripples at the sea surface. The condition for constructive interference of the back-scattered radar waves is

$$2\lambda_s \sin\theta_i = n\lambda_R$$

where λ_s and λ_R are sea surface and radar wavelengths respectively and θ_i is the incidence angle measured from the vertical. Thus for $\lambda_R \sim 23$ cm (Seasat) then $\lambda_s \sim 30$ cm, assuming first order scattering, i.e. n=1. For the C-band SAR on ERS-1 then $\lambda_R \sim 6$cm and $\lambda_s \sim$ 9cm.

Imaging of the larger waves must involve modulation of those surface ripples. Three mechanisms have been proposed for this modulation and are shown on Fig. 17 :

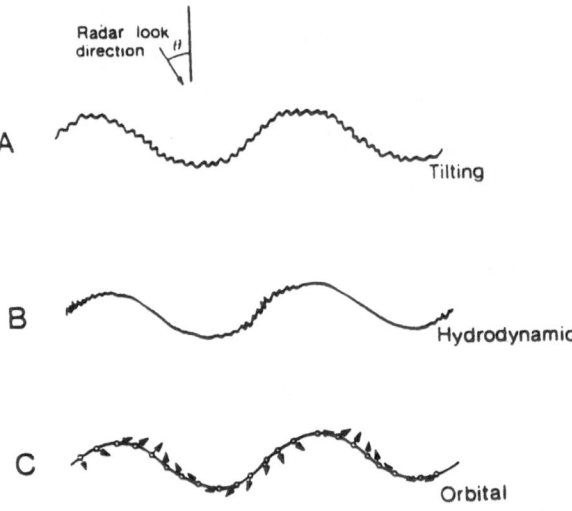

Fig. 17 Three effects which are believed to play a part in the
interaction of SAR with a moving sea surface.

a. The long waves could tilt the ripples as shown so that the effective angle of incidence changes with respect to target plane (facet) thus modifying the radar return by changing the Bragg back-scattering coefficient.

b. Hydrodynamic interactions between the scattering ripples and longer gravity waves could produce a concentration of ripples on the face of the waves which would modulate the energy returned.

c. The facet parameters could be changed by orbital motions during the time the wave remains in the beam. Thus, the facets across the wave profile would move in and out of phase with the radar look direction.

There is still a degree of controversy surrounding the dominant mechanism. Consideration of the effect of the orbital motion will show that one part of a wave should be displaced in one direction while another part is displaced in the opposite direction since the radial velocity of the facets changes sign across the wave profile. From the relationship between wavelength λ_s and wave frequency f (Eq. 3, remembering that frequency is 1/T).

$$\lambda_s = g / (2\pi\ f^2)$$

where g is the acceleration due to gravity. A wave of wavelength 200m has a wave frequency of $0.09s^{-1}$ (or a wave period of 11.1s). Its orbital velocity is given by:

$$u_r = \pi\ H\ f \tag{7}$$

where H is the waveheight. Then for H ~ 5m, u_r ~ 1.4 m s^{-1}. With the Seasat (ERS-1) R/V ratio of 130 s then one part of the wave could be displaced by 180m in one direction with another part shifted 180 m in the other direction. Such large displacements (referred to as velocity bunching) could lead to smearing so severe as to destroy the image. However, if the waves were travelling across-track (range waves) the displacement would be along the direction of the wave train so that the dominant wavelengths might be retained in the image. This would not, of course, be true for azimuthal travelling waves where some distortion should be expected.

Beal (1987) has pointed out that because of the smaller R/V ratio for shuttle imaging radar, the distortion is reduced. He argues that the higher-flying satellites such as Seasat and ERS-1 could fail to image North Atlantic storms which would be correctly imaged by Shuttle. His argument follows the line:

If the crest of a wave is displaced in the image plane by $+\lambda_s/4$ (while the trough is displaced $-\lambda_s/4$) then cancellation will occur. To avoid destructive interference (from Eqs. 6 and 7)

$$R\ \pi\ H\ f\ /\ V < \lambda_s\ /4$$

For a fully developed spectrum the significant waveheight, $H_s = 0.04\ T^2$ so that

$$\lambda_{min}\ > (R\ 4\pi\ H_s\ /\ V)(0.04/H_s)^{1/2}$$

$$> (2R/V)\ H_s^{1/2} \qquad \text{(in S.I. units)} \qquad (8)$$

(or in terms of period
$$T_{min}\ > (4\pi/g)^{1/2}\ (R/V)^{1/2}\ H_s^{1/4}\ )$$

For Seasat and ERS-1 R/V~ 130 s, then for H_s ~ 5 m

$$\lambda_{min}\quad \sim 550\ m$$

A graphical representation of Beal's argument is shown in Fig. 18. From this discussion a cut-off in the azimuthal direction of Seasat SAR spectra might be anticipated at

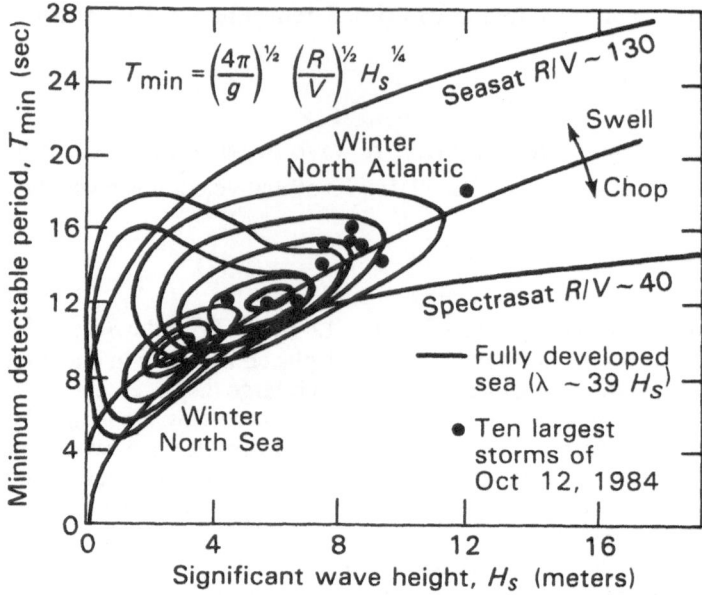

Fig. 18 Joint probability densities of H_s and dominant wave period (contoured at the 0.1,0.3,0.5,0.7 levels) for the North Sea and North Atlantic in winter, superimposed on predicted SAR azimuth limits for both a low (Spectrasat) and high (Seasat) orbit (Beal, 1987).

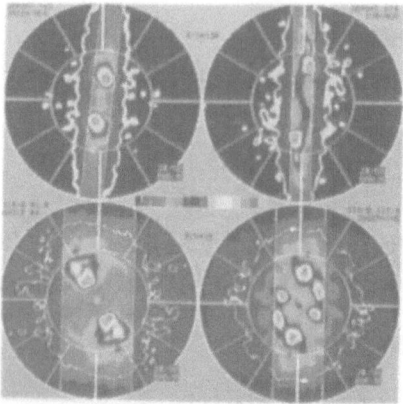

Fig. 19 Examples of SAR wavenumber spectra showing a cut-off in the azimuthal direction (horizontal). Top pair: Seasat, bottom pair: SIR-B. Azimuthal wavenumber limits calculated from Eq. 8 are shown by the pairs of vertical parallel lines (Beal, 1987).

high frequencies and one example is given in Fig. 19. It should be noted that the discussion has centred on the frequency and directional distribution of image intensity (which we refer to as the directional image spectrum) and not on the ability to recover amplitude information which is necessary to derive the directional wave spectrum. Thomas (1982) and Vesecky et al. (1986) have made some attempts to retrieve waveheight.

3.4 VALIDATION OF SAR WAVE MEASUREMENTS

Due to the coverage of the Seasat and Shuttle missions and the brevity of the latter the scope of SAR wave validation is more limited than with the altimeter. Moreover, even after ten years not all of the Seasat data has been digitally processed or analysed because no real-time SAR processor has been available. The tremendous amount of data produced by SAR, together with the current lack of an efficient processor means that it has been difficult to analyse the information in a systematic fashion and studies have tended to be ad hoc in nature.

The first point to note in considering SAR imaging of waves is that the fundamental limitation is the resolution of the SAR system. The Seasat SAR and SIR-A had identical resolution in range and azimuth, being respectively 25 m x 25 m and 40 m x 40 m. In contrast, SIR-B had a resolution of 25 m in azimuth and 17-58 m in range, the latter being dependent on the look angle of the radar, which was variable. The effect of the limited resolution is to restrict the imaging of surface waves to waves of wavelength greater than twice the resolution of the SAR. This means that for the SARs flown to-date only waves of wavelength greater than about 50 m could be imaged. Additionally, the azimuthal distortion effects, discussed in 3.3, tend to limit the ability of the SAR to image shorter waves (that is, roughly those of wavelength less than 100m equivalent to a period of ~ 8s). This means that information on a significant portion of the wave spectrum cannot be obtained from the SAR images and we are restricted to considering the "swell" part of the spectrum. Clearly, this has implications for any studies of waves using SAR data.

A second point that needs to be noted is that there is evidence to suggest that, above a certain wind speed, SAR cannot image waves at all (Allan & Guymer, 1984). Such effects may be dependent on various factors, including the effect of spray and wave breaking, and are not well understood. The implication for waves studies using SAR is that there may be occasions when no wave information can be retrieved from the image, even though waves are present on the sea surface. Possible examples of this type of phenomenon may be seen in a SIR-A image of the western Mediterranean (Fig. 51 in Ford et al., 1983) and in a Seasat image taken during the JASIN experiment (Fig. 24 in Allan & Guymer, 1984). The results of Vesecky & Stewart (1982) suggest that waves will also not be imaged if the significant waveheight is less than 1.4m.

With these provisos some encouraging results have nevertheless been obtained. Vesecky & Stewart (1982) compared the dominant wavelength and direction inferred from Seasat SAR images with those derived from buoy data. SAR wavelengths were biased high by about 12% and directions agreed to within 15° (Figs. 20 & 21). Sometimes two swells are imaged as in Fig. 22. On this occasion the two directions were corroborated by visual observations from a nearby ship. Regarding extraction of H_s, Thomas (1982) obtained agreement to 20%, based on only two comparisons with buoys; Vesecky et al. (1986) made 9 comparisons which exhibited a scatter of ± 1m.

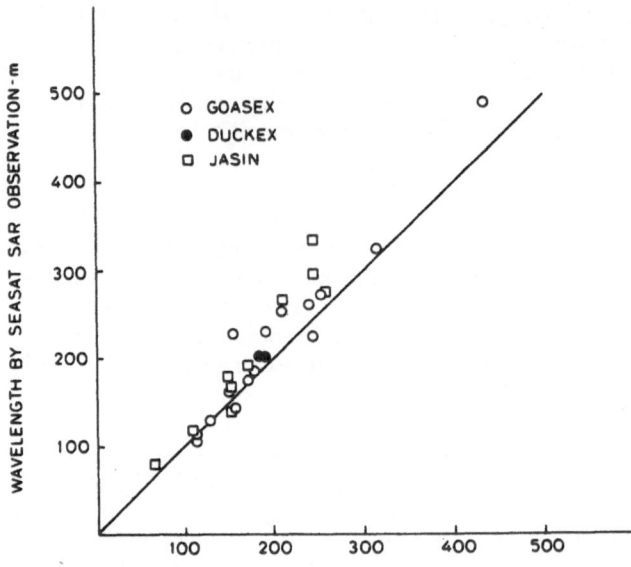

Fig. 20 Comparison of dominant wavelength from SAR with that measured by a pitch-roll buoy (Vesecky & Stewart, 1982).

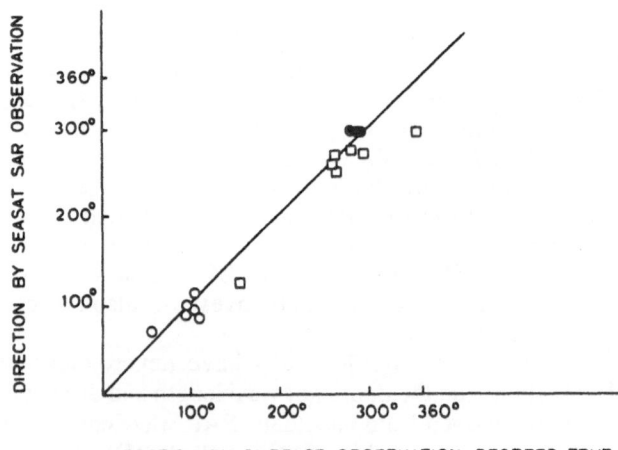

Fig. 21 Comparison of dominant direction from SAR with that measured by a pitch-roll buoy (Vesecky & Stewart, 1982).

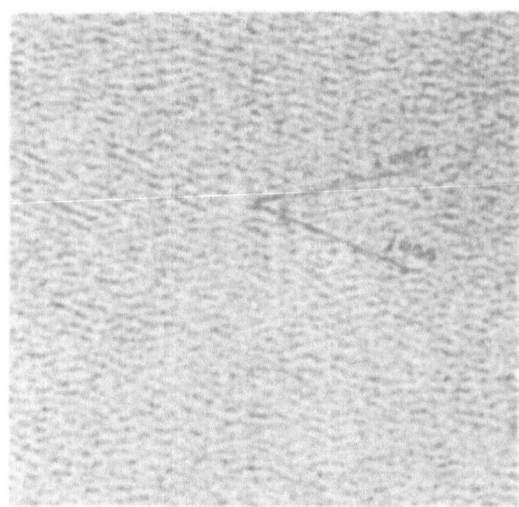

Fig. 22   Example of two swells observed by Seasat SAR in the
NE Atlantic on Rev. 762 (Allan & Guymer, 1984).

3.5 APPLICATIONS

A number of studies have been made of the spatial variation of the 'wave' spectrum as
given by SAR. Two types are described briefly below: wave refraction and the estimation
of the source and age of hurricane-generated swell.

3.5.1 *Wave refraction*
As discussed in 1.1 the propagation of waves generated in the deep ocean is modified as
they reach shallow water and refraction of the wavefronts takes place. A good example of
this is seen in Fig. 23 in which an Atlantic swell is approaching the Portuguese coast.
There are several other examples, particularly off the north coast of Scotland, in which
wave diffraction due to headlands is also evident. It is possible to model the change of
direction and wavelength in terms of the water depth if the direction and wavelength of the
waves in deep water is known. Schuchman et al. (1981) calculated the modification for an
actual case of swell approaching Cape Hatteras. The open sea values of dominant
wavelength and direction were taken from a SAR image and the depths were obtained from
a bathymetric chart. Predicted values were then compared with the wavelengths and
directions observed by the SAR at a number of locations (Fig. 24). Agreement was
generally good despite the assumptions involved in accepting the depths as correct and that
no azimuthal distortion effects were allowed for. In addition to swell SAR sometimes
images surface roughness manifestations of submerged sandbanks (Kenyon, 1983). By
analysing sequences of images over the same area it may prove possible to obtain
information on uncharted bathymetric features and to monitor any temporal changes due to
the effects of major storms.

Fig. 23   Seasat SAR image (Rev. 785) of swell undergoing
refraction as it approached the Portuguese coast.

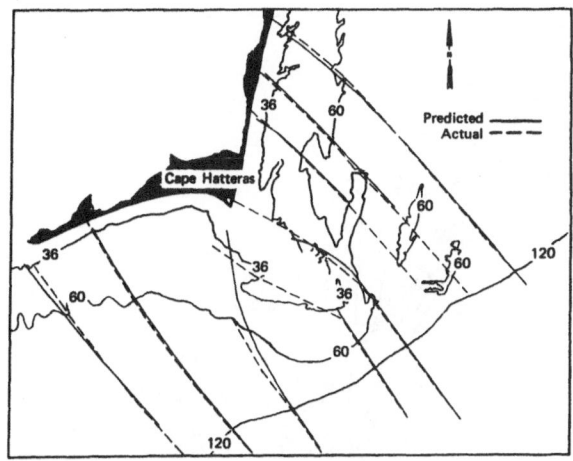

Fig. 24   Wave fronts in shallow water derived from SAR imagery
compared with predictions from a wave refraction model
(Schuchman & Kasischke, 1981).

Wave fields also experience changes as they encounter strong currents such as coastal tidal streams. However, even in deep water such effects may be noticeable and Beal et al. (1986) found changes in SAR-derived wavelength as waves propagated across a mesoscale eddy and across the Gulf Stream. One of the reasons for the study of wave-current interactions is the cause of unexpectedly high waves in some regions such as the Agulhas Current.

3.5.2 *Swell generation*

As mentioned earlier swell waves from intense storms may propagate rapidly away from their generation regions as relatively simple wave trains. Gonzales et al. (1987) made use of this in a case study of hurricane-generated waves. Analysis of the variation of directional spectra along a SAR track showed systematic variations in wavelength and direction which were postulated to be due to the swell at each location having been generated at different points in the storm. From the dominant wavelength the group speed of the waves was calculated and by tracing back along the direction of propagation the path of each wave train could be established as a function of time. In order to find out the location of the source region a further piece of information was used - the position of the storm centre at various times. By making an assumption about the inflow angle of the surface wind it was possible to identify the points which satisfied all the constraints necessary to be a source region and hence determine the age of the swell reaching the SAR swath. Fig. 25 shows the results for one value of inflow angle. The oldest waves (7-9 hr) occurred at each end of the 600 km swath, travelling towards the NE at the southern end and towards the NW (i.e. parallel to the track) in the north. Waves observed in the middle of the strip were 2-3 hr old with generation regions lying only 200 km from the hurricane centre. Such analyses may yield new information on wave generation, particularly when combined with satellite winds.

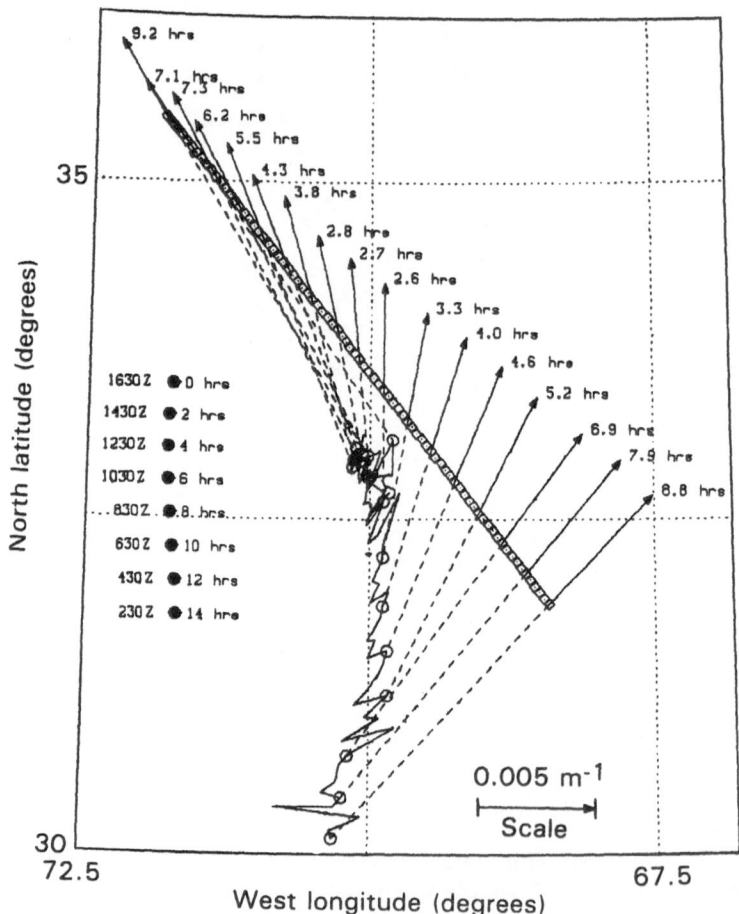

Fig. 25 Inferred source regions (open circles) and wave age (at head of arrows) calculated from a hurricane swell kinematic model with an inflow wind angle of 20° (Gonzales et al., 1987).

4. Concluding remarks

This survey of wave data from satellite missions has shown that valuable information can be obtained from radar altimeters and high resolution imaging radars. The parameter in which we have most confidence is significant waveheight from the altimeter since it is a measurement whose theoretical basis is understood and which can be easily validated against conventional data.The retrieval of other wave parameters from the altimeter is more speculative and new techniques may have to be developed to ensure proper validation. The greatest disadvantage of the altimeter is the narrowness of its swath and resulting limitations in sampling. The swath of a synthetic aperture radar is about ten times broader and potentially gives information on wavelength and direction. However, there are many problems to be overcome:-

(1) Its high data rate and power consumption preclude global coverage in a high resolution mode.
(2) There is uncertainty in wave imaging mechanisms. Sometimes waves are not observed at all even though significant wave activity is clearly present. The shorter wind waves (< 50 m) are never resolved so an important portion of the spectrum is omitted.
(3) Even when waves are imaged there may be significant distortion for certain combinations of viewing geometries and environmental conditions. Whether the full wave spectrum can ever be recovered in such circumstances is a subject of debate.
(4) The ERS-1 Active Microwave Instrument, of which the SAR is one component, will operate at C-band. Since no spaceborne C-band SAR has been flown before there is uncertainty over the conditions under which useful oceanographic information will be obtained.

On ERS-1 an attempt is being made to overcome (1) by having the AMI operating in wave mode, in which it will be possible to image a 5 km x 5 km scene every 200-300 km along the track, storing the data via the onboard tape recorder. Thus coverage will not be constrained by the position dedicated SAR receiving stations and will be quasi-global. Wave models may help to overcome sampling limitations and correct image spectra to wave spectra and schemes for assimilating satellite wave data into models are being developed (e.g. Thomas, 1987). An advantage of ERS-1 over Seasat is that some data products will be disseminated in real-time which allows the possibility of incorporating them in coupled wind-wave numerical forecasting models, such as that developed at the European Centre for Medium Range Weather Forecasts (Hasselmann, 1985).

5. Acknowledgements

I am grateful to Peter Challenor and David Carter (both at the Institute of Oceanographic Sciences Deacon Laboratory), Meric Srokosz (Remote Sensing Applications Development Unit, British National Space Centre) and Tom Allan (now of Satellite Observing Systems, Ltd.) for providing much of the material on which this paper is based.

6. References

Allan T.D. & Guymer T.H. 1984 SEASAT measurements of wind and waves on selected passes over JASIN. Int. J. Remote Sensing, 5, 379-408.
Barnett T.P & Wilkerson J.C. 1967 On the generation of ocean wind waves as inferred from airborne radar measurements of fetch-limited spectra. J. Mar. Res., 25, 292-321.
Barrick D.E. & Lipa B.J. 1985 Analysis and interpretation of altimeter sea echo. Adv. Geophys., 27, 60-99.

Beal R.C. 1987 Spectrasat: a hybrid ROWS/SAR approach to monitor waves from space. Johns Hopkins APL Tech. Digest, **8**(1), 107-115.

Beal R.C., Gerling T.W., Irvine D.E., Monaldo F.M., and Tilley D.G. 1986 Spatial variations of ocean wave directional spectra from the Seasat Synthetic Aperture Radar. J. Geophys. Res., **91**, 2433-2449.

Brown G.S. 1977 The average impulse response of a rough surface and its applications. IEEE Trans. Ant. Propag., **AP-25**, 67-74.

Carter D.J.T. & Draper L. 1988 Has the north-east Atlantic become rougher? Nature, **332**, 494.

Carter D.J.T., Foale S. and Webb D.J. 1989 Variations in global wave climate throughout the year. Proc. of Seasat 10th Anniversary Meeting, London, June 1988 (to be published).

Challenor P.G. 1983 Spatial variation of significant wave-height. pp451-460 in *Satellite microwave remote sensing* (ed. T.D. Allan), Ellis Horwood.

Challenor P.G., Guymer T.H. and Srokosz M.A. 1986 The influence of spatial and temporal scales on calibration/validation. pp 11-16 in Proc. Workshop on ERS-1 Wind and Wave Calibration, ESA SP-262.

Challenor P.G., Foale S. and Webb D.J. 1989 Seasonal changes in the global wave climate measured by the Geosat altimeter. Int. J. Remote Sensing (in press).

Challenor P.G. & Srokosz M.A. 1989 Wave studies with the radar altimeter. Proc. of Seasat 10th Anniversary Meeting, London, June 1988 (to be published).

Chelton D.B., Hussey K.J. and Parke M.E. 1981 Global satellite measurements of water vapour, wind speed and waveheight. Nature, **294**, 529-532.

Chelton D.B. & McCabe P.J. 1985 A review of satellite altimeter measurements of sea surface wind speed: with a proposed new algorithm. J. Geophys. Res., **90**, 4707-4720.

Deacon G.E.R., Darbyshire J. and Smith N.D. 1949 Use of the airborne sea and swell recorder to measure changes in the wave spectrum from west to east across the Irish Sea. Admiralty Research Laboratory Report ARL/R1/103-18/W.

Dobson E.B., Monaldo F., Goldhirsh J. and Wilkerson J. 1987 Validation of Geosat altimeter derived wind speeds and significant wave heights using buoy data. J. Geophys. Res., **92**, 10,719-10,732.

Fedor L.S.& Brown G.S. 1982 Waveheight and windspeed measurements from the Seasat radar altimeter. J. Geophys. Res.,**87**, 3254-3260.

Ford J.P., Cimino J.B. and Elachi C. 1983 Space Shuttle Columbia views the world with imaging radar: The SIR-A experiment. JPL Publication 82-95, Jet Propulsion Laboratory, 179pp.

Griffiths H.D., Wingham D.J., Challenor P.G., Guymer T.H. and Srokosz M.A. 1987 A study of mode switching and fast-delivery product algorithms for the ERS-1 altimeter. ESA Contract Rep. 6375/85/NL/BI, 204pp.

Gonzalez F.I., Holt B.M. and Tilley D.G. 1987 The age and source of ocean swell observed in hurricane Josephine. Johns Hopkins APL Technical Digest, **8**, 94-99.

Guymer T.H., Challenor P.G., Srokosz M.A., Rapley C.G., Queffeulou P., Carter D.J.T., Griffiths H.D., McIntyre N.F., Scott R.F. and Tabor A.R. 1984 A study of ERS-1 radar altimeter data processing requirements. ESA Contract Report 5681/83/NL/BI.

Guymer T.H. & Srokosz M.A. 1986 The determination of sea-state bias and non-linear wave parameters from satellite altimeter data. pp115-120 in Proc. EARSeL Symp. on Europe from Space, ESA SP-258.

Hasselmann K. 1985 Assimilation of microwave data in atmospheric and wave models. pp47-52 in Proc. of a Conf. on the Use of Satellite Data in Climate Models, ESA SP-244.

Kenyon N.H. Tidal current bedforms investigated by SEASAT. pp261-270 in *Satellite microwave remote sensing* (ed. T.D. Allan), Ellis Horwood.

Mognard N. 1984 Ocean wave parameter extraction using satellite short-pulse radar altimeters. pp37-41 in Proc. Workshop on ERS-1 Radar Altimeter Data Products, ESA SP-221.

Monaldo F. 1988 Expected differences between buoy and radar altimeter estimates of wind speed and significant wave height and their implications on buoy-altimeter comparisons. J. Geophys. Res., **84**, 3979-3986.

Phillips O.M. 1977 *The dynamics of the upper ocean.* (2nd edn.), C.U.P, 336pp.

Pierson W.J. & Moskowitz L. 1964 A proposed spectral form for fully developed wind seas based on the similarity theory of S.A. Kitaigorodskii. J. Geophys. Res., **69**, 5181-5190.

Queffeulou P. 1983 Seasat waveheight measurement: A comparison with sea-truth data and a wave forecasting model-application to the geographic distribution of strong sea states in storms. J. Geophys. Res.,**89**, 2041-2051.

Robinson I.S. 1985 *Satellite oceanography.* Ellis Horwood Ltd, 455pp.

Schuchman R.A. & Kasischke E.S. 1981 Refraction of coastal ocean waves, in Spaceborne Synthetic Aperture Radar for Oceanography, Johns Hopkins Oceanographic Studies, Number 7, 128-135.

Srokosz M.A. 1986a On the probability of wave breaking in deep water. J. Phys. Ocean., **16**, 382-385.

Srokosz M.A. 1986b On the joint distribution of surface elevation and slopes for a nonlinear random sea, with an application to radar altimetry. J. Geophys. Res., **91**, 995-1006.

Thomas M.H.B. 1982 The estimation of wave height from digitally processed SAR imagery. Int J. Remote Sensing, **3**, 63-68.

Thomas J.P. 1988 Retrieval of energy spectra from measured data for assimilation into a wave model. Quart. J. R. Met. Soc., **114**, 781-800.

Vesecky J.F. & Stewart R.H. 1982 The observation of ocean surface phenomena using imagery from the SEASAT synthetic aperture radar: An assessment. J. Geophys. Res., **87**, 3397-3430.

Vesecky J.F., Stewart R.H., Shuchman R.A., Assal H.M., Kasischke E.S. and Lyden J.D. 1986 On the ability of synthetic aperture radar to measure ocean waves. pp403-421 in Wave dynamics and radio probing of the ocean surface. ed. O.M. Phillips & K. Hasselmann, Plenum Press, 694pp.

Webb D.J. 1981 A comparison of Seasat altimeter measurements of waveheight with measurements made by a pitch-roll buoy. J. Geophys. Res., **86**, 6394-6398.

THE SCATTEROMETER: DATA AND APPLICATIONS

R.A. Brown
University of Washington
Department of Atmospheric Science
Seattle, Wash. 98195   USA

ABSTRACT: The application of a new class of data such as that
furnished by a satellite scatterometer requires a basic understanding of
the nature of the data.   Here are examined the assumptions and techniques
required to translate backscatter cross-section into surface stress, air-
sea fluxes, wind profiles and pressure fields.   Although these are
extensive, the results are convincing and rewarding.

1. Scatterometer Characteristics

Satellite scatterometers furnish a new spectrum of data for geophysical studies.   Because the very nature of the data is new, it is worthwhile to review the basic physics behind the final data product.   The
characteristics presented here are for the Seasat scatterometer (SASS),
flown in 1978, and for the ERS-1 and NASA (NSCAT) scatterometers to be
flown in the early 1990s.

1.1. THE SIGNAL

The basic premise of a surface wind measuring device several hundred
kilometers removed from the wind is straight-forward and intuitively
reasonable.   The microwave radar emits energy in wavelengths which are
Bragg-scattered from the centimeter scale ocean waves.   The strength of
the return signal is related to the amplitude and density of the waves.
These waves are in turn related to the wind.   It is common practice for
sailors to follow the passage of a gust of wind over the water by watch-
ing for the wind's subtle impact on the interface, like a shadow passing
over the surface of the water.   The wind evidently generates the very
short waves with nearly instantaneous response times, reducing the
ocean's reflecting capacity.   Since the shape of the waves depends on the
wind direction, their ability to reflect energy varies with the angle of
observation related to the wind direction.   Thus, two looks by a radar at
different angles can produce different measurements which together give
information on the wind direction.

Although it is convenient and practical to discuss the wind in
relation to the wave field, the latter is more directly related to the
momentum flux, or surface stress.   This quantity is the mechanical energy
transferred from the wind to the waves.   The correlation of this stress
to the wind at any height is not unique.   It depends on surface

R. A. Vaughan (ed.), Microwave Remote Sensing for Oceanographic and Marine Weather-Forecase Models, 99–123.

roughness, stratification (which is closely related to air-sea temperature difference), and wind speed magnitude. The theory relating wind speed to wave generation and equilibrium spectrum is not well developed, so an analytic algorithm is not likely to replace the empirical correlation between wind and radar backscatter cross-section, σ_o. However, there are a number of caveats to this relation which must be kept in mind when applying the algorithm winds to a specific geophysical problem. They include:

1. The short waves correspond to small-scale stress, sometimes called skin friction. Some percentage of energy must go directly into longer waves due to dynamic pressure variations, called form drag. For the wind correlation with σ_o to work, this percentage must be fairly constant.
2. The short wave generation and equilibrium spectrum is attenuated by the long waves. (e.g. Phillips, 1978).
3. Long waves in the form of swell can travel thousands of kilometers and be unrelated to the local winds. (e.g. Mognard, 1984).
4. Capillaries can be mechanically generated by long waves, and interact in the wave breaking process (e.g. Cox, 1958).
5. The role of breaking waves and white water is unclear, but likely significant (Donelan, 1978; Monahan, 1986; Banner and Melville, 1976).
6. There is an expected saturation of the capillary waves at high wind speeds.
7. The effect of rain on the surface and capillary spectra can be significant. (e.g. Guymer et al., 1981).
8. The sea surface temperature (SST), through surface tension, influences capillary development (Woiceshyn et al. 1986).
9. Oceanic and atmospheric mixed layer dynamics influence capillary development, providing variable convergence regions (Brown, 1980).
10. The momentum flux can be transferred to, and depends on, the ocean currents. For instance, the generation mechanism depends on the difference between wind speed and currents. Wave-current interaction can affect the wave shapes. (e.g. Longuet-Higgins, 1978).

Many of the problems in the wind algorithm are pertinent to winds only. The stress-σ_o correlation would be more direct and fundamental to oceanic needs. The only reason to use winds in the algorithms is their relative ease of measurement and abundance. Marine stress measurements are practically non-existent.

1.1.1. *Wind Speed*. The possibility of inferring the winds by measuring the wave density with radars originated in the 1950s and was the subject of conferences and early measurements in the 1960s. The correspondence between microwave scatterometer backscatter cross-section, σ_o, and measured surface winds was established in the 1970s using an airplane-borne scatterometer and buoys and ships. This was extended to satellite data when NASA launched SEASAT in 1978. (e.g. Jones et al., 1982).

There is an arbitrary curve fit, called a model function, or algorithm, for parameterizing the wind speed σ_o.

$$\text{Log } \sigma_0 = G(\theta, \chi, \epsilon) + H(\theta, \chi, \epsilon) \log U \qquad (1)$$

where θ is incidence angle, χ is antenna azimuth-wind direction angle, ϵ is the beam polarization, and G and H are curve-fitting coefficients.

Evaluation of the parameters in this relation evolved during SEASAT using buoy, ship and area averaged winds from several oceanographic experiments, described in Brown et al. (1982) and Jones et al. (1982).

1.1.2. *Direction*. The unique capability of the microwave scatterometer is the direction measurement. The anisotropy in the wave field is related to the wind direction. If a scatterometer were to look at a given wind field from every angle, there would be peaks in σ_0 in the upstream and downstream directions, with minima at cross-wind. The model function produces several possible wind directions for each σ_0. This ambiguity in direction had to be resolved with ancillary data for SASS. When three looks are available the model function will have at most a 180° ambiguity.

1.2. THE FIELD

The radar signal from the antenna is processed to yield a geophysical data record (GDR) of winds. The nature of this data product is influenced by the application requirements. There are compromises with respect to resolution versus accuracy and other tradeoffs in processing. The products available for geophysical analysis span considerable spatial and temporal scales.

The product generated by the sensor data systems engineers may not be directly useful for many geophysical studies. Most studies require regularly spaced data at specific times. Satellite scatterometer data are densely sampled within swaths, but are not spatially homogeneous and they are sequential in time. Thus, the first chore for a study is to develop accurate and efficient objective methods for interpolating and averaging the data, and establishing the error structures of the resulting fields. The latter will involve inherent instrument errors plus errors due to the analysis as it reconciles the irregular scatterometer sampling with the natural variability of the wind fields.

1.2.1. *Footprint*. Each radar beam is electronically subdivided into resolution cells; of approximately 18 X 70 km in SASS. This will be approximately 50 X 50 km for ERS-1 and 25 X 25 km for NSCAT. The swath of the antenna coverage varies from about 20° to 55° incidence. Each σ_0 data point is comprised of two or three looks which must be co-located. This process yields a somewhat irregular field of data points across and along the swath. The cross track and long track spacing between resolution cells is arbitrarily set, so that extra data points within the cells are averaged. This separation was 100 km for SASS, and will be 100 km for ERS-1 and 50 km for NSCAT.

102

1.2.2. *Swaths*. The SASS swaths were 475 km wide, on each side of the nadir with a 400 km separation. The sun-synchronous, polar orbiting satellite lays a progressive sinusoidal path over the global surface, returning to approximately the same region 24 hours later. There is a gap between adjacent northward or southward orbit segments of approximately 90 minutes and 0-500 km depending on latitude. These ascending/descending orbits can be paired and linked in various fashions through extrapolation and averaging, depending on the phenomena being investigated.

Swath data from a single sensor can be connected to provide a picture of the weather over the oceans in a 12 or 24 hr period. Using both ascending and descending modes, a 12 hour picture of global winds can be produced, as calculated by Woiceshyn et al. (1986). There are two patches in such a figure where adjacent winds differ in time by 12 hours. When a sequence of ascending or descending swaths are used, the juncture between the beginning and end of the calculation will exhibit a 24 hour gap. However, the rest of the map produces a fair synoptic picture as shown in figure 1 and plate 1 . If both ascending and descending portions of the orbit are used to form a 1-day average, then multiple values at different times will be obtained in the grid, and an averaging process must be introduced.

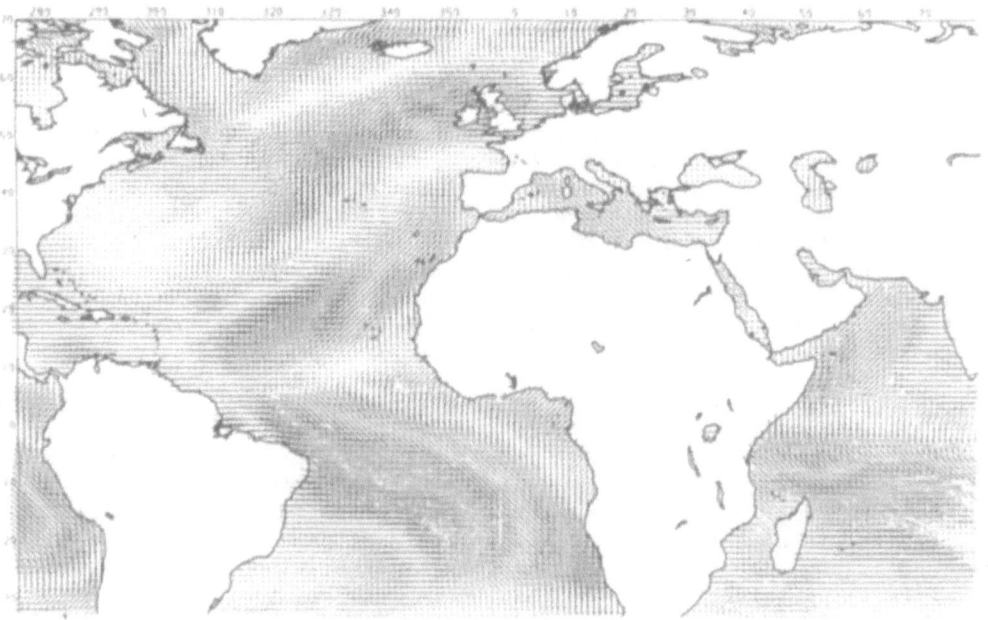

Figure 1. Atlantic and Indian ocean wind map derived from 12-hours of SASS swaths (Courtesy of S. Peteherych, M. Wurtele & P. Woiceshyn).

1.2.3. *Averages*. When enormous amounts of data are produced, as with satellite sensor data, then systematic averaging techniques must be employed. Since these will mask or eliminate sub-grid or shorter period phenomena, the averaging must be tailored to the phenomena to be studied. In many cases, the redundant data, such as extra resolution cells within the selected grid spacing, or ascending/descending data, have already been averaged simply to produce a better data point.

For regional studies, the swath data carries the maximum amount of data with minimum binning and averaging by the systems data processing. When a 1-day wind field is produced, the ascending and descending segments can be used separately, or combined to provide an average of data separated by about 12 hours. When global and climate phenomena are being studied, latitude-longitude cell averages can be compiled from daily to monthly (and hopefully, annually some day). The important application to numerical weather forecast models requires development of techniques for the accurate and efficient assimilation of scatterometer data into regional and global models. The many choices for time- and space-averaged fields (with corresponding error characteristics) make the development of appropriate interpolation and averaging techniques a prerequisite for most investigations.

Optimal interpolation schemes can be employed effectively to sparse and irregular data. Given the space/time covariance structure of the measurement errors, fields with prescribed mean square error minimized for space/time averages can be constructed. This kind of statistical error information is important to applications in numerical modeling. It requires substantial time and computational effort. Other operations ay wish to devise simpler methods, such as that in Levy and Brown (1986).

One motivation for careful evaluation of the data error statistics is obtained from the new quality of the scatterometer data. Data may vary due with unexpected parameters. For instance, studies have revealed unforeseen systematic variation in SASS data due to sea surface temperature, incidence angle, polarization, and wind magnitude.

2. The Wind Connection

The relation between the capillaries/short gravity waves and the wind involves some of the most difficult problems in geophysical fluid dynamics: stability, wave generation/propagation, and interaction theories. The mechanism of wind generation of water waves is a classic unsolved problem of fluid dynamics. One of the requisites for the problem is a model for, and an understanding of, the wind profile and the underlying water current profile. This is a classic boundary layer problem and much is known about the flow.

2.1. SURFACE LAYER DYNAMICS AND THE STRESS

The large quantity of land-based data taken in the surface layer is
assumed to apply over the water. An added consideration over the ocean
is the feedback due to the surface roughness variation as increasing
winds generate more roughness elements.

When surface observations are made, there exists a layer near the
surface where the turning of the wind due to Coriolis effects and the
variation in stress (momentum flux) appear insignificant (stress varies
less than 10% in the lowest 10-20 meters). We are interested in the
dynamics of this layer because this is where the measurements of wind and
stress are made. We wish to correlate these values to the surface σ_o
value. In this layer, the wind speed varies approximately as the natural
logarithm of the height.

$$U/u* = 1/k \ln(z/z_o) \qquad (2)$$

where $u* \equiv [\tau/\rho]^{\frac{1}{2}}$ is defined as the "friction velocity". In the boundary
layer, density, ρ, can be considered constant and $u*$ is a substitute for
stress, τ. k is von Karman's constant and z_o is a roughness parameter,
defined as the height where u = 0. The values of k and $u*$ can be found
from equation (2) by measuring u(z) to obtain a plot of u versus ln z,

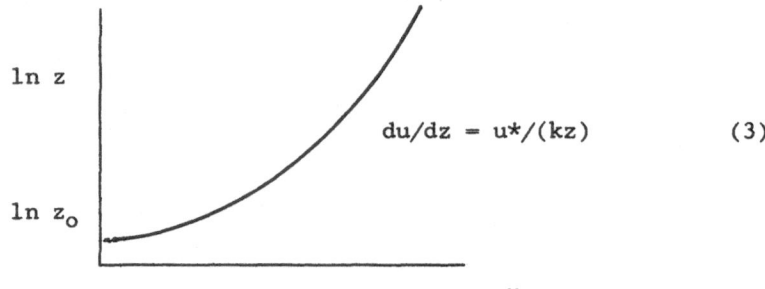

$$du/dz = u*/(kz) \qquad (3)$$

Figure 2. Plot of u(z) to determine z_o and k

In practice, z_o is very small and measurements near u = 0 are poor.
Extrapolation is necessary and values of z_o vary by orders of magnitude,
while k has been found to range from 0.32 to 0.45. Recently, in careful
experiments in the atmospheric surface layer, the value of k was
confirmed to be nearly constant at 0.4. Once k is known, measurement
plots such as figure 2 can be used to determine $u*$ from equation (3), in
what is called the profile method.

If $u*$ is measured with other methods, then z_o can be calculated from
a measurement of u at z using,

$$z_o = z /\{ \exp[ku/u*] \}. \qquad (4)$$

When stratification is important, a correction to this formula must be added,

$$u = u*/k \{ \ln[z/ z_o] - \psi \},$$ (5)

where ψ is an empirical correction for layer stratification (Paulson, 1970).

As measurement techniques improved, the capability to measure the eddy flux directly yielded "direct" stress measurements according to the eddy-correlation formula,

$$\tau = \rho \, u'w'.$$ (6)

where $u'w'$ is the average correlation of the turbulent components of the wind. It is equal to the momentum flux. Sonic anemometers provide this data and have been deployed on land, on the oceanic pack ice, and on towers over the ocean. However, stable alignment is crucial to this measurement, and inherent errors in the measurements suggest that large errors may generally be attached to these measurements (Blanc, 1987). Nevertheless, the drag coefficients established with respect to sonic anemometer surface measurements form the basis of flux parameterizations.

Finally, a technique for inferring stress from the turbulence spectra in the inertial subrange has evolved. This relies upon the hypothesis that the inertial subrange is a universal characteristic of the turbulent energy cascade in the dissipation regime. The stress (momentum flux) is a boundary condition for the spectral density and is related to the dissipation, which can be found from high-frequency velocity measurements. A hot film produces voltage variations proportional to the high frequency velocity fluctuations. Again, there is evidence (Blanc, 1987) that intrinsic errors in the dissipation method are comparable to the eddy-correlation method. However, recent results suggest that this technique can yield errors less than 30% (Guest and Davidson, 1987).

Most of these measurements have been made in the surface layer, which may extend to 40-50 meters in the atmosphere, 1-2 meters in the ocean. The instruments and methods are continually evolving. The physics have been the subject of several texts (e.g. Favre & Hasselman, 1978; Kraus, 1972). Most modeling has attempted to skip the smallest-scale details by concentrating on surface layer theories combined with semi-empirical parameterization of the fluxes. Thus there is a body of data relating the fluxes of momentum, heat, and humidity to surface layer winds. If these data were complete, then a measurement of wind, temperature and/or humidity anywhere in the surface layer (including at the surface) would be sufficient to determine all fluxes.

2.2. PBL DYNAMICS

The Planetary Boundary Layer (PBL) is defined as the region next to the surface wherein the influence of the surface is significant in calculating the flow (It is much deeper than the surface layer, where the surface stress is dominant). The influence can be due to surface roughness bringing the flow to a halt, surface temperature providing a heat sink or source, or the surface acting as a sink/source for any constituent (e.g. water vapor, CO_2). This region is tens of meters thick in the ocean and up to several kilometers in the atmosphere. We have seen that the scatterometer measures the surface boundary condition for the wind profile. A surface layer model then relates this value to the wind at 10 or 20 meters height (U_{10}, U_{20}). When a complete PBL wind profile is known, one might also say that the scatterometer implies the geostrophic flow, U_G. Since the pressure field is related to this flow, we have the new, exciting prospect that the scatterometer 'measures' the pressure field.

In 1904, V. Walfrid Ekman obtained the analytic solution for the geophysical flow due to a balance between Coriolis, pressure gradient and viscous forces. Published in 1905, Ekman's solution yielded a logarithmic spiral velocity profile in which the surface effect decayed exponentially away from the surface. It implies that the geostrophic balance between Coriolis and pressure gradient forces is a good approximation a short distance away from the surface, where viscous forces are negligible. Since the PBL is thin, the pressure field can be assumed to be impressed upon the layer---it is the same throughout. These concepts are important in the use of scatterometer winds to determine surface pressure fields, discussed below.

Ekman's solution successfully explained the pack ice drift and net transport in the oceanic mixed layer. However, the predicted logarithmic spiral is seldom, if ever, seen. Nevertheless, it is a touchstone for PBL solutions in the atmosphere and the ocean. The velocity profile is frequently plotted in U-V coordinates as the locus of velocity vectors at various heights (or depths) in the PBL. The continuous velocity profile for linked atmospheric-oceanic PBLs is shown in figure 3.

In the 1960s, Ekman's solution was found to be unstable to infinitesimal perturbations, but stable to finite perturbations (Brown, 1970). There are modified Ekman solutions which include secondary flows whenever wind speeds exceed 7-8 ms-1 in neutral stratification or 5 ms-1 in unstable stratification. The velocity profiles for most flow situations now contain large, organized eddies. These cause the wind profiles to vary considerably on horizontal scales comparable to the PBL height, with modified mean hodographs which are only qualitatively similar to Ekman's solution (Brown, 1980). One very important consequence of this flow solution is that point profiles, such as those taken with a radiosonde or dropsonde, do not measure the average velocity profile. This will vary point to point within the large eddies. Similar variations can be expected in the oceanic PBL (where the large eddies are called Langmuir circulations). One consequence of the large organized eddies for scatterometer correlations is that there are alternate convergence and divergence zones in the wind and the current fields. The latter produce the familiar wind rows---linear regions of suppressed capillaries on the oceanic surface. There is evidence from synthetic aperture microwave radar data that capillary intensity also varies on the atmospheric convergence-divergence scale (2-4 km).

Some knowledge of the flow in the PBL is needed to apply the surface
stress measurements from satellite microwave sensors to geophysical problems. Weather analyses involve wind dynamics and prediction. Climate
analyses use winds and surface pressures as fundamental parameters.

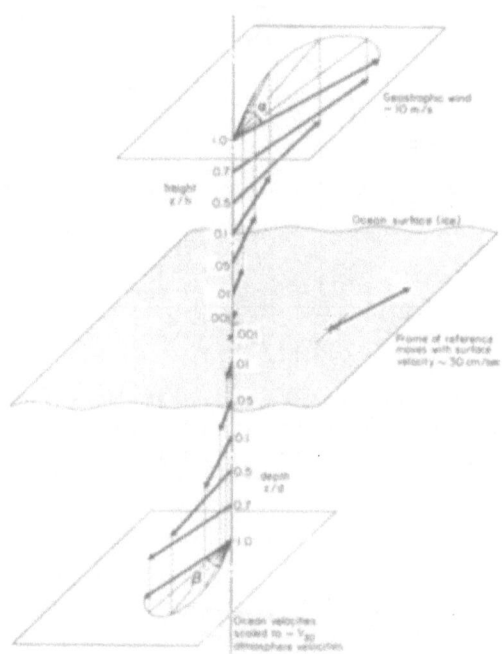

Figure 3. Linked hodographs of velocity vectors in oceanic and
atmospheric PBLs.

2.3. U_{10}, U_{20}, or U_G

The solutions for the surface layer and Ekman layer have been combined in
a two-layer model of the PBL dynamics as described by Brown and Liu
(1982). The results show the dramatic influence of stratification on the
ratio of surface stress or 10-meter wind to geostrophic, as shown in figure 4. The influence of both stratification and wind magnitude over the
ocean are shown in the variation of neutral drag coefficient in figure 5.

Since the wind profile depends on stratification and surface roughness, the scatterometer measurement of surface stress (or related neutral
layer winds) can correspond to different values of actual $U10$, $U20$, or
UG. Some idea of the magnitude of this variation due simply to surface
layer variations from equation (5) are shown in figure 6.

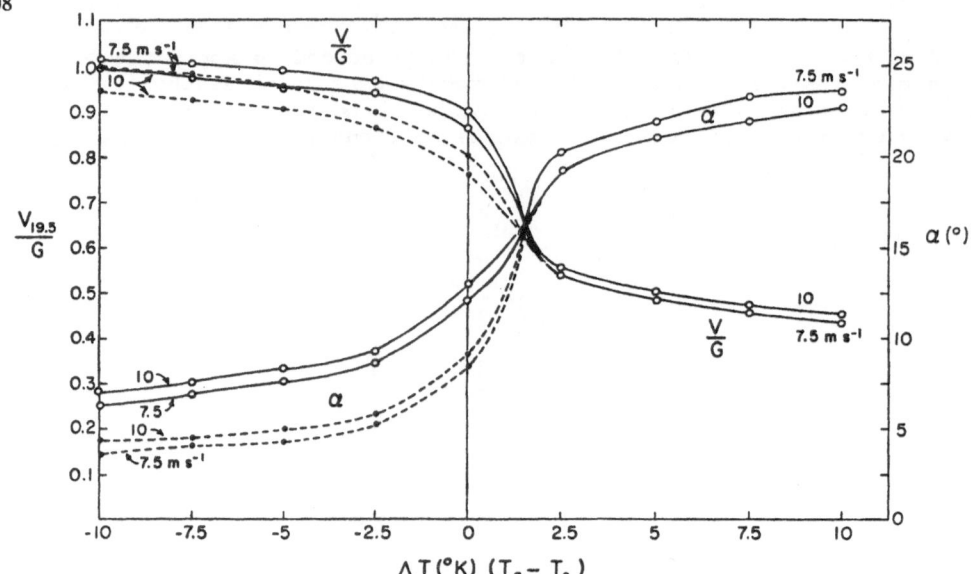

Figure 4. Model results for the ratio of wind at 19.5 meters to geostrophic (solid), and turning angle, α, (dashed), versus air-sea temperature difference, for G = 7.5 and 10 ms-1.

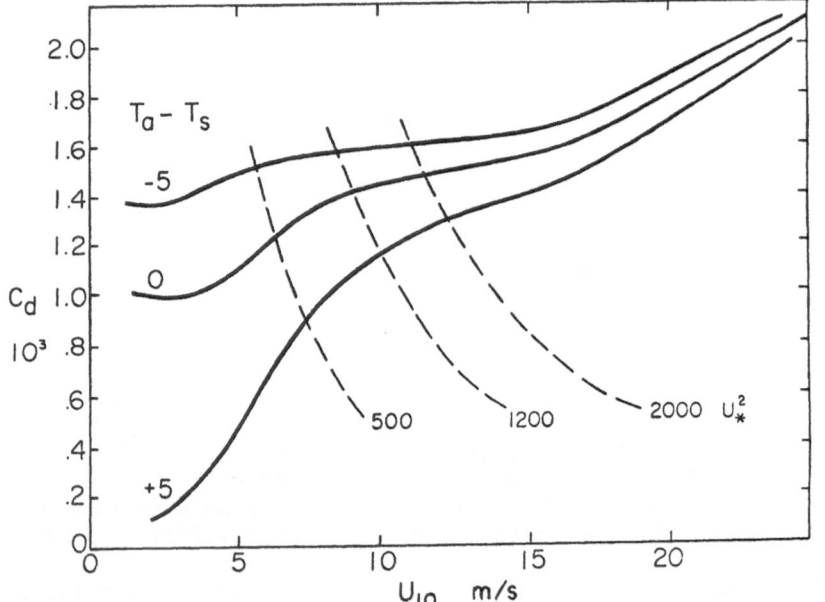

Figure 5. The model function neutral drag coefficient (proportional to surface roughness) versus windspeed, for various stratifications.

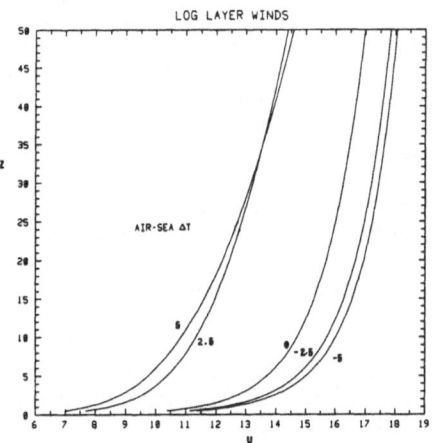

Figure 6. The velocity profile in the surface layer for various stratifications.

The problems of variable roughness and stratification, PBL instability with consequent large eddies, baroclinicity, and an inversion capping the PBL have all been addressed in observational experiments and theory. Using the flow solutions, the measurement made by scatterometers can be used as a boundary condition for the flow throughout the PBL. There is a large body of work modeling air-sea interaction and PBL dynamics, from bulk coefficient methods, to simple two-layer similarity parameterizations, to complete numerical simulations of the PBL flow. These solutions greatly enhance application opportunities for scatterometer data.

2.4. AVERAGING

The establishment of the algorithms relating σ_0 to wind or surface stress requires a good geophysical measurement under an appropriate average. The satellite instantaneous average over a 900-1500 km^2 region is difficult to compare to a point measurement, or even several point measurements, within the area. The basic algorithm was established with respect to buoy winds. These are recorded as 8 minute averages, and in SASS comparisons, a 60 minute average was used. For 10-ms^{-1} winds, this would represent a pie-shaped sample of a steady-state wind field with radius 36 km and arc segment about 4 km, or about 150 km^2. However, averages on all scales will encounter variations due to turbulent or organized variations on the same time and space scale as the measurement. This produces a basic difference in comparisons between satellite and surface data which is intrinsic to the different nature of the measurements. In other words, a true comparison measurement for even the 25 X 25 km footprint cannot be made short of a budget-busting investment in buoys.

Practical considerations determine that comparisons be made with data taken within 50 or 100 km of the cell centers. Representative mean gradients are about 1-2 ms^{-1} per 100 km or per hour, and contribute to the error. When storms are present, non-uniform, non-stationary and even discontinuous (at fronts) wind fields are present, and comparison data are most suspect (and sparse).

Aircraft measurements offer a somewhat broader average. They can sample a 60 km leg in 10 minutes, providing a more compatible average for comparison. However, they fly at a minimum of 30 meters height, and the error in extrapolating to the surface is incurred.

Apparently, in view of these concerns, the averaging problem sets a limit on the accuracy to which the scatterometer data can be verified. This accuracy is close to the projected specifications for the scatterometer---±2 ms^{-1} or 10%, and ±20°.

2.5. THE GEOPHYSICAL PHENOMENA

The success of a satellite sensor depends to a large extent on the ingenuity of the investigators using the product. It should be clear that the scatterometer data product is quite different from previous available data. The basic information---spectral density of the short waves---is of narrow geophysical interest. However, the fact that these numbers correlate to the surface stress is of emphatic interest to oceanographers and atmospheric scientists. This connection provides information on the clouds and the deep water. The surface stress serves as the boundary condition to drive the ocean currents, generate the waves and swell, affect the sea surface temperature, and force the mixed layer. The same stress anchors the atmospheric PBL solution, allowing calculation of wind profiles, fluxes, free stream (geostrophic) velocities and surface pressure fields.

We have seen that the σ_o is an average, and correlates to a wind speed or surface stress average, unlike any previously available data. Derivation of wind direction from multiple 'looks' at the same region yields another new set of data, a two directional wave spectrum. They are presumedly the same waves, but looking different from different directions. The mechanics of selecting the correct solution are discussed elsewhere, here we are interested in using the available data for geophysical interpretation. The multiple vector solutions are particularly sensitive to sea state. When the wind turns sharply, as it does through an atmospheric front, the individual multiple vector footprints yield close-in information.

While the flux coefficients for heat, moisture and CO2 need only wind speed to determine their magnitudes, the important physics relating regional characteristics to general circulation requires wind directions. The wind convergence/divergence and vorticity provide the essential weather and climatic features which enable analysis and prediction of important flux situations. The El Nino/Southern Oscillation phenomena provide a dramatic example of this. The extreme importance of latent heat flux through evaporation and precipitation in the tropical circulation is another application requiring the wind direction information.

The fact that the scatterometer (and other microwave sensors) are sampling the interface between the atmosphere and the ocean emphasizes the importance of the study of air-sea interaction. To use the data in either medium, it is necessary to understand the characteristics of the PBL and the means of communication between the surface and the free stream. The latter acts on larger scales, seemingly independent of the surface, yet ultimately depending completely on the transfer of energy from this interfacial surface.

As larger scales are considered, the selection of wind direction becomes easier. Fronts disappear, and with several day averages, storms are no longer evident. However, on the larger scales, the ocean's role in absorbing CO_2 will determine the greenhouse heating rate. The importance of CO_2 in the glacial/interglacial cycle has been demonstrated, and the role of ocean and atmosphere storage and interchange of CO_2 may be crucial to ice age cycles. While wind speed statistics determine fluxes, determination of weather patterns and their long term interaction with land, pack ice, and SST anomalies require directional information which is uniquely supplied by scatterometers.

2.6. FLUXES

Fluxes are the transport of momentum, sensible or latent heat, humidity, CO_2, or any other transportable entity. Their measurement has always been quite indirect to varying degrees. For instance, the momentum flux is related to the mean velocity profile or to the correlation between the vertical turbulent velocity component and the horizontal velocity, or to the turbulent dissipation rates. (Other fluxes are related to the correlation of the vertical turbulent velocity component and temperature, or moisture etc.) All of these methods are questionable, with significant error (Blanc, 1987). In addition, the flux may be greatly influenced by the organized large-eddies, or rolls in the PBL. These are present in neutral to unstable stratification above a very low threshold flow velocity. When the rolls are present, or the flow switches from simple eddy-diffusion to large eddy advection, the large-eddies must be accounted for explicitly, as part of the mean flow.

Thus it seems extraordinary to expect that microwave remote measurements will determine fluxes. However, the hypothesis is supported by two observations: the fluxes are accomplished by the turbulence which correlates to the wind profile (especially the shear); and the short waves on the water surface correlate to the surface stress, which constitutes the lower boundary condition on the wind profile. In fact a direct link between σ_o and the flux is made when one notes that the small reflecting waves are being generated by, and are proportional to, the momentum flux itself in a direct energy transfer. Add the logical speculation that the same (or nearly the same) turbulence elements which move momentum also produce flux of all entities. We then have available from the microwave backscatter signal the momentum flux (possibly via the wind as an intermediary parameter) and in some proportion to this, the remaining fluxes. There are some data over land which relate the various fluxes through their respective aerodynamic coefficients.

The surface stress can be related to the geostrophic flow in the geostrophic drag coefficient, $u*/U_G$. Ekman's solution predicts a Newtonian type surface stress at $z = 0$,

$$\tau_o = \rho \ K_E \ du/dz \ |_o. \tag{7}$$

where τ_o is surface stress, and K_E is eddy viscosity.
Similar comments apply to eddy diffusive modeling of sensible heat, H, and latent heat E, with respect to gradients in mean temperature θ, and moisture Q, in analogy to the molecular fluxes,

$$\tau = \rho \ \nu \ \partial u/\partial z; \quad H = -\rho C_p \ k_h \ \partial\theta/z; \quad E = - \ \rho \ k_e \ \partial Q/\partial z \tag{8}$$

where ν, k_h, and k_e are the respective molecular diffusivity coefficients.

2.7. PRESSURE FIELDS

Pressures are the most widely, and accurately, available field data on synoptic and large scales. Thus, surface pressure fields are a basic tool of the operational weather forecaster, the synoptician, and the climatologist.

The geostrophic balance between Coriolis force and pressure gradient force provides a simple relation between wind speed and pressure gradient which depends on latitude only (density assumed constant),

$$\rho \ \mathcal{F} \ [U_G, \ V_G] = [- \ \partial P/\partial y, \ \partial P/\partial x \ ]. \tag{9}$$

where $\mathcal{F} = 2 \ \Omega \ \sin$ (latitude); Ω is the earth rotation rate.

Endlich et. al. (1983) used the SASS data together with GOES-2 cloud motion data and the balance equation to calculate pressure fields with good success. The PBL model discussed above was developed to calculate surface winds and fluxes based on input geostrophic flow, or equivalently, surface pressure fields. When this model is inverted, scatterometer winds or stress serve as an input to obtain the geostrophic winds and thus to calculate surface pressure fields. Since the relation between surface winds and geostrophic winds is quite variable with stratification, the PBL model is essential to obtaining accurate pressures. In a typical marine storm system, stratification will vary from unstable to stable in various sectors. Variation in these parameters will yield significantly different results from a calculation employing constants for these parameters.

Independently deduced surface pressure fields for the Storms
Response Experiment (STREX) storms by Nuss and Brown (1987) were in
substantial agreement (0.5 mb) for fairly uniform synoptic fields but
exhibited errors of several millibars in the vicinity of a front. Since
the storms regime and fronts are an important weather phenomena and
provide maximum challenge for pressure field variation, the analyses
concentrated on these cases.

In the inverse PBL model (pressure fields from surface winds) the
input winds and thermodynamic data must be appropriately gridded to latitude-longitude coordinates. In the case of conventional data this requires elaborate interpolation and extrapolation to account for generally sparse data sets. While this is still true for the thermodynamic
variables in this study, the SASS wind data present different problems.
The inter-swath region was filled in with an interpolation scheme. Subsequent adjacent revolutions produced strips separated by 0 -1000km and
about 90 minutes depending on latitude. At the mid-latitudes of our
study, this yielded a gap of 0-400 km between subsequent orbits. The
famous storm of 9-10 September 1978 which struck the QE II in the west
Atlantic provided an example of such rapid cyclogenesis that it did not
appear in the National Weather Service (NWS) prognosis. Subsequent analysis has been done by Gyakum (1983) and the SASS derived pressure fields
are compared with this analysis in Figure 7. There is a time difference
of 55 minutes between Gyakum's analysis (solid) and the SASS swath used
for the model analysis (dashed), however the surface low was still quite
shallow and the changes during that time are not expected to be large.
The SASS derived pressure solution presented in Figure 7 represents a
basic solution that could be improved by inclusion of thermodynamic and
secondary flow corrections. A notable distortion of the 1008mb isobar
occurs along the coastline of Long Island due to the proximity of land.

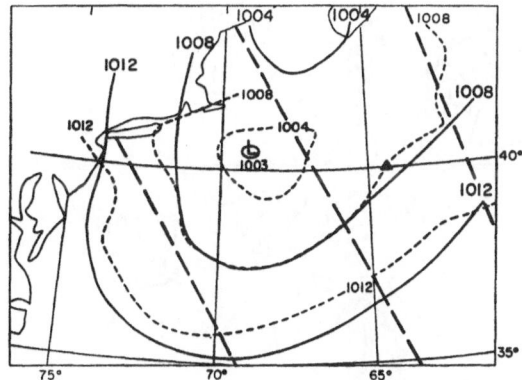

Figure 7. Conventional (solid) (Gyakum, 1983) and SASS/model (dashed)
pressure fields for 1200 GMT 9 Sept. 1978. Bold dashed lines mark SASS
swaths (between 63° and 69°m and NE corner). Triangle denotes base
pressure measurement.

114

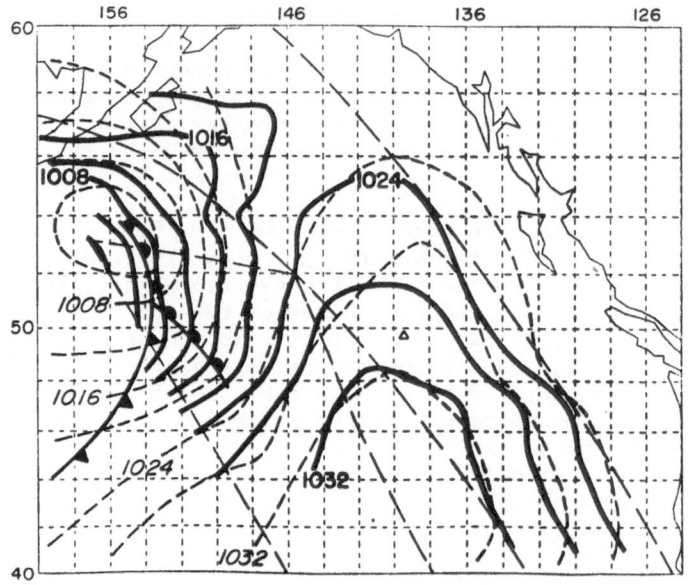

Figure 8. NWS (dashed) and model (solid) pressure fields for 17 Sept. 1978. SASS swaths are bounded by long dashes.

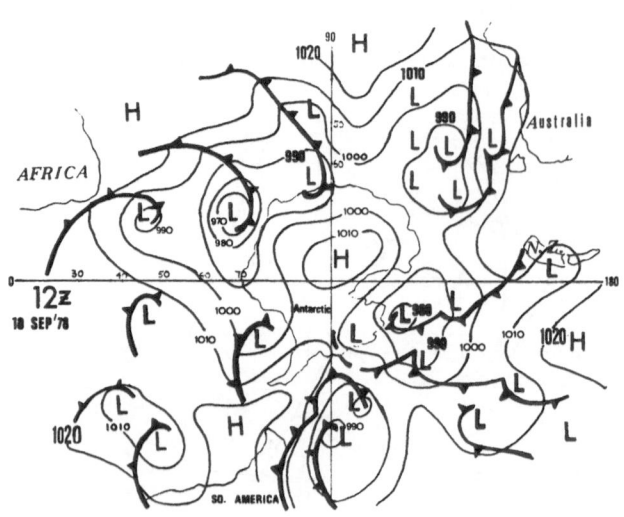

Figure 9. Typical day in the southern hemisphere ocean. Surface pressure analysis from Australian Met office.

On September 17 1978, the Seasat viewed a typical 'blocking' situation with a high pressure center to the east and the eastern edge of a
stationary front (part of a 'double low' system). The separation in time
between the swaths was approximately 90 minutes and the analysis is based
on these two orbits. The model and the NWS surface pressure analysis at
1800 GMT are shown in Figure 8. The region is dense in SASS winds from
the low winds in the ridge to the high winds associated with the fronts.
The geographic gap in data is smaller in this case. However, due to
missing data in one of the orbits there is no region of overlap. The
short wave variations in the model pressure fields can be a result of the
SASS directional selection procedure and/or curve fitting procedures in
the model; or they can be geophysical.

The air to SST difference based on ship and buoy reports was used to
correct for PBL stratification. The stratification gradients produce
significant changes in the model pressure map. The low was deepened
compared to the neutral analysis by about 4 mb. There are no reports at
the center of the low that can independently support either analysis. In
connection with the increased horizontal air temperature gradients, there
is a thermal wind component which can effect the U_{10}/U_G by 0-20%.

When the SASS/PBL model results were compared to the NWS maps, there
was basic qualitative agreement even for the model with minimum physics.
However, in the limited cases studied, the addition of secondary flow,
thermal wind, and stratification each tended to improve the agreement
with NWS analyses. In some cases, the less than 2 mb changes are not
greater than the inherent accuracy of the best fit fields, which can have
errors of ±2 mb due to instrument accuracy limitations. Short waves
appear in the model isobars under different circumstances. Sensitivity
tests indicate that these are sometimes due to the gridding and contour-
ing procedures wherein polynomials are fit to the data. However careful
attention to this procedure minimized this effect in the cases given.
They may also be due to the nature of the de-aliasing scheme which allows
lines of relatively abrupt direction change where one branch of the four
possible wind vectors is switched to another. The algorithm also experiences this abrupt transition in number of aliases from two to three or
four. These changes can be smoothed out, however in some areas the smooth
variation of the SASS wind vectors can produce similar wavelength variations. These might reflect actual flow dynamics. Only the examination
of more detailed pressure fields can reveal which phenomena are model
generated and which are real physical characteristics of these mesoscale
wind and pressure fields.

The choice of one central pressure point as an integration constant
was sufficient to carry out the integration to areas as large as 2000km X
2000km with a few mb error even in very dynamic regions. This combined
with the low sensitivity of the pressures to the thermodynamic fields
gives merit to the idea that in areas where such data is unavailable, an
assumed central pressure and thermodynamic field (based on scatterometer
measurements and forecast fields) would result in qualitatively good
analyses.

It is evident from these initial results that this process applied
to global scatterometer data can produce marine surface pressure maps of
quality similar to that of the global ship and buoy collection procedure.
This comparison was done in relatively high density ship report regions,
and one might expect a significant improvement in sparse report areas
such as the southern hemisphere.

Data is very scarce in the southern hemisphere oceans. The Australian Bureau of meteorology relies heavily on the visual and IR cloud
images to locate storms and fronts. Figure 9 shows a 'typical' day in
September southern seas. The SASS pressure analysis for orbit segments
within ± 2 hours of the synoptic time are shown. This comparison suggests that lows may be systematically 10-30 mb lower than predicted by
the ABM. Either that, or the SASS winds are too high, a conclusion
contrary to studies in the northern hemisphere. That is where it is at
the moment.

2.8. STORMS

The study of marine storms entered a new era with the advent of SASS wind
fields. There are no routine observations of these storms, ships conven-
tionally steer clear of them and measurements suffer in quality and quantity during them. There have been a few modestly instrumented studies of
storms (e.g. STREX; Ocean Storms; and GALE). Wind direction shifts, convergence and vorticity are important characteristics of storm behavior.
They are all available with scatterometer data. When a multi-frequency
radiometer is simultaneously available, unprecedented information on
storm dynamics results.

It is clear that oceanic mixed layer dynamics is strongly altered by
the episodic storms events. Mixing, current forcing, and wave generation
are singularly important in storms events. In the practical need for CO_2
transfer to the ocean, only a few flux parameterizations for steady,
uniform, conditions exist. It is likely that this transfer is uniquely
important within the high wind regimes of storms. Only the simultaneous
measurements of CO_2 flux and mean winds will establish this quantity.
Only the dense scatterometer wind fields will allow calculation of the
total transport associated with the storms regimes.

2.8.1. *Convergence/Vorticity.* Scatterometer wind vectors allow
calculation of the convergence/divergence and vorticity fields. These new
parameters are used routinely in weather and climate analysis to give
information on the vertical coherence of the atmosphere. Convergence
regions coincide with Lows, storm centers, mesoscale cumulus production
or simply upward movement in the wind field. Divergence associates with
Highs, clear weather and downdrafts. The statistics of convergence
patterns gives information on storm centers, high wind regions, maximum
flux potential, oceanic mixed layer forcing, up and down welling, wave
generation and other phenomena. When these are correlated with larger
scale forcing parameters, analysis and prediction models for weather,
fluxes, wave fields, SST and ocean dynamics can result. The correlation

between anomalous SST and regionally anomalous wind divergences during El Nino illustrates the importance of this parameter in defining the air-sea interaction.

2.8.2. *Fronts.* The marine front, delineating the line between two disparate air masses, with different wind vectors, temperatures, and moisture, has little documentary evidence. Models assume similarity to the relatively well measured land-based fronts. However, moisture, surface roughness, or stratification are seldom similar. The scatterometer provides unprecedented data in the vicinity of a front.

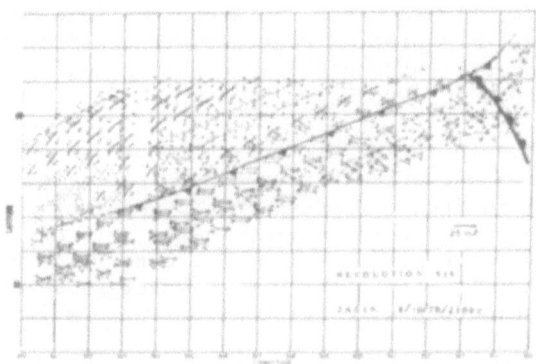

Figure 10. Seasat ambiguous wind vectors during JASIN.

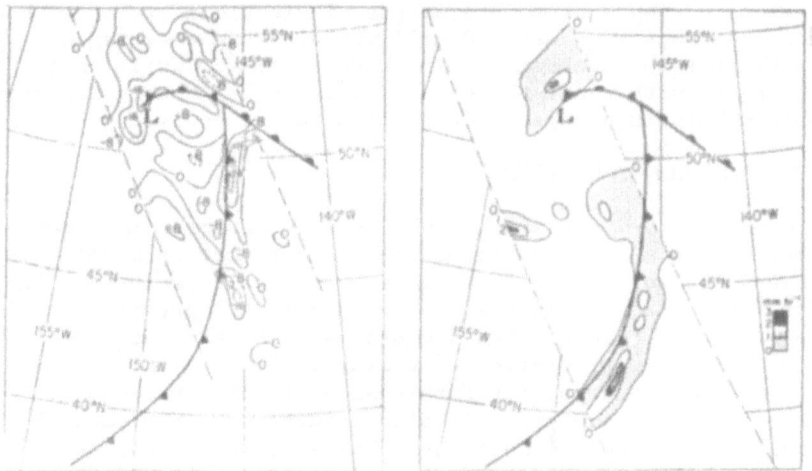

Figure 11. QEII storm analysed with scatterometer winds (convergence 10^{-5} s^{-1}) and SMMR integrated water vapor, kg m^{-2}.

The quality of the sea-state that is shown in the sensitivity of the ambiguous velocity signatures can define the frontal location within a distance approaching the scatterometer footprint scale (e.g. Brown, 1983). Figure 10 shows such a pattern in the vicinity of a Joint Air-Sea Interaction (JASIN) front.

The extraordinary data density from SASS suggest that fronts are not smoothly curved, but contain kinks or waves. These may indicate insta-bilities, which spawn smaller scale storms. They may also simply be irregularities in the wind direction selection process in this difficult region. SEASAT included a nearly coincident multifrequency radiometer, SMMR, which provided total liquid water, moisture and rain rate informa-tion. When these contours are plotted together with SASS divergences, corresponding patterns occur---strong evidence that a real geophysical phenomena is being observed (figure 11).

2.8.3. *Precipitation*. Precipitation hitting the ocean surface affects the scatterometer signal. One effect is to damp capillaries and smooth the surface. Another is to create ripples and roughen the surface. The net effect will depend on wind speed and rain rates. Insufficient data currently exists for any calibration of these effects. The usual proce-dure is to identify high rain rate areas with a multi-frequency radiometer and exclude them from wind calculations.

2.8.4. *Saturation, White Water*. When high winds are present, the most important air-sea interaction processes are strongest. There was concern that at higher winds the sea surface would 'saturate' with short waves and the scatterometer would peak-out in wind magnitude. Although the high wind regimes during Seasat were sparse, indications are that the σ_o-wind relation continues even near 30 ms^{-1} winds. Wave breaking and white water must influence the backscatter in these regimes. While these parameters increase with wind, there is no backscatter theory supporting a relationship. This is a topic in need of data and study.

2.9. SYNOPTIC SCALE ANALYSIS

Synoptic scale weather is the staple of national weather services, and enters the domain of the numerical general circulation model. There exists an immense network of land stations, ships of opportunity, buoys and satellite visual and IR pictures to be used in the creation of the world weather analysis. These data are used as initial conditions in numerical models for the prediction of weather. It is clear that over marine surfaces, the scatterometer furnishes an important boundary condition for atmospheric models. It also produces the fundamental boundary condition for oceanic modeling. The atmosphere respose is immediate, whereas the ocean responds in days to seasons.

The numerical analysis requires the data field to be redefined on a grid. The scatterometer data is very dense within the swaths with significant gaps, which include temporal gaps when joining different orbits. This nature of the data precludes some interpolation techniques.

The desire to retain smaller-scale features and to minimize processing time restricts other methods. Generally, an objective analysis method must be designed for the data and goals to be met. One such method was devised for SASS data by Levy and Brown (1986).

We can assume that the directional ambiguity is resolved by the application of ancillary data or the additional antenna information obtained by future scatterometers. With SASS data, directional selection can be done by using a first guess wind field and minimizing vector differences to select the correct direction. By scanning in a radius around selected grid points, assigning a weight function to each point (dependent on distance), and calculating a weighted average of all observations, an objective gridded wind field is produced. The error contributed by this analysis procedure can amount to 30% of the instrument error in extremely dynamic regions. In normal regions the analysis error becomes negligible compared to instrument error.

2.9.1. *Extrapolation (the gap)*. The 400 km gap between SASS swaths is not easily closed by extrapolation and interpolation. Within the gap an objective gridding scheme must reach the swaths to obtain data points. Weighting functions will act as filters depending on the density of the local data. To preserve subsynoptic features, the 400 km gap (for SASS), and 2.5 hours in time are probably upper limits.

The gap between successive orbits depends on the latitude. At mid-latitudes, it is about 100 km. The typical pass interval is 90 min. These characteristics allow reasonable steady-state assumptions and interpolation formulas. Divergence and vorticity patterns on the 50 km scale are well preserved.

2.9.2. *Temporal lapse*. Although subsequent orbit intervals were about 90 min for SASS, the repeat orbit is separated in time by 24 hours. For constructing global averages, this can be reduced to two gaps of 12 hours when both ascending and descending segments are used. However, the variable gaps and overlap regions at higher latitudes complicates the analysis and produces a less satisfactory result.

When longer averages are calculated, all data is collected in latitude-longitude bins and averaged. These data will have different times and densities, and variable gaps or holes depending on latitude. The averaging will filter shorter scale phenomena, but the new long term averaged data will be comprised of myriad more data points than any previous such calculation.

2.9.3. *Weather prediction*. Satellite scatterometer data offer great opportunities for improving numerical weather prediction models by furnishing more complete initial conditions. The lack of sufficient conventional data, particularly over the oceans, and especially over the southern hemisphere oceans, has been a major factor in limiting numerical weather forecasts. While attempts to evaluate SASS wind impact on forecasts have had mixed results (e.g. Anderson et al., 1986; Duffy and

Atlas, 1986), prospects remain bright. Consideration must be given to the coarse resolution of the numerical models, the scatterometer errors, the proper averaging and interpolation, and extension of the surface data into the vertical. The last task is closely linked to the dynamics of the inverse PBL model solution for the vertical wind profile. These factors are all subjects of research proposed for ERS-1 and NSCAT data.

2.10. CLIMATE

The applications to climate analysis remain theoretical since only 99 days of SASS data exist. Eventually, carefully analyzed and archived data will obviously offer abundant opportunities for correlations between oceanic thermodynamic and weather regimes and global ocean and atmospheric dynamics and climate.

NSCAT will have 97% coverage every 2-days at $\pm 70°$ latitude. Using the flux and boundary layer modeling methods, this will establish a global marine data record for surface stress, heat, moisture, CO_2, and any other passive variable. When these parameters are combined with SST, humidity, and other remotely sensed data, the opportunity for insightful correlations is immense.

2.11. OCEAN DYNAMICS

Surface stress is a fundamental boundary condition for many oceanic studies. The much slower response time of the ocean precludes many studies with presently available data. Certainly long term scatterometer data will become important in ocean circulation modeling. However, one can anticipate that improved surface stress fields will have immediate impact on regional current and mixed layer models (e.g. de Szoeke and Richman, 1984) and on ocean wave modeling and prediction.

2.11.1. *Currents*. The wind stress is a major driving force for a variety of oceanic motions. Present stress calculations are generally made with bulk parameterizations on surface layer or geostrophic winds. These incur the liabilities of a once-removed measurement---errors in boundary layer modeling are attached to the stress values. Since scatterometer values are associated with the surface, these liabilities do not exist. The problem is; there do not exist surface stress measurements with which to establish the correlation with σ_o. As more scatterometer data and the corresponding stress values are accumulated, a better algorithm for stress-σ_o will be developed.

The Antarctic Circumpolar Current (ACC) is one of the major current systems in the world. It plays a critical role in the transfer of heat across latitudes, and links each of the major ocean basins. The wind forcing is strong and direct due to the relatively unbounded geometry of the ACC. Yet the winds are poorly known, so that stress and fluxes are only crudely approximated. The dramatic changes in forcing suggested by the SASS results discussed above will have large influences on relevant ocean and atmosphere models.

The midlatitude ocean circulation depends on the wind forcing in an important, but unknown degree. Available wind data has large and unquantified errors that prevent ocean models from examining other constraints. There is consensus that wind forcing produces the net meridional water movement in the vast ocean interior, and that wind-forced currents on mesoscales are significant. The advent of a complete scatterometer data set offers the first opportunity to investigate the manner in which spatial and temporal variability in wind forcing is felt by the midlatitude ocean.

Finally, there are many regional and sub-basin scale phenomena which are known to exist from historical or satellite SST data. These phenomena invariably involve wind-forcing, and suffer from inadequate wind information. On these scales, the curl of the wind stress is important, and scatterometer data must be evaluated for error in this derived parameter. Local events such as upwelling, coastal sea level fluctuations and SST anomalies can correlate with either regional wind patterns, basin average atmospheric averages, or distant atmospheric forcing.

2.11.2. *Ocean waves*. Wave forecasting is closely allied to wind analysis and forecasting. When other microwave satellite sensors are used with a scatterometer, the complicated relationship between surface stress, short waves, and the long waves may be examined. In addition to increased physical understanding of this problem, there exists the possibility of increased accuracy in wave forecasting through parameterizations.

There are fundamental questions regarding the relation between stress and the spectrum of wave characteristics, wave field equilibrium requirements, and the wind. The wave energy is proportional to the fourth power of the wind speed, so that feedback from wave information to stress values is strong. The same factor indicates large errors in wave predictions result from stress errors. The stress-σ_0 algorithm must be tested in various long-wave environments, in conjunction with a good wave model.

2.11.3. *Tropical Oceans*. The tropical oceans present unique characteristics for the application of scatterometer data and modeling techniques. Important surface phenomena take place on scales suited to satellite surveillance. Upper-ocean motions are accompanied by strong, time-varying temperature and topography signals. There is strong correlation with wind forcing and convergence/divergence patterns. The average winds are light (5-8 ms^{-1}), with episodes associated with strong convection events. There are inadequate conventional data to define these phenomena.

The PBL models do not extend to the tropics, being dependent on Coriolis forces. While the important physics in the tropics seems to be a direct ocean response to atmospheric forcing, analyses show that wind field inaccuracies are the largest source of error in model predictions. With the planned ten-year conventional data set from the TOGA project, and concurrent scatterometer data, the evolution of a PBL wind model across the tropics will be greatly facilitated. The evaluation of current flux models such as that of Cane and Zebiak (1985) require better surface wind data, as will be supplied by satellite scatterometers.

3. Future Scatterometers

Plans for the future include more than one scatterometer, providing the capability to define and monitor storms from the sudden small-scale explosive lows (bombs, and polar lows) to hurricanes. More than one scatterometer will provide frequent large-scale coverage for weather and ocean circulation models and complete data for climate studies. These planned instruments include:
▷ ERS-1: scheduled to fly in 1990 by ESA. This will be a basic, one-side swath, three antenna, polar orbiting, microwave scatterometer.
▷ NSCAT: scheduled to fly in 1993 by NASA. It will be a SASS derivative, six-antenna, polar orbiter. It will provide over 80% coverage of the global oceans each day.
▷ SCANSCAT: proposed for EOS (the Earth Observing System proposed for the late 1990s). This will be a scanning, tropical (±30° from the space station) scatterometer designed especially to capture the low winds and high resolution needed in the tropics. It will have an 800 km swath, no gaps, 4-looks, and 25 km resolution.
▷ EOS scatterometer: At least two are planned for the EOS platforms, scheduled for liftoff in 1997.

It is evident that the 99 days of Seasat data have demonstrated that scatterometer data, as surface waves, stress, winds, pressure fields or basic flux parameters, are a tremendously exciting milestone for the understanding of many geophysical phenomena.

Acknowledgments: The author received support from NASA through JPL NSCAT contract 957648.

References
Anderson, D., A. Hollingsworth, S. Uppala and P. Woiceshyn, 1987. A study of the Feasibility of using sea and wind information from the ERS-1 Satellite. *ECMWF Contract report to ESA.*
Banner, M.L. and W.K. Melville, 1976: On the Separation of Air Flow Over Water Waves. *J. Fluid Mech.*, **77**, 825-842.
Blanc, R.V., 1987. Accuracy of Bulk-method-determined Flux, Stability, and Sea Surface Roughness, *J. Geophys. Res.*, **92**, C4, 3867-3876.
Brown, R.A., 1983. On a Scatterometer as an Anemometer. *J. Geophys. Res.*, **88**, C3, 1663-1673.
Brown, R.A., 1980. Longitudinal Instabilities and Secondary Flows in the Planetary Boundary Layer: A Review; *Rev. of Geophysics and Space Physics*, **18** (3), 683-697.
Brown, R.A., 1970. A Secondary Flow Model for the Planetary Boundary Layer. *J. Atmos. Sciences*, **27**, 742-757.
Brown, R.A., W.L. Jones, D.H. Boggs, E.M. Bracalente, T.H. Guymer, D. Shelton and L.C. Schroeder, 1982. Surface Wind Analyses for SEASAT, *J. of Geophys. Res.*, **87**, C3, 3355-3364.

Brown, R.A. and T. Liu, 1982. An Operational Large-scale Marine Planetary Boundary Layer Model, *J. Applied Meteor.* **21**, 3, 261-269.

Cane, M.A. and S.E. Zebiak, 1985: A theory for El Nino and the southern oscillation, *Science*, **228**, 1085-1087.

Cox, C.S., 1958: Measurements of Slopes of High-Frequency Wind Waves. *J. Mar. Res.*, **16**, 199-225.

Donelan, M., 1978: Whitecaps and Momentum Transfer, in *Turbulent Fluxes Through the Sea Surface, Wave Dynamics and Prediction*, edited A. Favre and K. Hasselmann, pp. 273-288.

Duffy, D. and R. Atlas, 1986: The Impact of Seasat-A Scatterometer Data on the Numerical Prediction of the QEII Storm. *J. of Geophys. Res.*, **91**, 2241-2248.

Ekman, V.W., 1905. On the influence of the earth's rotation on ocean currents, *Arkiv. Math. Astron. Fysik* . **2**, 11, 1-52.

Endlich, R.M., D.E. Wolf, C.T. Carlson and J.W. Maresca Jr, 1981. Oceanic wind and balanced pressure-height fields derived from satellite measurements, *Mon. Wea. Rev.*, **111**, 2009-2016.

Favre, A. and K. Hasselman, ed., 1978: *Turbulent Fluxes Through the Sea Surface, Wave Dynamics and Prediction*, Plenum Press, N.Y., 665pp.

Guest, P.S. and K. L. Davidson, 1987: Wind Stress Measurements Over rough Ice During the 1984 MIZ Experiment, *J. Geophys. Res.*, **92**, C7, 6933-6941.

Guymer, T.H., J.A. Businger, W.L. Jones, and R.H. Stewart, 1981: Anomalous Wind Estimates from the Seasat Scatterometer. *Nature*, 294,735-737.

Jones, W.L., L.C. Schroeder, D.H. Boggs, E.M. Bracalente, R.A. Brown, G.J. Dome, W.J. Pierson and F.J. Wentz, 1982. The SEASAT-A Satellite Scatterometer: The Geophysical Evaluation of Remotely Sensed Wind Vector. *J. of Geophys. Res.*, **87**, C3, 3297-3317.

Kraus, E.B., 1972: *Atmosphere-Ocean Interaction*, Claredon Press, Oxford, 275pp.

Levy, G. and R.A. Brown, 1986: A Simple Objective Analysis Scheme for Scatterometer Data, *Jn. Geophys. Res. Oceans*, **91**. C4. 5153-5158.

Longuet-Higgins, M.S., 1978: Dynamics of steep gravity waves in deep water, in *Turbulent Fluxes Through the Sea Surface, Wave Dynamics and Prediction*, edited by A. Favre and K. Hasselmann, pp. 199-218, Plenum, N.Y.

Mognard, N., 1984: Swell in the Pacific ocean observed by Seasat radar altimeter. *Mar. Geod.*, **8**, 183-210.

Monahan, E.C. and G.M. Niocaill Eds. 1986: *Oceanic Whitecaps and their role in air-sea exchange processes*, D. Reidel Pub. Co. 304 pp.

Nuss, W. and R.A. Brown, 1987: Evaluation of Surface Winds and Flux Analysis in Mid-Latitude Marine Cyclones, *J. Dynamics of Atmospheres/Oceans*, **10**, 291-315.

Paulson, C.A., 1970. The Mathematical Representation of Wind Speed, and Temperature Profiles in the Unstable Atmospheric Surface Layer, *J. Applied Math.*, **9**, 857-861,

Phillips, O.M., 1978: Strong Interaction in Wind-Wave Fields. *Through the Sea Surface, Wave Dynamics and Prediction*, edited by A. Favre and K. Hasselmann, 665pp. Plenum, New York.

Woiceshyn, P.M., M.G. Wurtele, D.H. Boggs, L.S. McGoldrick, and S. Peteherych, 1986: The Necessity for a New Parameterization of an Empirical Model for Wind/Ocean Scatterometry, *J. Geophys. Res.*, **91**.

SUNGLINT AND THE STUDY OF NEAR-SURFACE WINDSPEEDS OVER THE OCEANS

Arthur P. Cracknell,
Department of Applied Physics and Electronic
  & Manufacturing Engineering,
University of Dundee,
DUNDEE DD1 4HN.
Scotland, U.K.

ABSTRACT. Sunglint is a phenomenon that is widely neglected or ignored in the use of satellite-remote-sensing techniques to study the oceans. This lecture is concerned with a description of the effect of near-surface winds on sunglint patterns and of the established theory of Cox and Munk for establishing a quantitative relation between near-surface windspeed and the intensity distribution in an area of sunglint. The results from some of our own calculations based on using this theory with AVHRR data for the Mediterranean are presented. Correlations between near-surface windspeeds and the warming of the surface layer, as manifested in thermal infrared channel data from the AVHRR, are also discussed. The point is made that sunglint-derived near-surface windspeeds could provide an alternative to in-situ measurements of windspeeds for the validation of algorithms for the derivation of windspeeds from the microwave instruments on ERS-1.

1. INTRODUCTION

Perhaps I should explain that in taking on the role of a director of a summer school I do not programme myself in to give any of the lectures. However, I do stand by in reserve in case there should happen to be an empty slot in the programme. I had, however, for some time been preparing a lecture on sunglint for other purposes. Since it is relevant to winds and waves and since no one else was scheduled to talk about this at this summer school I thought I would inflict this lecture on sunglint on the summer school participants. I should hasten to add that

- much of what I shall say is well-established; I am not propounding anything really new. The ideas are all to be found already published in the literature. One of the best descriptions of sunglint is to be found in the report by Fett (1977) which is not very easy to obtain.

125

R. A. Vaughan (ed.), Microwave Remote Sensing for Oceanographic and Marine Weather-Forecase Models, 125–139.
© 1990 Kluwer Academic Publishers.

- I do have two PhD students working on this subject, Mr S. Khattak and Mr W.G. Huang. I shall refer to their results at a later stage.
- I shall cite some important references to some major work on the subject.
- The subject has some potential relevance to ERS-1 and to the validation of ERS-1-derived data.

2. SUNGLINT

In most applications of scanner data one is either dealing with diffusely reflected radiation, in the visible or near-infrared bands, or with emitted radiation, in the thermal infrared bands. In such cases sunglint, that is the direct specular reflection of sunlight at the surface of the sea, is something that people usually regard as a nuisance and they try to avoid it. Thus, for example, on the NIMBUS-7 satellite there was special provision made to introduce a tilt with a mirror so as deliberately to avoid observing any areas of sunglint. That was reasonable enough in the context of what CZCS was concerned with, i.e. studying ocean colour and water quality. We take a different view and try to see whether we can positively use sunglint to obtain some useful oceanographic or meteorological information. Basically, it is possible to study the area and intensity of a region of sunglint and to extract information about the surface roughness, i.e. the waveheight, and therefore the near-surface windspeed. The theory is largely due to Cox and Munk (1954, 1956).

Let us consider what is involved. Suppose that the surface of the sea is perfectly flat, i.e. it behaves as a plane mirror. Then if the satellite's scanner (the AVHRR for instance) is looking in the right direction for specular reflection there will be an enormously bright signal, see Figure 1.

One can do a simple calculation to estimate the area of very high brightness in the image. The angle subtended at the Earth, or at the scanner, by the Sun's diameter is D/H where D is the diameter of the Sun and H is the distance from the Sun to the Earth. The area of intense illumination at the surface is then an ellipse, the minor axis of which is of length $(h/\cos \theta) \times (D/H)$ where θ is, typically, an angle of about $30°$ or $40°$. Putting in values of these quantities gives a minor axis of this ellipse of about

$$\frac{900 \times 10^3 \text{ m}}{\cos 30°} \times \frac{2 \times 6.95 \times 10^8 \text{ m}}{1.49 \times 10^{11} \text{m}}$$

$$\simeq \frac{2 \times 2}{\sqrt{3}} \times \frac{900 \times 6.95}{1.49} \text{ m} \simeq 9.7 \text{ km}$$

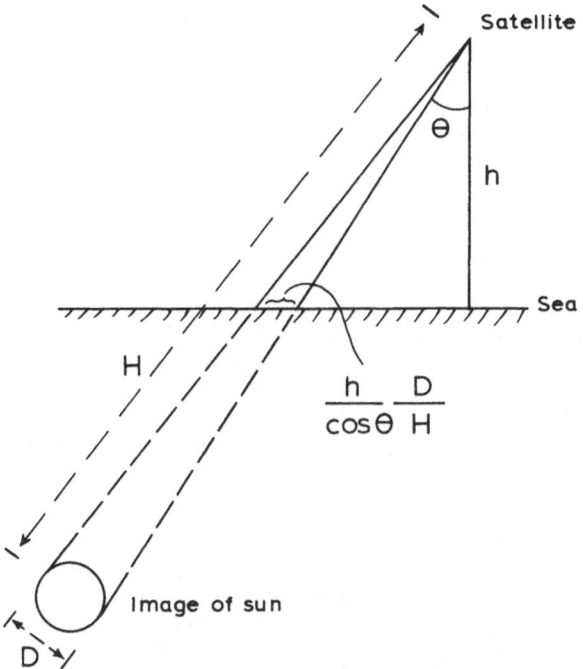

$$\frac{h}{\cos\theta} \quad \frac{D}{H}$$

Fig. 1. Sketch to illustrate specular reflection from the Sun at a smooth surface (not to scale).

Therefore, from a perfectly calm sea one would expect to obtain a dazzling white ellipse in the image with a minor axis of about 9 AVHRR pixels.

   There is one complication that ought to be mentioned. Most satellite-derived images are obtained these days from scanners not from cameras and the argument given above only really applies to the case of a photograph in which the whole scene is imaged virtually instantaneously. With a scanner such as the AVHRR a complete scene is imaged over a period of several minutes, during which time the satellite travels a large distance. The details of the geometry of the imaging process become much more complicated and the general consequence is that the sunglint pattern becomes extended in a direction parallel to the satellite's motion, which is approximately north-south (or vice versa). Fig. 2 illustrates the geometry and the procedure for using this diagram to calculate the primary specular point (PSP), the centre of the sunglint pattern, is as follows:

1.   Plot the satellite subpoint track on a gnomonic projection base chart.
2.   Locate the position of the solar subpoint on the base chart.

128

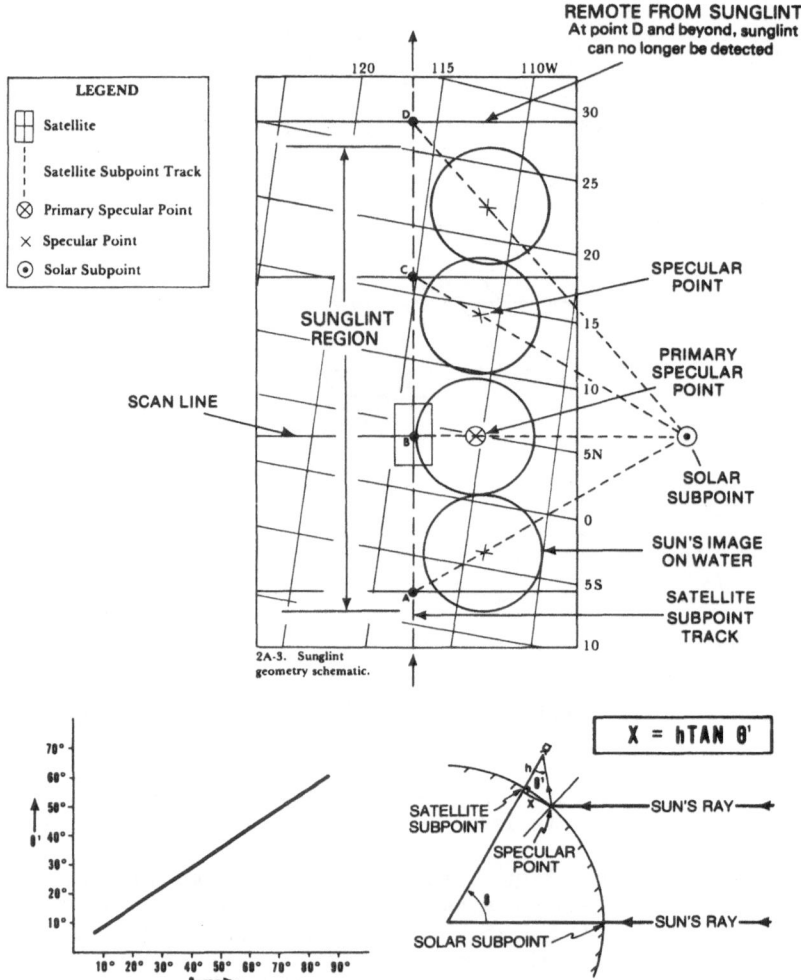

**Fig. 2.   Diagram to illustrate the geometry of the formation of a
sunglint pattern (after Fett 1977).**

3.   From the solar subpoint draw a line perpendicular to the satellite
subpoint track.
4.   Measure the distance from the solar subpoint to the intersection
of the perpendicular line to the satellite subpoint track in n mi.
Convert to degrees of latitude (1 degree latitude = 60 n mi) to
obtain θ.
5.   Enter θ in the nomogram (Fig. 2) and find θ'.
6.   Use formula $x = h\tan\theta'$ to calculate the distance of the PSP
from the satellite subpoint along the perpendicular line to the
solar subpoint.

One can obviously just look manually through an archive of satellite data, such as the one that we have in Dundee, and search for examples of scenes with sunglint data. However, there are a very large number of such scenes and one thing that one of my students, Mr. S . Khattak, has done was to write a FORTRAN program using the orbital parameters and the laws of physics to predict which orbits of NOAA-series satellites can actually lead to the possibility of sunglint in the AVHRR data. It is necessary to satisfy the laws of refection, namely:

(i)   the incident ray (from the Sun), the reflected ray (to the AVHRR) and the normal at the surface of the Earth must be in the same plane and

(ii) the angle of incidence must be equal to the angle of reflection. The details of satisfying these conditions involves some quite complicated geometry and spherical trigonometry.   The testing of his computer program is described in a paper we presented at the Remote Sensing Society's annual conference in 1987 at Nottingham (Cracknell et al. 1987).

So far we have assumed that the surface of the sea is perfectly smooth and occasionally that may be very nearly true.   However, waves are nearly always present and we need to examine that situation in some detail.

Suppose that the sea surface is not smooth but has a single wave present, see Figure 3.   In this figure the scale of the waves,

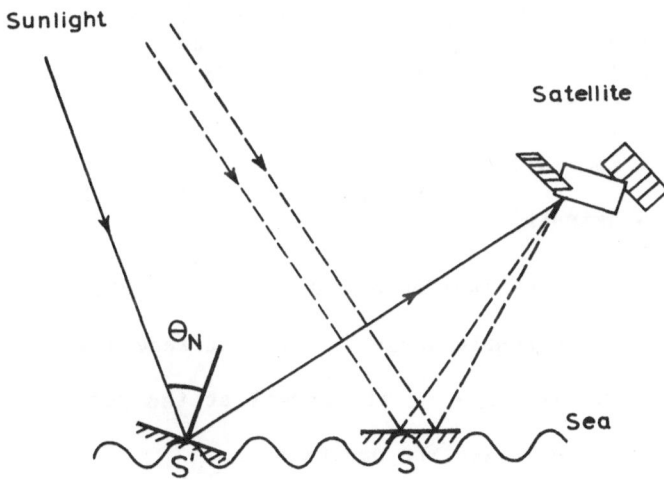

Fig. 3.   Sketch to illustrate specular reflection from the Sun at a surface with the shape of a sinusoidal wave.

relative to the IFOV (instantaneous field of view), of the scanner is wrong. Consider a wavecrest (or trough) in the IFOV; the original geometry still survives because the tangent plane is horizontal. For parts of the wave near to a trough or crest, where the inclination of the tangent plane is quite small, the specular reflection conditions will still lead to the reflected ray reaching the scanner on the satellite. But for other parts of the wave, where the inclination of the tangent plane is greater, the geometrical conditions required for sunglint no longer apply. Thus, the intensity received from the area S is reduced. The greater the amplitude (height) of the waves then, qualitatively, the greater the reduction in the sunglint intensity that can be expected.

However, if we consider an area outside S then there will be some part of the wave for which the tangent is now sloping at an angle for which it happens to be in the right orientation to satisfy the conditions for specular reflection. Thus, whereas for a calm sea we would get no specular reflection except from the small area S, now we get some specular reflection from an area outside S.

Thus we see, in qualitative terms that if the sea is not calm, the sunglint will be (a) enlarged in area and (b) reduced in intensity.

3. THEORY OF COX AND MUNK

The theory which was developed by Cox and Munk (1954) was published long before anyone could reasonably foresee the vast amount of relevant satellite data that would become available. This theory is concerned with obtaining a quantitative relation between the sunglint pattern and the near-surface windspeed.

I should like to summarise the theory of Cox and Munk. We consider a piece of wave (not in S) from which specular reflection occurs, then

$$\left. \begin{array}{l} \theta = \text{zenith} \\ \alpha = \text{azimuth} \end{array} \right\} \text{of reflected ray relative to Sun}$$

β = angle of incidence = angle of reflection

ξ = solar zenith angle at ocean surface and

θ_N = zenith angle of the normal at the point on the wave at which reflection occurs.

Then, from some spherical trigonometry, one finds that

$$\cos \theta_N = \frac{\cos \theta + \cos \xi}{2 \cos \beta} .$$

Then Cox and Munk showed empirically that for uniform ocean surface roughness there was a Gaussian distribution of wave slopes with a distribution function

$$P(\theta_N, V) = \frac{1}{2\pi\sigma^2} \left(\exp - \frac{\tan^2\theta_N}{\sigma^2} \right)$$

where V is the near-surface windspeed. The standard deviation was found to be related (empirically) to the windspeed V by

$$\sigma^2 = 0.00512 \ V + 0.003$$

where V is expressed in ms^{-1}. Then we have to relate the satellite received radiance, $L_s(\lambda)$, to the probability distribution of the slopes, i.e. of the θ_N. $L_s(\lambda)$ can be written in the form

$$L_s(\lambda) = L_w(\lambda) + L_{sk}(\lambda) + L_g(\lambda)$$

where

$L_w(\lambda)$ = diffuse radiance reflected from the water surface

$L_{sk}(\lambda)$ = diffuse sky radiance

and

$L_g(\lambda)$ = sunglint radiance.

If we consider an area that is far away from the sunglint area then only $L_w(\lambda)$ and $L_{sk}(\lambda)$ are present. Therefore we take $L_s(\lambda)$ from the satellite data for the sunglint area and we assume that we can take $L_w(\lambda)$ and $L_{sk}(\lambda)$ from the satellite data from elsewhere in the scene and thus determine $L_g(\lambda)$. This does assume that $L_w(\lambda)$ and $L_{sk}(\lambda)$ do not vary significantly between the sunglint area and elsewhere. Then we are able to make use of the expression

$$\frac{L_g(\lambda) \ \cos^4\theta_N \ \cos\theta}{L_s(\lambda) \ r(\omega)} = P(\theta_N, V)$$

where $r(\omega)$ is the Fresnel reflectivity, to determine $P(\theta_N, V)$. Thus for a given pixel in the sunglint area we get $L_g(\lambda)$ and $L_s(\lambda)$ from the satellite data, do all the geometry to find θ_N for this pixel and then use this equation to determine $P(\theta_N, V)$. The equations given previously can then be used to determine σ from $P(\theta_N, V)$ and then, from this value of σ, to determine V, the windspeed corresponding to this pixel. Thus it is possible to determine a pixel-by-pixel map of the near-surface windspeed throughout the sunglint area.

4. SOME WINDSPEED RESULTS

Various people have used the theory of Cox and Munk to calculate windspeeds from satellite data. We claim no uniqueness but it is easier to demonstrate this with our own results. What we have done is to consider a scene from 15 June 1982 and to obtain windspeed data for the Mediterranean for that day from the U.K. Meteorological Office. Two points should be noted, first, the Meteorological Office's windspeeds were not given at exactly the locations used in the sunglint-derived windspeed calculations and we had to do some weighting to account for this. Secondly there was a time difference, of the order of 60-90 minutes, between the acquisition of the meteorological data and of the satellite data. The results obtained by Cracknell et al. (1988) for 34 points in the sunglint area in the data from 15 June 1982 are shown in Table 1.

The differences between the satellite-derived and Meteorological Office-derived windspeeds in table 1 vary from +1.0 to -0.8 ms^{-1}. This agreement is quite gratifying but I would make the usual caveat that several other people have made, namely that we do not really know how much of the error is attributable to the satellite-derived values and how much is attributable to the Meteorological Office-derived values.

5. NON-UNIFORM WINDFIELDS

What we have described so far in our qualitative description of sunglint, or in terms of the theory of Cox and Munk, has assumed a uniform windfield. However, this is often not the case in practice. It is common to find dark areas within a sunglint pattern and there are various important cases to consider.

(a) There may be dark areas within the region near the centre of the sunglint pattern. These correspond to calm areas and are illustrated schematically in Fig. 4 which is taken from some calculations of McClain and Strong (1969). This figure is based on calculations of the sunglint intensity for a uniform wind speed of 5 ms^{-1} except in the shaded area, which would appear dark on an image, for which a wind speed of zero was assumed.

An example of such a feature in an AVHRR image is shown in Fig. 5. The large dark patch SE of Mallorca and the smaller dark patches north of Mallorca and Menorca in Fig. 5(a) (visible) and Fig. 5(b) (near infrared) are all indicative of calm areas. In Fig. 5(d) (thermal infrared) these calm areas show up as exhibiting local heating of the surface layer (see Robinson et al. (1984) for example) and fit quite

TABLE 1: Wind speeds

No.	WINDSPEED Weighted Met.Data	Computed	Difference	No.	WINDSPEED Weighted Met.Data	Computed	Difference
	ms^{-1}	ms^{-1}	ms^{-1}		ms^{-1}	ms^{-1}	ms^{-1}
1	4.7	4.0	0.7	18	5.0	4.7	0.3
2	4.3	4.1	0.7	19	4.7	4.9	-0.2
3	4.9	4.1	0.8	20	4.6	5.4	-0.8
4	4.9	4.1	0.8	21	4.8	5.4	-0.6
5	4.9	4.2	0.7	22	5.1	5.7	-0.6
6	5.0	4.2	0.8	23	4.6	3.7	0.9
7	5.0	4.8	0.2	24	5.0	4.4	0.6
8	5.0	4.4	0.6	25	5.0	4.7	0.3
9	5.0	4.5	0.5	26	4.9	5.2	-0.3
10	5.0	4.8	0.2	27	4.9	4.4	0.5
11	5.0	5.0	0.0	28	5.1	4.3	0.8
12	4.8	5.2	-0.4	29	4.9	5.3	-0.4
13	4.6	4.6	0.0	30	4.8	5.2	-0.4
14	4.2	4.9	-0.7	31	6.2	5.2	1.0
15	4.2	4.9	-0.7	32	5.0	4.4	0.6
16	4.7	4.9	-0.2	33	5.5	4.5	1.0
17	4.7	4.7	0.0	34	5.0	4.8	0.2

exactly with the areas shown in bands 1 and 2 (Figs. 5(a) and (b)).
This is printed, as band 4 in Fig. 5(d), with dark corresponding to a
high intensity of radiation and light, e.g. the clouds in the
north-west corner of the scene, corresponding to a low intensity of
radiation.   For band-3 data (3.5 - 3.9 μm) shown in Fig. 5(c) the
sunglint area is dominated by reflected radiation at this wavelength;
this is printed, like Fig. 5(d), with dark corresponding to a high
intensity of radiation and light corresponding to a low intensity.
The calm areas noted from the other bands reflect less radiation to the
scanner (as in Figs. 5(a) and (b)) and hence appear light in Fig. 5(c)
too;   they are also very well co-registered with the band-1 and band-2
data shown in Figs. 5(a) and (b).

Another example is given in Fig. 6 where there are two calm areas
apparent.   One is south-west of Sicily and one is north-east of Sicily
just south of the edge of the clouds;   these are seen as dark in
Figs. 6(a), (b) and (d) and light in Fig. 6(c).   Similar correlations
between areas of low windspeed and local heating of the sea surface
have been noted, for example, by Saunders et al. (1981) in data from
the North Sea.

134

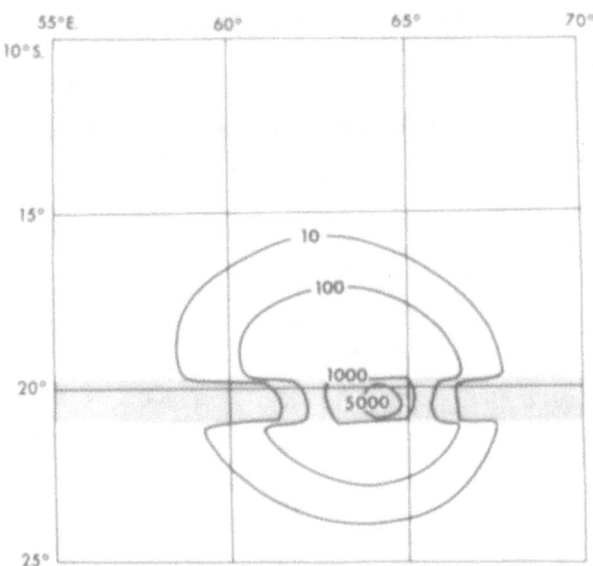

Fig. 4. Theoretical sea-surface sunglint for a composite pattern. Insopleths are relative reflected intensity per 10^4 steradian incident flux; satellite height 722 km (from McClain and Strong 1969).

(b) There may be dark areas well away from the PSP (primary specular point) and in the large sunglint area of low intensity indicating calm conditions as illustrated schematically in Fig. 7. This, like Fig. 4, is taken from McClain and Strong (1969) and is based on a uniform windspeed of 5 ms^{-1} everywhere except in the shaded area, for which a windspeed of zero was assumed.

(c) If, however, the windspeed becomes quite large, then the sunglint pattern becomes very much reduced in intensity. An example is shown in Fig. 8 where a high wind rushing out through the Straits of Gibraltar from the Mediterranean destroys the sunglint in an area to the west of the Straits see, for example, Scorer (1986). According to Strong and Ruff (1970) this occurs at windspeeds of about 10 ms^{-1}. That is, dark patches may occur or, if the windspeed is large over the whole area, the whole sunglint pattern will disappear completely. Thus with dark patches, particularly in the extended sunglint area, one needs to be a little careful because they may arise from very calm conditions or they may arise from rather rough sea conditions and high windspeeds; ambiguities may need to be resolved by reference to other data.

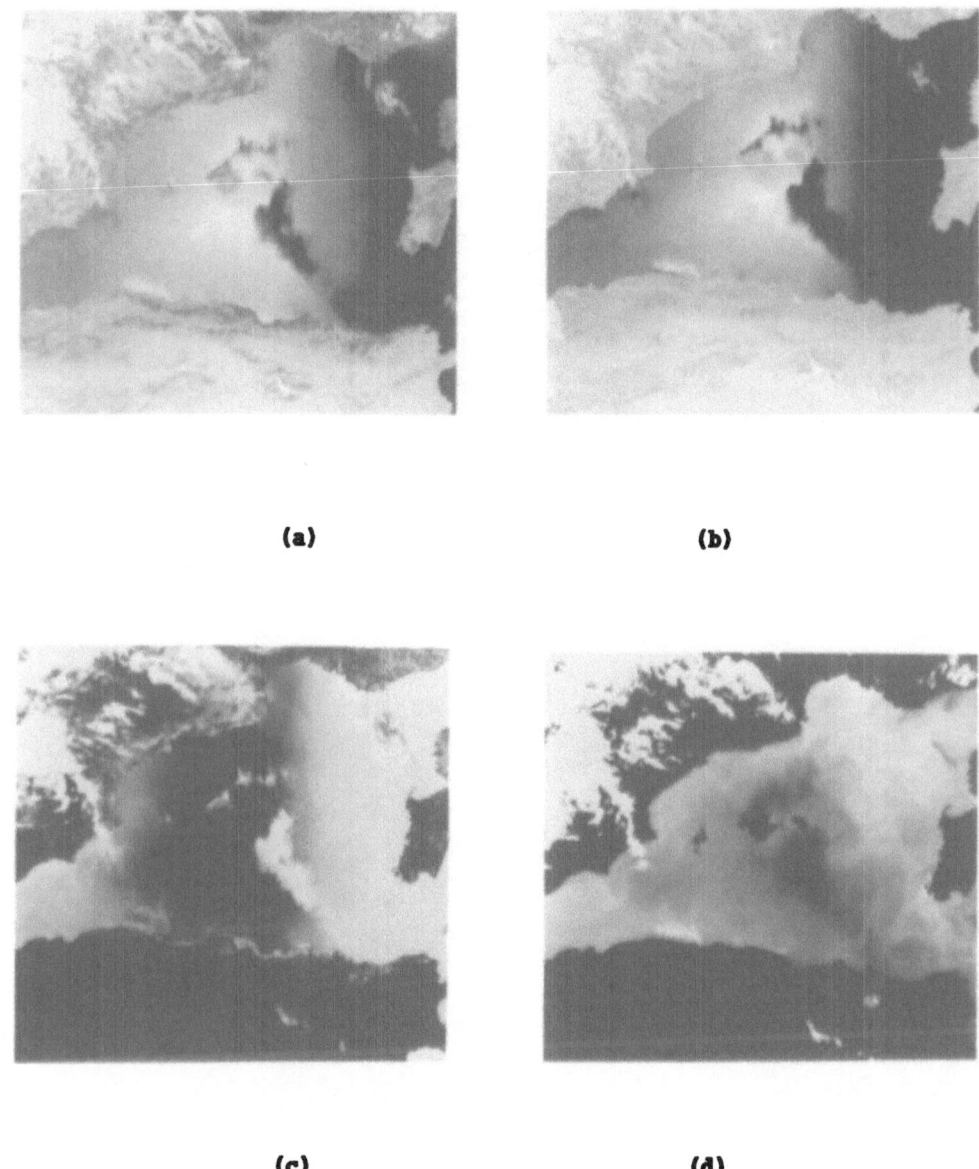

(a)   (b)

(c)   (d)

Fig. 5. NOAA-7 images, orbit no. 5266, 1348 GMT on July 1982. (a) band 1, (b) band 2, (c) band 3 and (d) band 4 (copyright Dundee University).

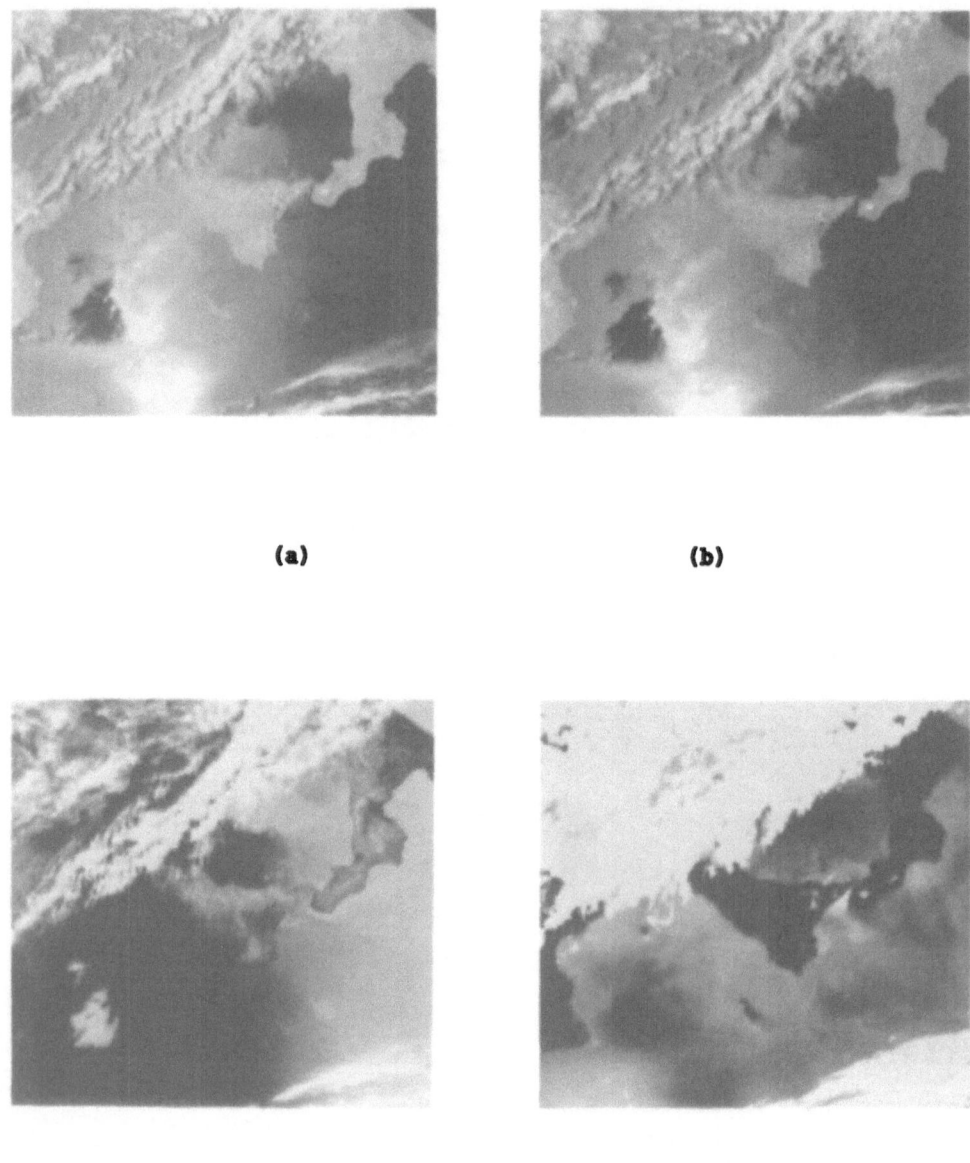

(a)   (b)

(c)   (d)

Fig. 6   NOAA-7 images, orbit no. 4715, 1310 GMT on 23 May 1982
         (a) band 1,   (b) band 2,   (c) band 3 and   (d) band 4
         (copyright Dundee University).

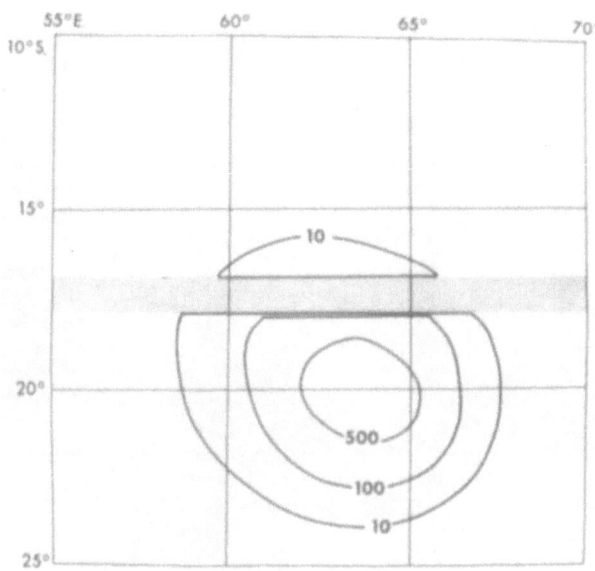

Fig. 7.   Theoretical sea-surface sunglint for a composite pattern,
with isopleths and satellite height as in Fig. 4 (from
McClain and Strong 1969).

6.   RELEVANCE TO ERS-1

One might wonder what is the relevance to ERS-1 and this summer
school.   Generally one thinks in terms of validating ERS-1-derived
windspeeds with in-situ measurements.   I do not dispute the value of
that.   But what I would also like to suggest is that if AVHRR sunglint
data and ERS-1 data can be obtained simultaneously for a given area
then we could use the sunglint-derived wind speeds as an alternative to
in-situ data for ERS-1 validation purposes.   It may not work - for
example the orbits may not fit - but it is worth thinking about.
Alternatively, it may be possible to obtain sunglint data from the
ATSR/M on ERS-1.

ACKNOWLEDGEMENTS

I am grateful to Mr P.E. Baylis for the supply of the AVHRR data
and the loan of the manual by Fett (1977).

138

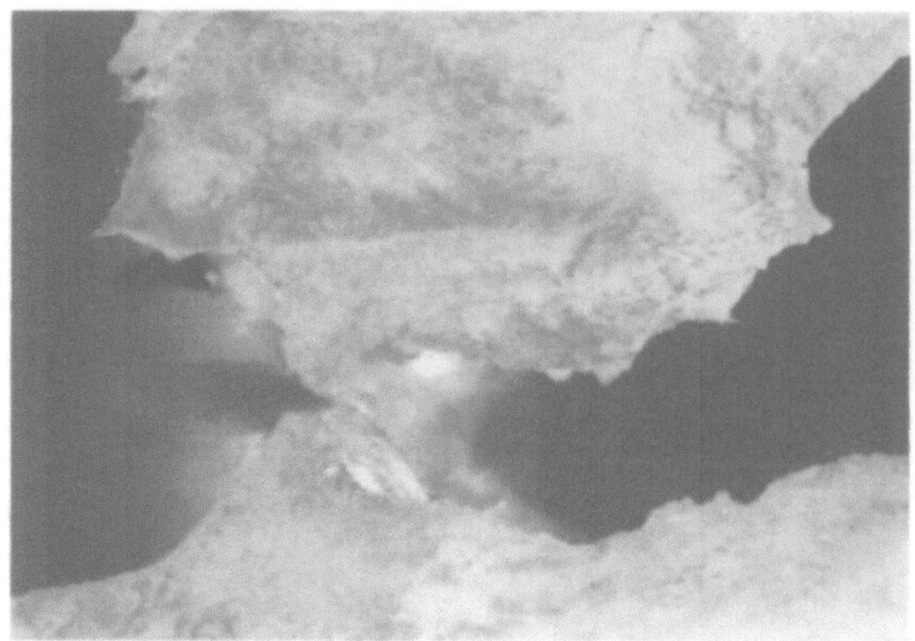

Fig. 8.   Effect of wind through Straights of Gibraltar, NOAA-7, orbit
           no. 5224 GMT on 28 June 1982 (copyright Dundee University).

REFERENCES

Cox, C. and Munk W., 1954, 'Measurement of the roughness of the sea
   surface from photographs of the sun glitter' Journal of the Optical
   Society of America, **44**, 838-850.
Cox, C. and Munk W., 1956, 'Slopes of the sea surface deduced from
   photographs of sun glitter', Bulletin of the Scripps Institute of
   Oceanography, **6**, 401-488.
Cracknell, A.P., Khattak, S. and Vaughan, R.A., 1987, 'The automatic
   identification of areas of sunglint in AVHRR data; in Advances in
   Digital Image Processing, Proceedings of the Annual Conference of the
   Remote Sensing Society, Nottingham, September 1987 (Nottingham:
   Remote Sensing Society) 403-415.
Cracknell, A.P., Khattak S. and Vaughan, R.A., 1988, 'Sunglint in AVHRR
   data and the determination of near surface windspeeds' International
   Archives of Photogrammetry and Remote Sensing, **27**, B7, 118-124.

Fett, R.W., 1977, 'Navy Tactical Applications Guide, Volume 1, Techniques and Applications of Image Analysis'. (Park Ridge, Illinois: Walter A. Bohan).

McClain, E.P. and Strong, A.E., 1969, 'On anomalous dark patches in satellite-viewed sunglint areas' Monthly Weather Review, 97, 875-884.

Robinson, I.S., Wells, N.C. and Charnock, H., 1984, 'The sea surface boundary layer and its relevance to the measurement of sea surface temperature by airborne and spaceborne radiometer', International Journal of Remote Sensing, 5, 19-45.

Saunders, R.W., Ward, N.R., England, C.F. and Hunt, G.E. 1981, 'Sea surface temperature measurements around the U.K. derived from Meteosat and TIROS-N data' in Matching Remote Sensing Technologies and their Applications, Proceedings of an International Conference held in London in December 1981 (Nottingham: Remote Sensing Society) pp 191-8.

Scorer, R.S. 1986 'Cloud investigation by Satellites' (Chichester: Ellis Horwood).

Strong, A.E. and Ruff, I.S., 1970, 'Utilizing satellite-observed solar reflections from the sea surface as an indicator of surface wind speeds' Remote Sensing of Environment, 1, 181-185.

SATELLITE INFRARED SCANNING RADIOMETERS – AVHRR AND ATSR/M

P. J. MINNETT
Applied Oceanography Group
SACLANT Undersea Research Centre
Viale San Bartolomeo 400
I-19026 San Bartolomeo
La Spezia
Italy

ABSTRACT. By providing data of higher spatial resolution and, in some cases, of higher absolute accuracy, satellite infrared scanning radiometers complement those detecting microwave radiation. Infrared measurements are, however, much more susceptible to contamination by the presence of cloud. In this paper the physical principles of the emission of infrared radiation from the sea-surface and its transfer through the atmosphere are presented. There follows descriptions of two satellite radiometers: the operational AVHRR and experimental ATSR/M. The method of retrieving quantitative measurements of SST is given and applications of infrared radiometer data are discussed.

1. Introduction

Imaging infrared radiometers have now been flown on many meteorological and earth observation satellites, both in geosynchronous and low, near-polar orbits. The images from these infrared scanning radiometers can provide information at much higher spatial resolution than can those from microwave radiometers. As far as the measurement of sea-surface temperature (SST) is concerned, the infrared radiometers also produce higher absolute accuracy, but the data are subject to contamination by cloud cover. This cloud cover, however, is the information, rather than the contaminant, sought by some meteorologists. Microwave radiometers, on the other hand, can provide measurements of the ocean surface in all weather conditions. The infrared and microwave satellite radiometers therefore provide data that are complementary, both in form and applications.

This paper provides an overview of satellite infrared radiometers, beginning with the basic principles of radiation emission and transfer through the atmosphere. Two instruments are described in some detail: the Advanced Very High Resolution Radiometer (AVHRR), a series of which have been flown on US operational near-polar orbiting meteorological satellites for decade now; and the Along Track Scanning Radiometer with Microwave Sounder (ATSR/M) which is an experimental device being developed for the earth-observation satellite, ERS-1, of the European Space Agency. The derivation of algorithms to derive SST, the

141

R. A. Vaughan (ed.), Microwave Remote Sensing for Oceanographic and Marine Weather-Forecase Models, 141–163.
© *1990 Kluwer Academic Publishers.*

accuracies of the SST measurements and some examples of applications are also discussed.

Other scanning radiometers, such as those on geosynchronous satellites, such as the METEOSAT radiometer (Campbell, 1982) and the Visible Infrared Spin-Scan Radiometer (VISSR), are not dealt with, nor are radiometers designed for deriving atmospheric profiles, e.g. HIRS, the High Resolution Infrared Sounder (Schwalb, 1978; Smith *et al.*, 1979) and VAS, the VISSR Atmospheric Sounder (Menzel *et al.*, 1981; Smith, 1983).

2. Physical principles

The satellite infrared radiometers are passive instruments that measure the electromagnetic energy falling on their detectors. The source of the energy is whatever is in the instantaneous field of view (IFOV) defined by the instrument's optical system.

The energy in an infinitesimal wavelength interval, $d\lambda$, radiated by a perfectly emitting surface is described by the Planck function, B:

$$B(T,\lambda)d\lambda = \frac{2\pi hc^2 d\lambda}{\lambda^5(e^{ch/\lambda\kappa T} - 1)}, \tag{1}$$

where B is the spectral radiance at wavelength λ μm in units of W m^{-2} sr$^{-1}\mu$m^{-1}, i.e. it is the energy per unit wavelength interval radiated per unit area of the source into unit solid angle; h is Planck's constant, κ is Boltzmann's constant and c is the speed of light in vacuum. The form of the Planck function is shown in Fig. 1a for the infrared part of the spectrum $\lambda = 2$ to 22 μm for temperatures of $0°C$, $10°C$, $20°C$ and $30°C$, which largely encompass the range of temperatures of the sea-surface.

The peak of the emitted radiance is in the 9 to 11 μm interval, going to shorter wavelengths with increasing temperatures (Wien's Displacement Law). Further, the temperature dependence of the emitted radiance (Fig. 1b) is high in this part of the spectrum, and, when expressed as a proportional change $B^{-1}\partial B/\partial T$, it increases dramatically with decreasing wavelength ($\sim T^5$ at $\lambda = 10$ μm, $\sim T^{13}$ at $\lambda = 3.7$ μm).

At these infrared wavelengths the sea-surface is not quite a perfect emitter. The ratio of the radiance emitted by the sea to that of a black-body, the sea-surface emissivity, ϵ_λ, is dependent on wavelength and emission angle. These dependences are weak, being only a few percent (Friedman, 1969), but are important nevertheless. The temperature dependence of ϵ is very weak in the infrared.

In contrast, the physics governing the emission of microwave radiation leads to a much different situation. At longer wavelengths, well away from its peak value (say for $\lambda >\sim$ 100μm at $T \sim 300$ K; e.g. Schanda, 1986), the Planck equation approaches a linear function of temperature – the Rayleigh-Jeans approximation – and so the radiometric advantage of the infrared is lost. Additionally, the sea-surface emissivity in the microwave region is much less than unity (~ 0.39 at $\lambda = 3$ cm for an emission angle of $0°$and T$=20°C$), and has significant dependence on surface roughness (i.e. on wind and waves), on temperature and, for $\lambda >\sim 10$ cm, also on salinity.

Planck Function

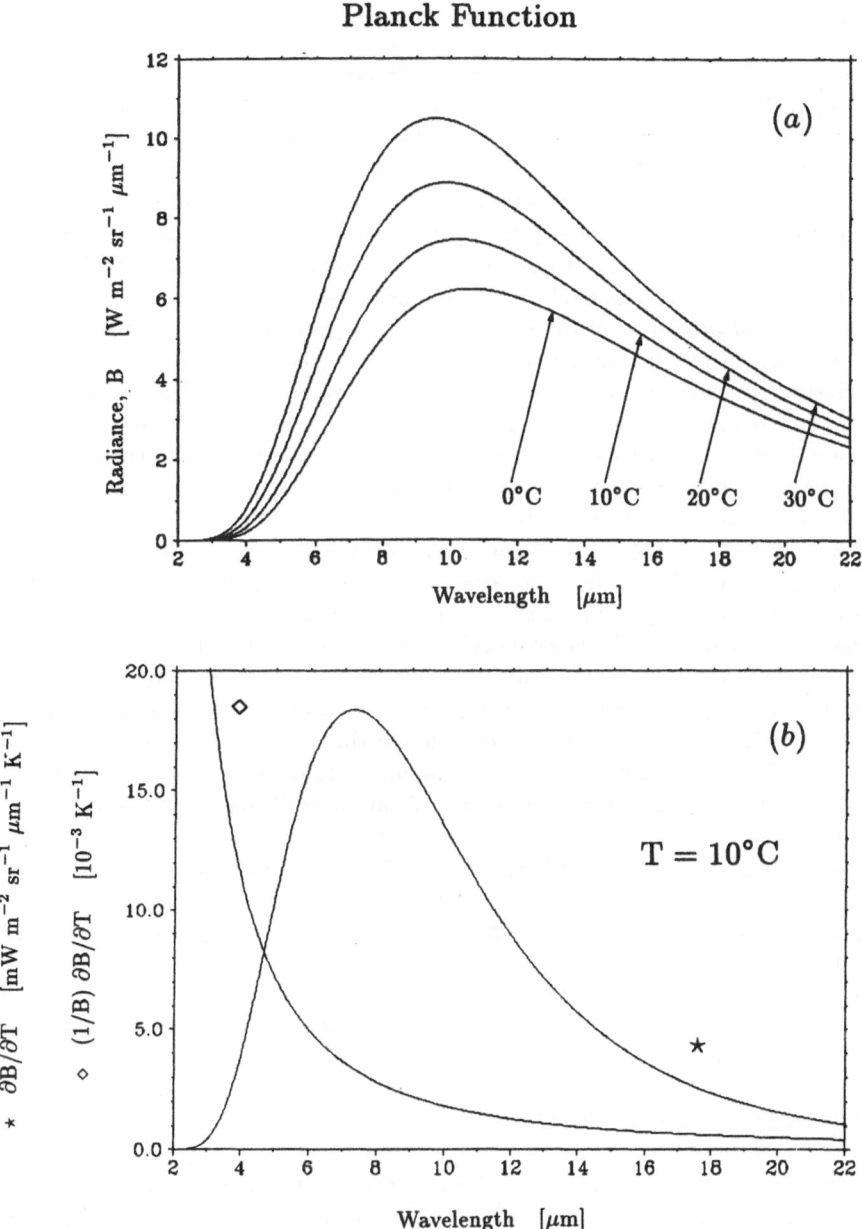

Fig. 1 The spectral infrared radiance of a perfectly emitting surface at temperatures of 0°C, 10°C, 20°C and 30°C, (a). This is the Planck function and increases at all wavelengths with increasing temperature. The temperature dependence of the black-body emission is shown in (b) as absolute and proportional values (both curves have the same numerical values).

3. Radiative transfer in the atmosphere

The energy received by a satellite radiometer directed towards the sea-surface has components originating from the sea, the atmosphere, and the sun (Fig. 2a). In the atmosphere, the processes of absorption, emission and scattering are caused by both molecules and suspended particles, such as dust, water-droplets, ice-crystals and sea-salt crystals. At the sea-surface emission and reflection of the incident radiation take place, and for radiation for which the sea-water is partially transparent, scattering from suspended material and reflection from the bottom in shallow seas can be important components of the measured signal. By confining ourselves to cloud-free conditions and to a relatively narrow wavelength interval in the infrared part of the electromagnetic spectrum, many of the components illustrated schematically in Fig. 2a can be safely neglected. The remaining components of the radiation received by the satellite radiometer are shown in Fig. 2b and are (a) the radiation emitted at the sea-surface, but modified by its passage through the atmosphere; (b) the radiation emitted by the atmosphere (including aerosols) into the radiometer field of view; and (c) the radiation emitted by the atmosphere (and aerosols) and reflected at the sea-surface into the radiometer field of view.

Throughout the atmosphere, emission and absorption are dependent on the local temperature, on the density of aerosols, on the density of molecular types composing the gaseous atmosphere, and on the spectral properties of those molecules, which in turn are temperature and pressure dependent. The atmospheric molecules of interest are water vapour (H_2O), carbon dioxide (CO_2), ozone (O_3), nitrous oxide (N_2O) and methane (CH_4). With the exception of water vapour, these can be considered 'well-mixed' throughout the atmosphere and not to exhibit significant seasonal or regional changes. Atmospheric water vapour, on the other hand, is extremely variable and, in addition to the line absorption and emission of infrared radiation by molecular rotation and vibration, it displays quasi-continuous absorption and emission which is dependent on the water vapour pressure. This has been termed the 'water vapour continuum' and is presumed to be caused by the presence of water vapour dimers, $(H_2O)_2$, and possibly larger water vapour polymers. The continuum effects form an important part of the total atmospheric contribution to the satellite radiometer signal, especially for moist atmospheres.

The density of aerosols also exhibits much spatial and temporal variability. The similarity in size between the wavelength of the infrared radiation and some aerosol particles means that they can be effective scatterers.

3.1. MATHEMATICAL FORMULATION

The change in spectral radiance $L(\lambda)$ of a non-divergent, monochromatic beam, at wavelength λ, when it traverses normally an infinitesimally thin layer of atmosphere, thickness dz at height z, and within which all properties are uniform and constant, is given by

$$dL(\lambda) = -\kappa(z, \lambda)L(\lambda)dz, \qquad (2)$$

where $\kappa(z, \lambda)$ is the absorption coefficient of that layer. This is Bouguet's Law. The

(a)

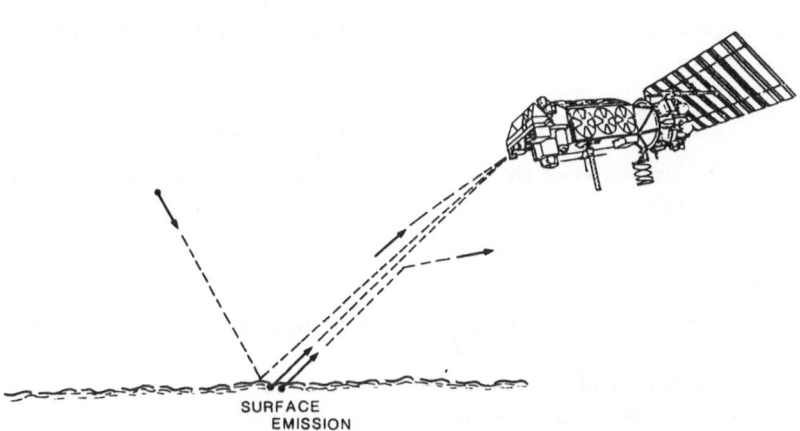

Fig. 2 Schematic diagram of the physical processes involved in the radiative transfer through the atmosphere. In the general case (a), the satellite radiometer signal has many components from the sun, atmosphere and ocean. In the limited wavelength range of the thermal infrared at ~ 10 to $\sim 13\mu$m, there are only three significant components (b).

atmosphere in the layer dz is also emitting radiation into the beam, $dL_e(\lambda)$, which, assuming local thermodynamic equilibrium, is equal to the local absorption, by Kirchoff's Law,

$$dL_e(\lambda) = \kappa(z, \lambda)B(z, \lambda, T_A)dz, \tag{3}$$

where $B(z, \lambda, T_A)$ is the blackbody spectral radiance at the local atmospheric temperature T_A, and B is given by Planck's Equation (Eq. 1).

Thus,

$$\frac{\mathrm{d}L}{\mathrm{d}z} = B(z, \lambda, T_A)\kappa(z, \lambda) - L(z, \lambda)\kappa(z, \lambda). \tag{4}$$

Solving this linear differential equation using the conditions at the sea-surface, $z = 0$, and the height of the satellite, $z = H$, results in

$$L_H(\lambda) = L_0(\lambda) \exp\left[-\tau(0, H)\right] + \int_0^H B(z, \lambda, T_A)\kappa(z, \lambda) \exp\left[-\tau(z, H)\right]\mathrm{d}z, \tag{5}$$

where $L_H(\lambda)$ is the spectral radiance measured at the satellite, $L_0(\lambda)$ is the upwelling spectral radiance at the bottom of the atmosphere and

$$\tau(z_1, z_2) = \int_{z_1}^{z_2} \kappa(z')\mathrm{d}z', \tag{6}$$

is the optical thickness of the layer $z_2 - z_1$. When $z_1 = 0$ and $z_2 = H$, τ is the optical depth of the atmosphere.

$L_0(\lambda)$ is the sum of the emission of the sea-surface at temperature T_s, and the reflected downwelling atmospheric radiation, $r(\lambda)L{\downarrow}(\lambda)$, where $r(\lambda)$ is the spectral reflectance of the sea-surface. Hence

$$L_0(\lambda) = \epsilon(\lambda)B(T_s, \lambda) + r(\lambda)L{\downarrow}(\lambda). \tag{7}$$

where $\epsilon(\lambda)$ is the emissivity of the sea-surface, which is related to the spectral reflectance through Kirchoff's Law,

$$r(\lambda) = 1 - \epsilon(\lambda). \tag{8}$$

Thus

$$r(\lambda)L{\downarrow}(\lambda) = (1 - \epsilon(\lambda)) \int_H^0 B(z, \lambda, T_A)\kappa(z, \lambda) \exp\left[-\tau(z, 0)\right]\mathrm{d}z. \tag{9}$$

In reality the channels of a satellite radiometer are not monochromatic, but have a finite width (Fig. 3). Thus it is necessary to integrate the spectral radiance at satellite heights, $L_H(\lambda)$ in Eq. (5), across the width of the channel, using $\phi_i(\lambda)$, the spectral response for channel i, to give the channel radiance L_i,

$$L_i = \int_0^\infty L_H(\lambda)\phi_i(\lambda)\mathrm{d}\lambda \Big/ \int_0^\infty \phi_i(\lambda)\mathrm{d}\lambda. \tag{10}$$

In the general case, when the beam does not propagate normally to the atmospheric stratification (which is assumed to be locally horizontal), but at a local zenith angle θ, the thickness of the infinitesimal layer $\mathrm{d}z$ must be replaced by $\sec\theta\,\mathrm{d}z$. This approximation, which neglects the changing refractive index through the atmospheric column, is valid for $\theta < 60°$; for larger zenith angles a more complicated expression should be used (e.g. Kondratyev, 1969).

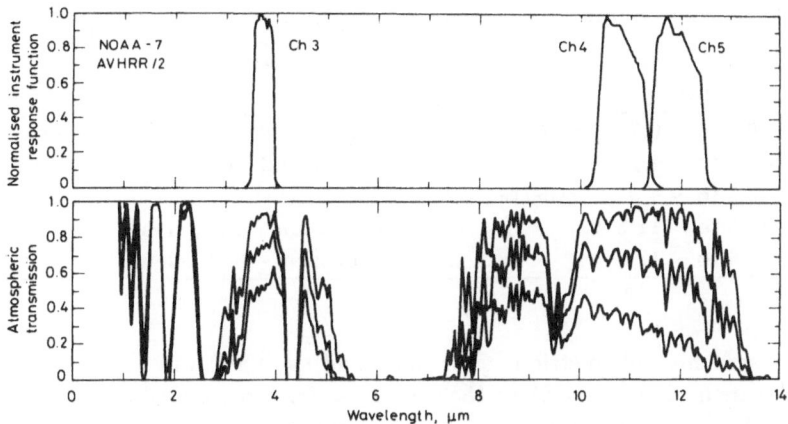

Fig. 3 Theoretical spectra of atmospheric transmission, at nadir, in the infrared region at 1 to 14 μm wavelength. The three spectra correspond to different amounts of precipitable water (polar: 7 mm; mid-latitude: 29 mm; tropical: 54 mm). The response functions of channel 3 ($\sim$3.7 μm), channel 4 ($\sim$11 μm) and channel 5 ($\sim$12 μm) of the AVHRR/2 on the NOAA-7 satellite are also shown. The different dependence of atmospheric transmission on water vapour amounts at each channel provides a correction for the atmospheric effect in SST measurement by multichannel methods. (After Llewellyn-Jones *et al.* 1984).

3.2. ATMOSPHERIC CORRECTIONS

Without detailed knowledge of the condition of the atmosphere, it is not possible to solve the radiative transfer problem (Eqs. (6), (7), (9) and (10)) and thereby derive the SST, T_s, from a radiance measurement of a single channel, L_i. However, the atmospheric transmission spectra for various atmospheric conditions (Fig. 3) have differing dependencies at different parts of the spectra. For example, the change in the atmospheric transmission in going from dry, cold, arctic conditions to warm, moist, tropical ones is more pronounced in the wavelength interval 11.5 to 12.5 μm (i.e. close to AVHRR/2 channel 5) than 10 to 11 μm (i.e. close to AVHRR/2 channel 4). This spectral dependence implies that coincident measurements at different channels can be used to correct for the effect of the atmosphere, and hence to measure the SST without recourse to ancillary measurements of the atmospheric profiles. This is the basis of the commonly-used multichannel correction for the effects of the atmosphere in SST measurements using data from AVHRR/2.

An alternative approach to making a correction for the atmospheric effects is to utilise its path-length dependence and make near-simultaneous measurements of the same piece of the sea-surface at different zenith angles. This was first done by Saunders (1967) using a radiometer mounted on an aircraft. He noted that the temperature deficit (i.e. the

Table 1. Channels of the Advanced Very High Resolution Radiometer, AVHRR

Channel	Wavelength (μm)	Signal
1	~ 0.6 to ~ 0.7	reflected solar energy – clouds, coastlines, vegetation
2	~ 0.7 to ~ 1.1	reflected solar energy – clouds, coastlines, vegetation
3	~ 3.5 to ~ 4.0	reflected solar and thermal emission – clouds, SST
4	~ 10.3 to ~ 11.3	thermal emission – SST, clouds
5	~ 11.5 to ~ 12.5	thermal emission – SST, clouds

difference between his radiometric measurement and the SST) was about twice as large for measurements inclined at $\sim 60°$ to the vertical as for those taken at nadir. A number of researchers have since followed the same approach by combining measurements from two satellites viewing the same area of the ocean at different zenith angles from polar-orbiters and geosynchronous satellites (Chedin *et al.*, 1982; Holyer, 1984) or from two geosynchronous satellites (Takayama *et al.*, 1983). The first purpose-designed satellite radiometer to measure SST using this technique is the Along Track Scanning Radiometer (ATSR/M), being developed for the ESA satellite ERS-1. Unlike the multichannel case, the multiangle atmospheric correction is based on a real measurement of the atmospheric effects.

4. The Advanced Very High Resolution Radiometer

The Advanced Very High Resolution Radiometer Mk. 2 (AVHRR/2) is a five-channel imaging device which forms part of the payload of the TIROS-N series of polar-orbiting weather satellites. There are two such satellites in operation at any time (currently NOAA-9 and NOAA-10) in sun-synchronous orbits; one overhead at about 0230 UTC and 1430 UTC, and the other at about 0730 UTC and 1930 UTC. The satellites are about 840 km above the sea-surface and have orbital periods of ~ 100 minutes. The channel characteristics are given in Table 1. However, it should be noted that the AVHRRs on TIROS-N, NOAA-6, NOAA-8 and NOAA-10 were without channel 5.

The instrument builds up images of the surface by scanning across the width of its swath, ~ 3150 km centred at nadir, using a rotating mirror inclined at $45°$ to the axis of rotation, which lies along the flight vector of the satellite (Fig. 4). The rotation rate is such that adjacent scan lines are contiguous, or in fact overlap, at the surface. The spatial resolution is 1.1 km at nadir but becomes poorer towards the edge of the swath due to geometrical effects.

The infrared channels are in the so-called 'atmospheric windows' where the atmosphere is relatively transparent (Fig. 3). The two longwave channels are at wavelengths close to the peak of the Planck function for emitted energy at marine surface temperatures (Fig. 1a), and are used primarily for SST determination. Unlike these, the 3.7 μm channel

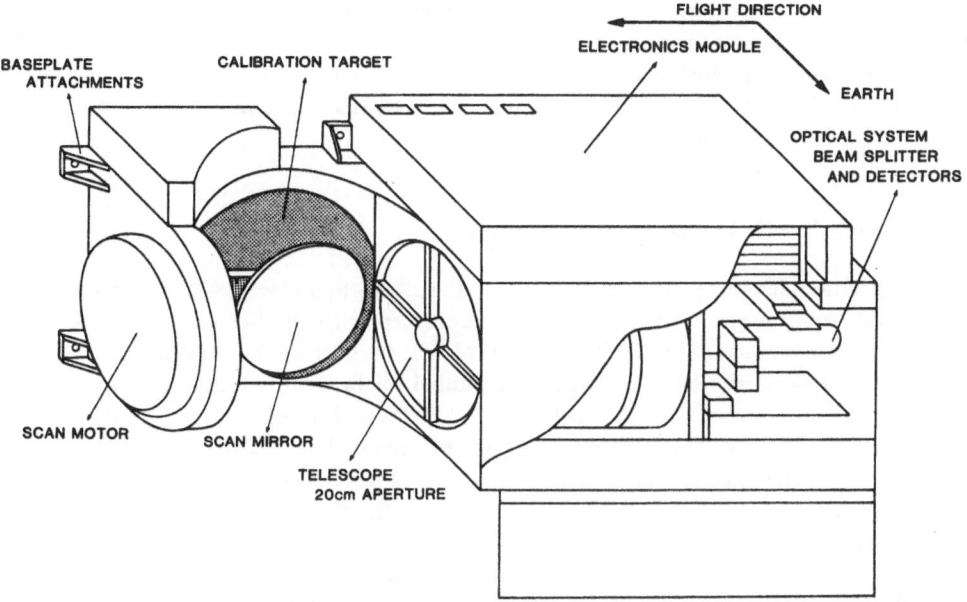

FLIGHT DIRECTION

ELECTRONICS MODULE

EARTH

OPTICAL SYSTEM
BEAM SPLITTER
AND DETECTORS

BASEPLATE
ATTACHMENTS

CALIBRATION TARGET

SCAN MOTOR

SCAN MIRROR

TELESCOPE
20cm APERTURE

ADVANCED VERY HIGH RESOLUTION RADIOMETER - NOAA POLAR - ORBITERS

Fig. 4 A schematic diagram of the Advanced Very High Resolution Radiometer, AVHRR. (After ITT, 1979).

is susceptible to contamination by reflected solar radiation during the day, and so it is used for SST measurements only when the viewing geometry is such that the solar contamination is impossible, or at least unlikely, or during the night. Data from this channel are useful in detecting and classifying clouds both during night and day. The two shortwave channels are also used for cloud-detection and for delineating coastlines. They are also used to determine the 'vegetation index' over land. The next generation of AVHRRs, planned to be introduced in about 1993, will have a sixth channel in the near-infrared at $\lambda \sim 1.6$ μm, which will be switchable with the channel at 3.7 μm. The new channel will be useful only during the daylight parts of each orbit, and will provide an improved measurement of marine aerosols and a better discrimination between snow, ice and clouds.

The AVHRR/2 data stream includes measurements in the infrared channels of an on-board black-body calibration target and of space (zero radiance measurement). The temperature of the black-body, which is close to terrestrial temperatures, is monitored by four platinum resistance thermometers. The noise levels in the two longwave channels is low, being given as 0.12 K for a 300 K target, but in reality they appear to be much lower. The

3.7 μm channels, however, have had much higher noise levels, often worsening with the age of the instrument. The data from each channel are digitised to 10-bit resolution (0–1023).

Information about the instrumental properties is published by NOAA as each new satellite becomes operational. Detailed descriptions of the AVHRR instrument and data stream are given by Schwalb (1978) and Lauritson et al. (1979).

5. The Along Track Scanning Radiometer

The Along Track Scanning Radiometer with Microwave Sounder* (ATSR/M) combines multichannel, multiangle, infrared imaging radiometry with dual-channel nadir-directed microwave measurements (Fig. 5). The ATSR/M will be part of the experimental component of the payload of the ERS-1 satellite, scheduled for launch in 1990 into a near-polar, sun-synchronous orbit at a nominal altitude of ∼780km and a local time of ∼1030 UTC.

In the case of ATSR/M, the rotating scan mirror is inclined so that it sweeps out a cone of half-angle 23.45°, the axis of which is tilted forwards, along the spacecraft's flight direction, by the same angle. As a consequence, the 500 km swath width of ATSR is narrower than that of the AVHRR, but it is scanned twice, at zenith angles close to and including 0° (the nadir scan), and at zenith angles up to 55° (the forward scan). The ground speed of the sub-satellite point is ∼6.67 km s^{-1}, and the distance of ∼905 km between the centres of the two scans is covered in ∼135 s. The ground resolution of the infrared channels is 1km square at nadir, increasing to ∼2 km × ∼4 km at the centre of the forward view. Towards the edges of the swaths, the contrast in zenith angles between the overlapping pixels of the two swaths becomes less. The scan geometry and relative dispositions of the two swaths are shown in Fig. 6. The relatively narrow swath width of the ATSR/M means that global coverage will not be possible for the shorter repeat orbit cycles. For example, complete coverage will be possible poleward of ∼44° for the 3-day repeat cycle planned for the start of the ERS-1 mission. The unsampled areas become smaller as the repeat cycle lengthens, vanishing for cycles of 6 days and more.

The infrared channels of the ATSR/M will be very similar to those of the next generation of AVHRR, i.e. ∼ 1.6μm, ∼ 3.7μm, ∼ 10.8μm, ∼ 11.9μm, with the ∼ 1.6μm, and ∼ 3.7μm, data being selectively telemetered.

---

* ATSR/M is being provided to the European Space Agency by a consortium consisting in the UK of Rutherford Appleton Laboratory, who are responsible for the overall design and manufacture of the instrument; Oxford University Department of Atmospheric, Oceanic and Planetary Physics, who are responsible for the conceptual design, together with the scientific calibration and testing of the instrument; Mullard Space Science Laboratory of University College London, who are principally responsible for the on-board black-bodies; the UK Meteorological Office, who are providing the infrared detector package; and in France, the Centre de Recherches en Physique de l'Environment Terrestre et Planétaire, Issy-les-Moulineaux, who are providing the microwave radiometer. There is also a significant contribution from Australia, from where the flight model digital electronics package will be provided, as well contributions to the data validation activities.

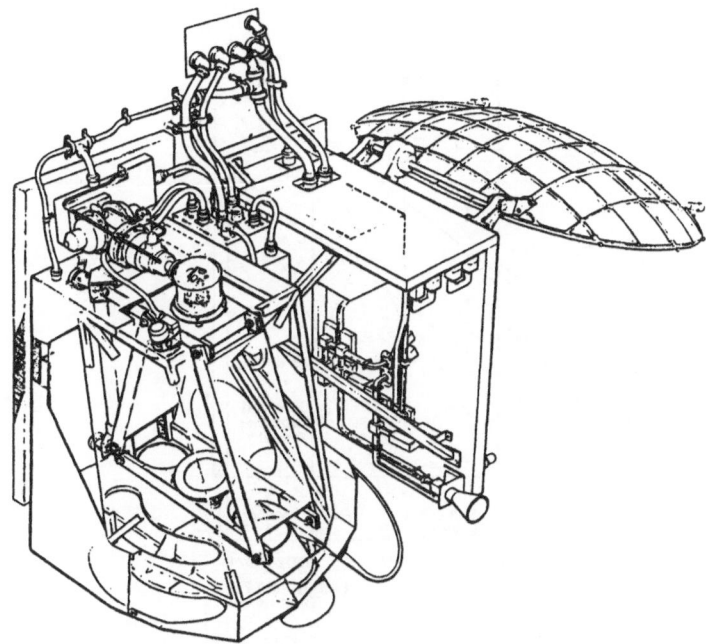

Fig. 5 A schematic diagram of the Along Track Radiometer with Microwave Sounder, ATSR/M. The infrared radiometer is to the left, in front of the microwave radiometer, the antenna of which is deployed in space.

The inherent noise levels of the ATSR/M detectors will be lower than those of AVHRR because they will operate at a lower temperature of $\sim$77 K compared with the $\sim$105 K. This is achieved by using new mechanical coolers instead of passive radiative coolers. The reduced swath width of ATSR also means that the sampling time of each pixel is longer than for AVHRR (75μs compared with 25μs). As a result the noise equivalent temperature difference (NEΔT) will be smaller than for AVHRR, e.g. $\sim$0.03 K for a 310 K scene. To exploit this, the data must be digitised to greater resolution and the on-board radiometric calibration must be very good. A number of selectable digitisation schemes will be programmed, including 11 or 12 bit resolution, as well as to the AVHRR-standard of 10 bits. The 11 and 12 bit data will be coded as "real words" using a mantissa and exponent. The on-board calibration will be achieved by using two purpose-designed black-body targets whose temperatures will be controlled to span the range of SST signals. These black-bodies are situated between the two apertures defining the nadir and forward scans and will be sampled during each mirror rotation (Figs. 5 and 6).

The microwave radiometer of ATSR/M will take measurements in the nadir at 23.8 and 36.6 GHz, with a spatial resolution of $\sim$20km. The channels will provide a precise

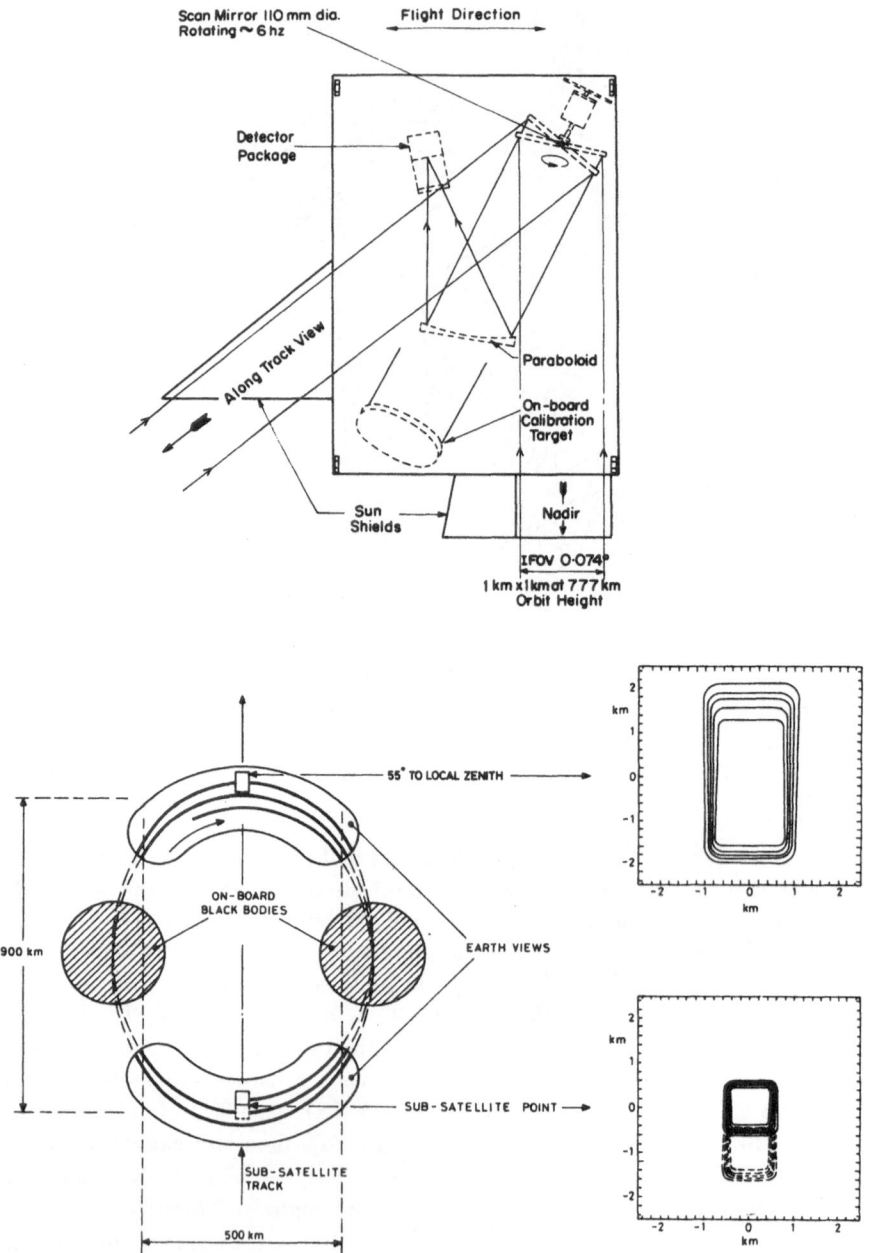

Fig. 6 The scan geometry of the infrared component of the ATSR/M, showing the relative positions of the forward and nadir scans and the on-board black-body calibration targets. The relative pixel sizes at the centres of the nadir and forward scans are also shown.

measurement of the total atmospheric water vapour content which will be used to improve the infrared SST retrieval, and to provide the correction for the effects of water vapour on the ERS-1 altimeter range measurement.

Further details of microwave radiometry are given by Bernard (this volume) and Pedersen (this volume), and of the ATSR/M in general by Llewellyn-Jones (1985) and Muirhead and Eccles (1987).

6. The SST retrieval algorithm

In practice the multichannel SST retrieval algorithm takes the form of a simple linear combination of the temperatures measured in the different channels, T_i:

$$T_s = a_0 + \sum_{i=1}^{n} a_i T_i, \tag{11}$$

where T_s is the SST measurement, a_i are dimensionless coefficients, n is the number of channels, and a_0 is a constant temperature. A full derivation of this expression is given elsewhere (e.g. McMillin and Crosby, 1984; Minnett, 1988). The minimisation of errors in the SST measurement is dependent on the correct choice of the coefficients a_0, a_i.

There are two approaches to determine the coefficients and these can be conveniently referred to as the empirical and the simulation methods. The empirical approach requires the collection of high-quality measurements from *in-situ* thermometers, such as on drifting meteorological buoys, that are coincident with satellite measurements. Then a regression analysis of the *in-situ* temperatures and the satellite data produces the coefficients. The alternative approach uses a computer model of atmospheric radiative transfer together with a large set of atmospheric profiles to simulate the satellite measurements in a range of conditions. The simulated measurements are then used with the associated SST values to derive the coefficients, again by regression analysis. In both methods, the regression analysis also provides an estimation of the accuracy with which the SST value can be derived.

The advantage of the empirical method is that it uses real satellite data and real *in-situ* data in real conditions. It does, however, have some disadvantages:

(a) The satellite measurement is of the radiation temperature, which, after correction for the atmospheric effect, is attributable to the surface 'skin' of the ocean, while the *in-situ* measurement is generally taken at a depth of a few centimetres to a few metres (the so-called 'bulk' temperature). Because of the heat exchange between ocean and atmosphere, the surface skin temperature is generally several tenths of a Kelvin colder than the bulk temperature (Robinson, Wells and Charnock, 1984).

(b) Spatial variability in the SST field introduces uncertainty into the comparison: the satellite measurement is a near-instantaneous spatial average whereas the *in-situ* measurement is not.

(c) The presence of undetected clouds in the satellite data is a likely source of significant error.

(d) The number of usable coincident sets of measurements is relatively small because many are invalidated by the effects of clouds in the satellite data. Generally, the

RADIATIVE TRANSFER MODELLING FOR SATELLITE DATA SIMULATION

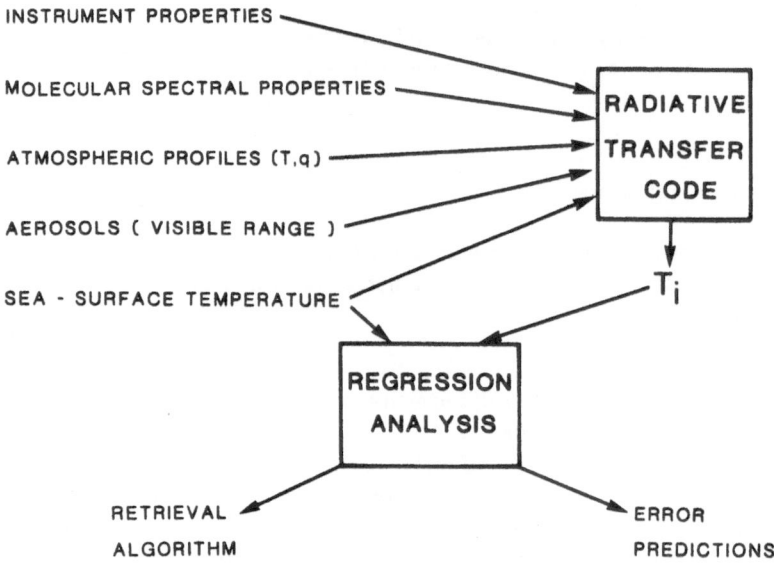

INSTRUMENT PROPERTIES

MOLECULAR SPECTRAL PROPERTIES

ATMOSPHERIC PROFILES (T,q)

AEROSOLS (VISIBLE RANGE)

SEA - SURFACE TEMPERATURE

RADIATIVE
TRANSFER
CODE

T_i

REGRESSION
ANALYSIS

RETRIEVAL
ALGORITHM

ERROR
PREDICTIONS

Fig. 7 Block diagram of the procedure for the numerical simulations of satellite radiometric measurements.

condition of coincidence must be relaxed to include *in-situ* temperatures taken within several hours of the satellite overpass, thus introducing the possibility of errors caused by temporal or advective temperature changes.

(e) Calibration errors in both the satellite and *in-situ* data are carried over into the SST retrieval expression. This could be considered an advantage if calibration errors in the satellite data are constant, as they would be compensated at each application of the retrieval expression.

The simulation approach is shown schematically in Fig. 7. It avoids all of the problems of the empirical approach, but has others:

(a) Errors can arise through inadequate knowledge of the spectral properties of the atmospheric constituents, and their dependences on temperature and pressure, which is especially relevant to the water vapour 'spectral continuum', and through assumptions necessary in the formulation of the numerical model, such as the effects of aerosols.

(b) Measurement errors in the atmospheric profiles will introduce errors in the simulated satellite data. Random measurement errors in individual profiles are not important, provided the profile remains physical, as the data are used to represent realistic distributions rather than the true state of the atmosphere at the time and location of each profile.

(c) Since the simulations are of skin temperature measurement, the predictions of the modelling, i.e. the coefficients of the retrieval expression and its accuracy, are difficult to verify properly. Discrepancies in predictions and validation data could be caused by several effects which may not be readily identifiable.

The main advantage of the simulation approach is that it permits controlled numerical experiments to investigate the behaviour of the predicted algorithm in response to variations in the input information. In particular, one can study its response to different atmospheric conditions or its dependence on particular variables, such as the satellite zenith angle and spectral response functions of different channels. The simulations may also be performed in advance of the launch of an instrument, and the retrieval algorithm can be derived ahead of time.

6.1. AVHRR SST RETRIEVAL ALGORITHM

The two AVHRR/2 channels most usually used for SST retrieval are 4 and 5, which share the atmospheric 'window' at 10 to 13 μm (Fig. 3). Thus they are sometimes called the 'split-window' channels. Their proximity supports the assumptions needed to linearise the radiative transfer problem and derive the multichannel retrieval expression, which can be written

$$T_s = a_0(\theta) + a_4(\theta)T_4(\theta) + a_5(\theta)T_5(\theta), \tag{12}$$

where the numerical subscripts 4 and 5 are the channel numbers, and θ is the local zenith angle to the satellite sensor, measured at the sea-surface. The zenith angle dependence is required not only because of the changes in atmospheric path length across the instrument swath, but also because of the dependence of the sea-surface emissivity, ϵ, on emission angle. The surface emissivity, weighted by the normalised channel transfer functions, is smaller for channel 5 than for channel 4 at all emission angles.

The atmosphere is generally cooler than the ocean surface below and as a result the brightness temperatures measured in space are usually lower than the sea-surface temperature. Further, since the atmosphere is less transparent for channel 5 than for channel 4, the atmospheric contribution to the channel 5 signal is greater and thus T_5 is generally cooler than T_4.

6.2. ATSR/M SST RETRIEVAL ALGORITHM

The algorithm for the retrieval of SST from measurements at different zenith angles results in an expression of the same form as the multichannel algorithm. Thus SSTs will be retrieved from the ATSR/M infrared data using Eq. 11, but where T_i now refers to temperatures measured at the forward and nadir views instead of, or as well as, at different channels. The correct combination of channels and views, and the values of the coefficients are a subject of continuing research (using numerical simulations), as is the optimum way of including the explicit water vapour information from the microwave sounder in the SST retrievals.

156

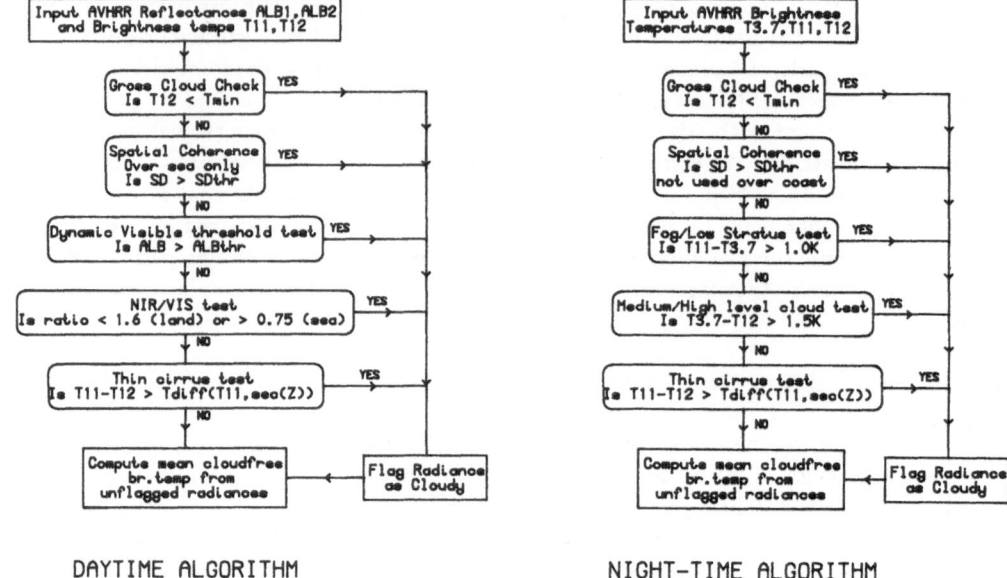

DAYTIME ALGORITHM   NIGHT-TIME ALGORITHM

Fig. 8 Flow diagram of the operations necessary for a scheme to detect clouds in a procedure for SST retrieval from AVHRR/2 data. 'ALB1' and 'VIS' refer to measurements from channel 1, and 'ALB2' and 'NIR' to those from channel 2; 'T3.7', 'T11' and 'T12' are the temperatures measured at channels 3, 4 and 5; 'SD' is the standard deviation of a small group of adjacent pixels; 'thr' means a threshold value, and 'Tdiff' is a temperature difference threshold that is a function of the satellite zenith angle 'Z'. (From Saunders and Kriebel, 1988b).

6.3. CLOUDS

The presence of clouds in the radiometer IFOV poses the biggest limitation on the use of infrared measurements for SST determination. Unlike microwaves, infrared radiation from the sea-surface does not pass through even very thin clouds, and it is enough for just a few percent of the area of a pixel to contain clouds for it to be a major source of error in the derived SST. Consequently, it is imperative to identify all pixels in which clouds obscure the sea-surface and remove them from the SST retrieval procedure. It is beyond the scope of this paper to describe the various techniques available for cloud identification in infrared line-scanner data, and this is done elsewhere (e.g. Bernstein, 1982; Coakley and Bretherton, 1982; Crosby and Glasser, 1978; Llewellyn-Jones *et al.*, 1984; McClain *et al.*, 1985; Minnett *et al.*, 1984; Phulpin *et al.*, 1983; Saunders, 1986; Saunders and Kriebel, 1988a and b). It is now generally realised that more than one single test is required to identify all types of cloud, and Fig. 8 shows a typical scheme developed by Saunders and Kriebel (1988b).

7. Accuracies

The results of some numerical simulations of AVHRR/2 measurements are given in Table 2 as SST retrieval coefficients for NOAA-7 split-window measurements for a number of satellite zenith angles. These coefficients are applicable to measurements over the north-eastern Atlantic Ocean or other areas with similar atmospheric conditions (Table 2a), or in the topics (Table 2b). The predicted rms uncertainties are also given, and these increase with increasing zenith angle, and with decreasing atmospheric transmission, i.e. in the topics (Llewellyn-Jones *et al.*, 1984; McMillin and Crosby, 1985). The predicted uncertainties are very small, and this may be due to a too restricted range of atmospheric conditions; subsequent results, using a much larger set of profiles for north Atlantic summer conditions, give an rms error of ~0.14 K for nadir measurements (Minnett, 1986).

Comparisons between *in situ* data and satellite retrievals fail to corroborate these predicted accuracies, and generally reveal rms errors of ~0.5 K at best, and mean errors of a few tenths of a degree (Llewellyn-Jones *et al.*, 1984; Schluessel *et al.*, 1987; Strong and McClain, 1984). However, much of this could result from the method of comparison (i.e. skin temperature *vs.* bulk temperature; spatial average *vs.* point measurement; lack of true coincidence of measurements etc. – as elaborated in section 6 – and it must be concluded that these comparisons do not constitute a critical test.

The improvements in accuracy that are anticipated with the ATSR/M are shown in Table 3, for a number of NEΔT values for the detectors, and a number of combinations of channels and zenith angles. The values are derived from simulations using a set of radiosondes describing global atmospheric variability, and it is assumed that the ATSR/M infrared channels have the same characteristics as those of the NOAA-7 AVHRR/2 (Zavody, 1982). By comparing the corresponding multiangle and multichannel values it can be seen that, provided the detector noise level is low, the multiangle approach gives about a twofold improvement in accuracy.

8. Applications

The use of satellite data in oceanographic and meteorological models is dealt with in other chapters of this volume, and elsewhere (e.g. Vaughan, 1987), but it is worthwhile to give examples here of some applications of infrared measurements.

The spectral peak of the incident solar energy arriving at the Earth is in the visible part of the spectrum, where the atmosphere is relatively transparent. So, over the major, oceanic portion of the globe, that part of the energy (~70%) that is not reflected directly back into space, from cloud tops for example, is absorbed in the main not by the atmosphere

Table 2. NOAA−7 AVHRR 'split window' SST retrieval coefficients, and predicted errors.

a) N. Atlantic conditions − 61 atmospheric profiles.

Zenith angle	Airmass	a_0 (K)	a_4	a_5	rms errors (K)
00.0°	1.00	−0.334	2.6710	−1.6689	0.07
36.9°	1.25	0.246	2.8478	−1.8479	0.08
48.2°	1.50	−0.017	2.9610	−1.9597	0.09
55.2°	1.75	−1.503	3.0011	−1.9932	0.11
60.0°	2.00	−5.595	2.9038	−1.8795	0.14

b) Tropical conditions − 39 atmospheric profiles.

Zenith angle	Airmass	a_0 (K)	a_4	a_5	rms errors (K)
00.0°	1.00	−17.182	3.9078	−2.8524	0.36
36.9°	1.25	−24.769	4.2469	−3.1667	0.48
48.2°	1.50	−33.612	4.5808	−3.4711	0.61
55.2°	1.75	−44.223	4.8849	−3.7391	0.74
60.0°	2.00	−54.367	5.1959	−4.0156	0.88

(After Llewellyn−Jones et al., 1984)

Table 3. ERS-1 ATSR/M predicted global SST retrieval errors.

Wavelengths (μm)	Multichannel			Multiangle		
	NEΔT=0.0	0.05	0.12	0.0	0.05	0.12
3.7	−	−	−	0.141	0.201	0.350
11	−	−	−	0.404	0.438	0.569
3.7,11	0.262	0.276	0.334	0.083	0.136	0.222
3.7,11,12	0.192	0.208	0.265	−	−	−
11,12	0.407	0.461	0.644	0.202	0.352	0.509

Predicted rms SST errors using multispectral and multiangle (zenith angles 0° and 60°) techniques for a range of detector noise levels. All temperatures are in K. (After Zavody, 1982).

but by the upper tens of metres of the sea. The major currents redistribute this heat around the oceans, and the subsequent release of the energy to the atmosphere drives the atmospheric circulation. The ocean-atmosphere heat exchange mechanisms, both turbulent and radiative, are strongly dependent on the SST. Thus SST fields derived from satellite measurements provide information to assist in the determination of the distribution of oceanic heat content and transport, and of the energy input to the marine atmosphere.

It is often the departures from the mean SST distribution that are important, rather than the distribution itself, and one of the most significant short-term climate fluctuations is associated with anomalous distributions in SST in the tropical Pacific Ocean. This is the 'El-Niño–Southern Oscillation' which occurs every few years and persists for about one year. The large-scale zonal SST gradient is largely removed with anomalies of up to 6 K appearing in the eastern Pacific (Plate 2). There is local ecological havoc and weather disturbances that can also reach higher latitudes, being caused by the response of the atmospheric circulation to changed SST patterns (e.g. Hoskins and Karoly, 1981).

Remaining in the tropical Pacific but looking to shorter time and spatial scales, sequences of infrared images have revealed cusped frontal waves at the boundary of the eastward flowing North Equatorial Countercurrent and the westward flowing South Equatorial Current (Legeckis, 1977). These waves have $\lambda \sim 1000$ km and periods of ~ 25 days, and although first discovered in infrared images, are such that their influence can be seen in weekly composites of microwave radiometer data (Bernstein and Morris, 1983).

Moving to higher latitudes, infrared images from AVHRR have disclosed the detailed structure of oceanic fronts, such as at the Gulf Stream in the Atlantic Ocean (Plate 3), the Kuroshio in the Pacific Ocean and the fronts in the Norwegian Sea (Plate 4). In the case of the Atlantic, monitoring of the Gulf Stream has revealed in detail the extensive variability of fronts, such as the evolution of meanders and the shedding of rings, which can be observed moving into the Sargasso Sea. More recently, the information of the position of the Gulf Stream and its rings has been used in ocean circulation modelling to predict the thermal variability of the area (Robinson, 1986). The two-dimensional images of the SST signals of Gulf Stream rings have also been instrumental in the interpretation of measurements from satellite altimeters of the sea-surface topography along the subsatellite track (Cheney and Marsh, 1981).

Turning to the atmosphere, the presence of clouds in AVHRR images, considered in section 6.3 as a contaminant, are also a valuable source of meteorological information. Weather forecasting is assisted by analysis of cloud patterns, their development and motion and the quantitative data from the AVHRR channels can be used to classify types of clouds, fog and haze (Liljas, 1987). The displacements of clouds between successive satellite images from geosynchronous satellites can be used to estimate winds (e.g. Warnecke et al., 1987), while their appearance and evolution can contribute significant information to the analysis of many meteorological situations: selections of 'case studies' are presented by Reynolds (1987) and by Scorer (1987, 1988).

Finally, it is noteworthy that some meteorological centres, such as NOAA and CMS, Lannion, routinely derive and distribute charts of SST, cloud cover and type, and ice cover from satellite images on an operational basis (eg. Dismachek et al., 1980; Castagne et al., 1987).

Further uses of AVHRR data are presented and discussed by Robinson (1985) and Stewart (1985).

9. Concluding comments

While some applications, discussed only briefly here, make use of the detailed spatial information of satellite infrared imagery, others require high absolute accuracy. One of the more demanding of these is the estimation of ocean atmosphere heat exchange, which needs accuracies of a few tenths of a degree. While this accuracy is currently not demonstrated a very encouraging study has been made by Gautier *et al.* (1988) in the case of the Indian Ocean Monsoon. Realistic estimates of the ocean-atmosphere heat fluxes were obtained using only satellite measurements from visible, infrared and microwave radiometers, although in this initial study the SST was derived from HIRS and not AVHRR. This successful example of the combination of data from several instruments on different satellites is an indication of some types of future applications of spaceborne sensors.

The multiangle, multichannel capability of the ATSR/M, coupled with the high-quality calibration targets and increased digital resolution, will provide opportunities for a detailed examination of the effects of the atmosphere on the SST measurements from space. The numerical models of the atmospheric radiative transfer, used to simulate the satellite measurements will be tested, and maybe revised, leading to more confidence in the retrieval algorithms they generate, including those for AVHRR.

The improved accuracy of the SST measurements will be especially useful for climate research, and although the anticipated three-year duration of the ERS-1 mission will be too short to establish climatic trends, the ATSR/M data should provide a high-quality "baseline" with which future global SST data sets could be compared.

Acknowledgements. The diagrams of the ATSR/M were supplied by Dr D. T. Llewellyn-Jones of the Rutherford Appleton Laboratory, Didcot, Oxfordshire, UK, from whom further details of the instrument may be requested. Dr R. W. Saunders of the Robert Hooke Institute, Oxford, UK, provided Fig. 8.

References

BERNSTEIN, R.L. and MORRIS, J.H. Tropical and mid-latitude North Pacific sea surface temperature variability from the Seasat SMMR. *J. Geophys. Res.*, **88**, 1983: 1877–1891.

CAMPBELL, S. Vicarious calibration of METEOSAT's infrared sensors. *ESA J.*, **6**, 1982: 151–162.

CASTAGNE, N., Le BORGNE, P., Le VOURCH, J. and OLRY, J.-P. Operational measurement of sea surface temperatures at CMS Lannion from NOAA-7 AVHRR data. *In*: CRACKNELL, A.P. and HAYES, L. *eds.* 'Remote Sensing Yearbook'. London, Taylor and Francis, 1987: 117–153.

CHEDIN, A., SCOTT, N.A., and BERRIOR, A. A single-channel, double-viewing angle method for sea surface temperature determination from METEOSAT and TIROS-N radiometric measurements. *J. Appl. Meteorol.*, **21**, 1982: 613–618.

CHENEY, R.E. and MARSH, J.G. Seasat altimeter observations of dynamic topography in the Gulf Stream region. *J. Geophys. Res.*. **86**, 1981: 473–483.

COAKLEY, J.A. and BRETHERTON. F.P. Cloud cover from high resolution scanner data: Detecting and allowing for partially filled fields of view. *J. Geophys. Res.*, **87**, 1982: 4917–4932.

CROSBY, D.S. and GLASSER, K.S. Radiance estimates from truncated observations. *J. Appl. Meteorol.*, **17**, 1978: 1712–1715.

DISMACHEK, D.C., BOOTH A.L. and LEESE, J.A. National Environmental Satellite Service catalog of products. NOAA Memo NESS 109. Washington, D.C., National Oceanic and Atmospheric Administration, 1980: 120pp.

FRIEDMAN, D. Infrared characteristics of ocean water $(1.5–15\mu)$. *Appl. Opt.*, **8**, 1969: 2073–2078.

GAUTIER, C., FROUIN, F. and SYMONOT, J.-Y. An attempt to remotely sense from space the surface heat budget over the Indian Ocean during the 1979 Monsoon. Submitted to *Geophys. Res. Lett.* 1988.

HOLYER, R.J. A two-satellite method for measurement of sea surface temperature. *Int. J. Remote Sens.*, **5**, 1982: 115–131.

HOSKINS, B.J. and KAROLY, D.K. The steady linear response of a spherical atmosphere to thermal and orographic forcing. *J. Atmos. Sci*, **38**, 1981: 1179–1196.

ITT. AVHRR/1-FM Advanced Very High Resolution Radiometer. Final Engineering Report. (NASA-CR-160059). Fort Wayne, Indiana, USA. ITT Aerospace/Optical Division, 1979: 311pp.

KONDRATYEV, K. Ya. Radiation in the Atmosphere. New York, NY, USA, Academic Press, 1969: pp. 161–171.

LAURITSON, L., NELSON, G.J. and PORTO, F.W. Data extraction and calibration of TIROS-N/NOAA radiometers, NOAA Memo NESS 107. Washington, D.C., National Oceanic and Atmospheric Administration, 1979.

LEGECKIS, R. Long waves in the eastern equatorial Pacific Ocean: a view from geostationary satellite. *Science*, **197**, 1977: 1179-1181.

LILJAS, E. Multispectral classification of cloud, fog and haze. *In*: VAUGHAN, R.A. ed. 'Remote Sensing Applications in Meteorology and Climatology'. Dordrecht, Holland, Reidel, 1987: 301–319.

LLEWELLYN-JONES, D.T., Precise SST data from satellites. Proc. Conf. on *'The use of satellite data in climate models'*, 10-12 June, Alpbach, Austria. ESA SP-244. 1985: 139–142.

LLEWELLYN-JONES, D.T., MINNETT, P.J., SAUNDERS, R.W. and ZAVODY, A.M. Satellite multichannel infrared measurements of sea-surface temperature of the N.E. Atlantic Ocean using AVHRR/2. *Q. J. R. Meteorol. Soc.*, **110**,1984: 613–631.

McCLAIN, E.P., PICHEL, W.G. and WALTON, C.C. Comparative performance of AVHRR-based multichannel sea-surface temperatures. *J. Geophys. Res.*, **90**, 1985: 11587–11601.

McMILLIN, L.M. and CROSBY, D.S. Theory and validation of the multiple window sea surface temperature technique. *J. Geophys. Res.* **89**, 1984: 3655–3661.

McMILLIN, L.M. and CROSBY, D.S. Some physical interpretations of statistically derived coefficients for split-window corrections to satellite derived sea surface temperatures. *Q. J. R. Meteorol. Soc.*, **111**, 1985: 867–871.

MENZEL, W.P, SMITH, W.L., and HERMAN, L.D. Visible and infrared spin-scan radiometer atmospheric sounder radiometric calibration: an inflight evaluation from intercomparisons with HIRS and radiosonde measurements. *Appl. Opt.*, **20**, 1981: 3641–3644.

MINNETT, P.J. A numerical study of the effects of anomalous North Atlantic atmospheric conditions on the infrared measurement of sea-surface temperature from space. *J. Geophys. Res.*, **91**, 1986: 8509–8521.

MINNETT, P.J. The numerical simulation of infrared satellite measurements over the Greenland-Iceland-Norwgian Sea. Report SR-137. La Spezia, Italy, SACLANT Undersea Research Centre, 1988.

MINNETT, P.J., ZAVODY, A.M. and LLEWELLYN-JONES, D.T. Satellite measurement of sea-surface temperature for climate research. *In*: GAUTIER, C. and FIEUX, M. *eds*. 'Large-scale oceanographic experiments and satellites'. Dordrecht, Holland, Reidel, 1984: 57–85.

MUIRHEAD, K. and ECCLES, D. The Along Track Scanning Radiometer with Microwave Sounder. *In*: VAUGHAN, R.A. *ed*. Remote Sensing Applications in Meteorology and Climatology. Dordrecht, Holland, Reidel, 1987: 411–423.

PHULPIN, T., DERRIEN, M. and BRARD, A. A two-dimensional histogram procedure to analyse cloud cover from NOAA satellite high-resolution imagery. *J. Clim. Appl. Meteorol.*, **22**, 1983: 1332–1345.

REYNOLDS, R. Studies of synoptic and mesoscale atmospheric features from satellites. *In*: VAUGHAN, R.A. *ed*. 'Remote Sensing Applications in Meteorology and Climatology'. Dordrecht, Holland, Reidel, 1987: 217–243.

ROBINSON, A.R. Data assimilation, mesoscale dynamics and dynamical forecasting. *In*: O'BRIEN, J.J. *ed*. 'Advanced Physical Oceanographic Numerical Modelling', Dordrecht, Holland, Reidel. 1986: 465–483.

ROBINSON, I.S. Satellite Oceanography. Chichester, UK, Ellis Horwood. 1985: 455pp.

ROBINSON, I.S., WELLS, N.C. and CHARNOCK, H. The sea surface thermal boundary layer and its relevance to the measurement of sea surface temperature by airborne and spaceborne radiometers. *Int. J. Remote Sens.*, **5**, 1984: 19–46.

SAUNDERS, P.M. Aerial measurements of sea-surface temperatures in the infrared. *J. Geophys. Res.*, **72**, 1967: 4109–4117.

SAUNDERS, R.W. An automated scheme for the removal of cloud contamination from AVHRR radiances over western Europe. *Int. J. Remote Sens.*, **7**, 1986: 867–886.

SAUNDERS, R.W. and KRIEBEL K.T. An improved method for detecting clear sky and cloudy radiances from AVHRR data. *Int. J. Remote Sens.*, **9**, 1988a: 123–150.

SAUNDERS, R.W. and KRIEBEL K.T. Correction to "An improved method for detecting clear sky and cloudy radiances from AVHRR data". *Int. J. Remote Sens.*, **9**, 1988b: *In the Press*.

SCHANDA, E. Physical fundamentals of remote sensing. Berlin, Springer-Verlag, 1986: 187pp.

SCHLUESSEL, P., SHIN, H.-Y., EMERY, W.J. and GRASSL, H. Comparison of satellite-derived sea-surface temperature with *in-situ* skin measurements. *J. Geophys. Res.*, **92**, 1987: 2859–2874.

SCHWALB, A. The TIROS-N/NOAA A-G satellite series. NOAA Memo NESS 95. Washington, D.C., National Oceanic and Atmospheric Administration, 1978: 80pp.

SCORER, R.S. Cloud formations seen by satellite. *In*: VAUGHAN, R.A. *ed*. 'Remote Sensing Applications in Meteorology and Climatology'. Dordrecht, Holland, Reidel, 1987: 1–18.

SCORER, R.S. Sunny Greenland *Q. J. R. Meteorol. Soc.*, **114**, 1988: 3–29.

SMITH, W.L. The retrieval of atmospheric profiles from VAS geostationary radiance observations. *J. Atmos. Sci*, **40**,1983: 2025–2035.

SMITH, W.L., WOOLF, H.M., HAYDEN, C.M., WARK, D., and McMILLIN, L.M. The TIROS-N operational vertical sounder. *Bull. Am. Meteorol. Soc.*, **60**, 1979: 1177–1187.

STEWART, R.H. Methods of satellite Oceanography. Berkeley, USA, University of California Press, 1985: 360pp.

STRONG, A.E. and McCLAIN, E.P. Improved ocean surface temperatures from space–comparisons with drifting buoys. *Bull. Am. Meteorol. Soc.*, **85**, 1984: 138–142.

TAKAYAMA, Y., TAKASHIMA, T., MATSUURA, K. and NAITO, K. Simultaneous measurements of sea surface temperature by GMS-1 and GMS-2. *Adv. Space Res.*, **2**, 1983: 165–172.

VAUGHAN, R.A. *ed.* Remote Sensing Applications in Meteorology and Climatology. Dordrecht, Holland, Reidel, 1987: 480pp.

WARNECKE, G., ZICK, C., CARUS, B., DORING, R., ERIKSSON, A., and VOELLGER, C. Information extraction from meteorological satellite image sequences. *In*: VAUGHAN, R.A. *ed.* 'Remote Sensing Applications in Meteorology and Climatology'. Dordrecht, Holland, Reidel, 1987: 259–283.

ZAVODY, A.M. Appendix J of 'The Along Track Scanning Radiometer with Microwave Sounder – ATSR/M'. Chilton, Oxfordshire, UK. Proposal to ESA by the Rutherford Appleton Laboratory *et al.* 1982.

HANDLING AIRBORNE THEMATIC MAPPER DATA

J M Anderson
Department of Applied Physics and Electronic and
Manufacturing Engineering
University of Dundee
Dundee DD1 4HN, Scotland, U K.

ABSTRACT.   Methods of handling the output data from an Airborne
Thematic Mapper system are discussed with particular reference to the
Daedalus AADS 1268 scanner.   The format of the data is described and
methods of determining the various necessary corrections such as
radiometric calibration, brightness temperature retrieval and
atmospheric correction outlined using the thermal infrared band as the
example.

1.   THE DAEDALUS AADS 1268 AIRBORNE THEMATIC MAPPER

1.1 Data Format

The format of the output data of the Daedalus Airborne Thematic Mapper
has been described fully by the operators in the U.K.   (Global Earth
Sciences Limited) and only a brief description will be given here.
    11 wavelength intervals are covered in 12 channels.   The first 5
channels cover the visible range 0.42 to 0.69 micron, channels 6 to 10
selected regions of the near infrared between 0.695 and 2.35 micron
and channels 11 and 12 both cover the thermal infrared between 8.5 and
13.0 micron the latter being set at half the gain setting of the
former.
    The data is recorded on computer industry standard 0.5 inch 9-
track magnetic tape at 1600 bpi.   A header record is included which
details client and site information, flight date and number, ground
clearance and over the ground speed and library information.   The
logical data records are of a constant size of 750 bytes and are
interleaved by channel.   Each logical data record contains in bytes
24 to 739 the video data for one line of data for one channel.   Bytes
13 and 14 represent the temperature of calibration source 1 and bytes
15 and 16 that of calibration source 2 while bytes 23 and 740
represent respectively the digitised detector output from the thermal
channel detector for these calibration sources.   This information is
used to calibrate the thermal channel data.   Byte 22 gives the gain
setting, channel number and tells whether or not the s-bend correction

165

R. A. Vaughan (ed.), Microwave Remote Sensing for Oceanographic and Marine Weather-Forecase Models, 165–175.
© 1990 Kluwer Academic Publishers.

is in operation. The gain setting varies from 8 to 0.5 and needs to by known if radiometric calibration for channels 1 to 10 is required. Byte 17 gives the scan speed in scans per second and bytes 5 to 8 detail sortie number and line count.

1.2 System Resolution

The instantaneous field of view of the instrument is 2.5 mrad. This gives 600 system resolution elements per scan line but since the data is resampled every 2.093 mrad the data consists of 716 pixels per scan line. Two track resolutions are therefore quoted, the nadir across track resolution of 0.002093 H and the physical across track resolution of 0.0025 H where H is the aircraft ground clearance. At a height above ground of 1000 m for instance the nadir across track resolution is therefore 2.093 m.

The system resolution is affected by the s-bend correction which is applied to the data within the instrument to reduce distortion towards the edge of the scene. When applied this means that the view angle, θ, is not linear with pixel number, P, but varies as

$$\theta = \arctan\left((P - 0.5) \times 0.75/357.5\right).$$

This gives a reduced effective field of view of 73.72 degrees.

2. HANDLING THE RAW DATA

2.1 Radiometric Calibration of Channels 1 to 10 Data

The data consists of an 8-bit value for each line pixel at each channel allowing a 0 to 255 intensity range. During the collection of the data however the gain setting on the instrument can be set to any one of five gain settings; 0.5, 1, 2, 4 or 8. The procedure for radiometric calibration of the data for channels 1 to 10 allowing for these gain settings and the details of the scanner calibration covering the instrument flown by Global Earth Sciences Ltd. during the period August 1982 to September 1987 have been detailed by Wilson. The procedure is as follows:

(a) The gain settings for each of channels 1 to 10 are extracted from byte 22 of each logical data record.

(b) The appropriate scanner calibration data is chosen and the corresponding system sensitivity performance sheet obtained.

(c) Detector output voltages are then extracted for each channel for zero output (V_o mV) and for the input from a known source (V_{cal} mV).

(d) Determine N, the average panel radiance for each channel from the calibration data for the standard source.

(e) Calculate $DN_{on} = (256.V_{cal})/4000$ and $DN_{off} = (256.V_o)/4000$.

(f) Calculate GAIN(1) = N/(DN$_{on}$ - DN$_{off}$) at gain setting 1 and BASE(1) = DN$_{off}$ at gain setting 1.

(g) Calculate GAIN(G) = GAIN(1)/gain setting used and BASE(G) = BASE(1)/gain setting used.

(h) Finally determine Radiance = GAIN(G).(DN$_{value}$ - BASE(G)).

2.2 Brightness Temperature Retrieval

Although channel 11 is of fixed gain there is a channel 12 which covers the same thermal band at one half of the gain. This can be used if channel 11 is saturated.

If the reference black bodies are at temperatures T_{bb1} and T_{bb2} with corresponding DN values of DN$_{bb1}$ and DN$_{bb2}$ then the brightness temperatures is usually calculated from a linear interpolation relating DN to T_b. This does introduce a small error since the sensor measures incident radiation and any interpolation ought to take into account the Planck function (Anderson and Wilson). The brightness temperature obtained in this way is the apparent temperature of the surface as determined by the detector which is sensitive only over the wavelength interval λ_1 to λ_2 and which has a sensitivity which varies within this range. Neglecting this latter effect at present we see that a surface at temperature T_s will emit radiation

$$M'_{\Delta\lambda} = \varepsilon_\lambda \int_{\lambda_1}^{\lambda_2} B(\lambda) \, d\lambda$$

where $B(\lambda)$ is the Planck function involving T_s. More conveniently for the range 8 μm to 14 μm and for T_s in the temperature range 250 K to 290 K Goldstein has shown that the relation

$$M'_{\Delta\lambda} = C\varepsilon_\lambda T_s^5$$

can be used where C = 7.26 x 10^{-11} W m^{-2} K^{-5}

Given no atmospheric effects this temperature T_s, allowing for the fact that the surface is not likely to be a perfect black body, could be directly linked with the brightness temperature T_b. Since there are atmospheric effects even for low flight paths however and since these are not negligible this is not advisable.

Nevertheless either of these equations can be used to calculate the radiance incident on the scanner originating from all sources, including the surface of interest, within the channel by equating T_s to T_b, T_s being an apparent surface temperature. For example

$$E' = \text{Radiance reaching the scanner} = CT_b^{5}$$

since the reference sources are black bodies with $\varepsilon = 1$. This is the first step in determining the thermal radiation leaving the viewed surface.

3. ATMOSPHERIC CORRECTION

Atmospheric corrections can be applied to all channels but here we will concern ourselves with some of the methods which can be used to correct channel 11 data. The first method described can be adapted to apply to all channels.

3.1 The Various Factors

The spectral irradiance at the scanner is given by

$$E' = \tau_\lambda \{M' + M_r'\} + M'_{atmos} \tag{1}$$

where

$$M' = C\varepsilon_\lambda T_s^{5} \text{ and } M_r' = (1 - \varepsilon_\lambda) E'_{sky}$$

M'_{atmos} is the radiance emitted by the atmosphere towards the scanner, τ_λ is the net transmittance of the atmosphere between the surface and the scanner for the given channel and E'_{sky} is the sky radiance incident on the surface.

E' is determined by the brightness temperature since the black bodies are adjacent to the detector and no corrections need be applied in this case. We have therefore

$$E' = CT_b^{5}$$

or E' can be determined by numerical integration over the channel wavelengths as mentioned earlier. Hence if E'_{sky}, ε_λ, τ_λ and M'_{atmos} can be determined M' can be calculated and T_s obtained either by numerical integration or from

$$M' = C\varepsilon_\lambda T_s^{5}.$$

3.2 The Atmospheric Transmission of Thermal Infrared Radiation

In general the atmosphere is opaque to thermal infrared radiation apart from the "windows" which exist in the bands 3 μm to 5 μm and 8 μm to 14 μm. Even in these windows however some of the thermal radiation will be absorbed by some of the atmospheric constituents. The second window is the one of relevance here.

Of the radiance leaving the surface M' an amount τ_λ(L) M' will reach the hemisphere of the scanner. τ_λ(L) is the wavelength dependent transmittance which increases with path length L. It can be written in terms of the absorption coefficient α_λ as

$$\tau_\lambda(L) = \exp(-\alpha_\lambda L).$$

Various processes contribute to α_λ. When radiation passes through the atmosphere it can interact with either molecules or aerosols (small particles of various types) and it can either be scattered or absorbed. Since in the case of thermal infrared radiation the wavelength is much greater than the size of the molecules comprising the atmosphere molecular scattering can be neglected and molecules need only be considered as absorbers of radiation in this case. The main absorbers turn out to be water vapour, ozone, nitric acid, CO_2, N_2O, CH_4 and CO. O_2 and N_2 have an indirect influence but do not themselves absorb in the 8 μm to 14 μm band. The most important absorbers are water vapour and carbon dioxide. Within the actual window water vapour has the biggest influence.

Aerosol scattering and absorption is difficult to allow for accurately. Fortunately it can be rendered negligible by suitable choice of experimental conditions. Use of an expensive and sophisticated scanner such as the Airborne Thematic Mapper is almost invariably confined to conditions of clear sky and good visibility. In this case reference to the relevant data included in the LOWTRAN computer code (Kneizys et al) indicates a value for the atmospheric transmittance of 0.999 over the thermal band for typical path lengths when the visibility is 20 km or more. Since this represents typical operating conditions the aerosol contribution can be discounted. The following procedure uses the data files listed in the LOWTRAN code. It does not necessarily use the actual programmes listed but these can either be adapted or written to suit the particular circumstances. Detailed examples of such programme listings are given eleswhere (Wilson).

(a) Given good visibility conditions the aerosol contribution is discounted.

(b) The atmospheric profile (variation of temperature, water vapour content and pressure with height) is determined from the fine structure Meteorological Office data taken at the closest meteorological station to the working area.

(c) Ozone content is assumed to be 2.5×10^{-3} atm-cm/km unless specific values for the area are known.

(d) Atmospheric nitric acid is assumed to be not present below the aircraft.

(e) Average values for the uniformly mixed gases (CO_2, N_2O, CH_4 and CO) densities are used to calculate LOWTRAN defined equivalent absorber densities (ω^*) for the prevailing conditions.

(f) A similar calculation is made for water vapour to cover line absorption.

(g) The LOWTRAN empirical formula

$$\tau_\lambda = f(C_\nu, \omega^*, L)$$

can then be used (where C_ν is an absorption coefficient which depends on the wavelength and the individual absorber and L is the atmospheric path length) and applied to ozone, the uniformly mixed gases and water vapour line absorption in turn.

(h) Water vapour continuum absorption is then calculated. This is due to collision broadening of spectral lines both by self broadening and by collisions with other atmospheric constituents. This is where the indirect importance of O_2 and N_2 referred to earlier comes in. This is achieved by using the LOWTRAN application of the work of Roberts. The contribution to the transmittance is then

$$\tau_\lambda = \exp\left(-\alpha(\lambda) L\right)$$

where $\alpha(\lambda) = C\left(\lambda, T(z)\right) \omega_{H_2O}(z) \left[P_{H_2O}(z) + \gamma\left(P(z) - P_{H_2O}(z)\right)\right]$, $P(z)$ and $T(z)$ are the pressure and temperature at height z, $\omega_{H_2O}(z)$ is the number of water molecules per cubic metre at height z, $\gamma = 0.002$ is the foreign broadening coefficient and $C\left(\lambda, T(z)\right)$ is the empirically derived self broadening coefficient.

Typical results are shown in figure 1. These were in fact derived for a particular October evening in the Dundee area.

3.3 Atmospheric Radiance

This can be calculated by using the absorber densities and absorbtion coefficients previously determined. Essentially the atmosphere is divided into a number of slabs (say each 10 m thick) and each slab is assigned the temperature corresponding to its height. By equating the absorption coefficient to the emissivity the amount of radiation emitted by each absorber in each slab can then be calculated. Knowing the transmittance of that part of the atmosphere between each slab and the scanner due to all the absorbers then enables the determination of the amount of radiation reaching the scanner from each slab due to each absorber in the slab. If this calculation is

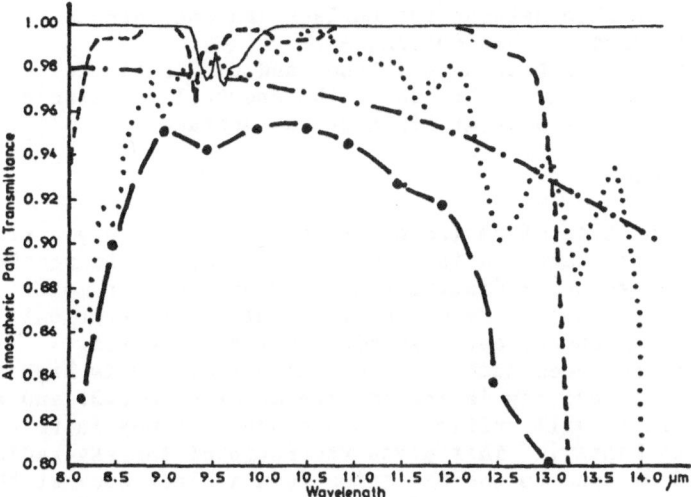

Figure 1. Atmospheric Path Transmittance
    ———— Ozone
    — — — Uniformly mixed gases
    · · · · · Water vapour line absorption
    —·—·— Water vapour continuum absorption
    ●———● Net transmittance

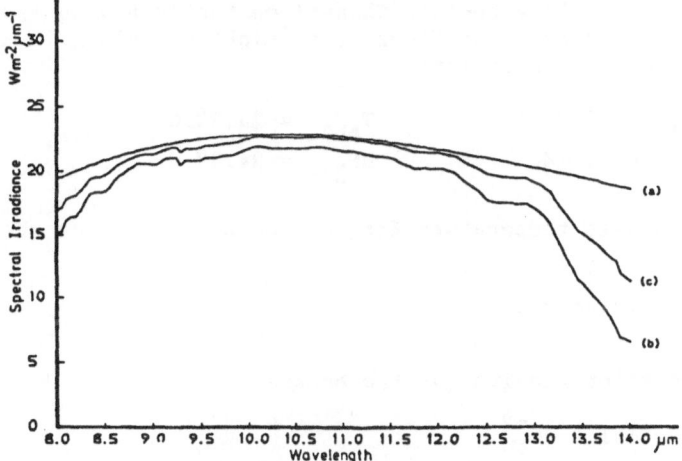

Figure 2. Irradiance at the scanner due to
    (a) a blackbody at temperature 282 K,
    (b) the blackbody allowing for path transmittance and
    (c) the blackbody allowing for path transmittance and
         atmospheric radiance

repeated for each slab between the surface and the scanner the total atmospheric radiance is then determined.

The net effect of path tranmittance and path radiance is illustrated in figure 2. This shows the spectral irradiance at the scanner due to a black body at 282 K at the surface.

3.4 Sky Radiance

The actual calculation of reflected sky radiation is very difficult. The radiance for instance varies from the horizon to the zenith both in magnitude and spectral distribution. Given that in most circumstances experimental determination would be impractical nevertheless some general determination will be required. Fortunately for surfaces such as water the emissivity is high and therefore the reflectivity is low (of the order of 0.002) and a reasonable estimate will suffice. One convenient way is to use the estimate due to Ramsay. This gives the ratio of the sky radiation in the 8 μm to 14 μm band to the radiation from a black body at the surface air temperature. This fraction varies with surface temperature and humidity. For instance if the surface air temperature is 21.5 C and the relative humidity is 54%, $E'_{sky} = 0.40 \ M'_{bb}$.

3.5 Sample Calculation

The following is an example of a calculation made for conditions prevailing in the western English Channel on the 19th June 1984 at 1300 GMT when the scanner was flown at a height of 4400 m. Scanner calibration data were as follows

$$T_{bb1} = 7.97 \ C \qquad T_{bb2} = 23.83 \ C$$
$$DN_1 = 27.14 \qquad DN_2 = 24.37 \ C$$

A typical brightness temperature for the channel water was then determined as

$$T_b = 17.95 \ C$$

giving the radiation received at the scanner as

$$E' = 151.2 \ Wm^{-2}.$$

Knowledge of the atmospheric profile enabled the transmittance to be calculated as

$$\tau_\lambda = 0.94 \text{ over the 8 } \mu\text{m to 14 } \mu\text{m band.}$$

The reflected sky radiation was estimated to be

$$M_r' = 1.4 \ W \ m^{-2}$$

and the atmospheric radiance towards the scanner from the intervening atmosphere as

$$M'_{atmos} = 3.6 \ W \ m^{-2}.$$

Substitution into the expression for E' then enables the ground leaving radiance to be calculated as

$$M' = 155.6 \ W \ m^{-2}.$$

Taking the emissivity of the sea surface as 0.98 gives a value for the sea surface temperature of $T_s = 21 \ C$.

4. OTHER ATMOSPHERIC CORRECTION METHODS

Other techniques for the determination of the net attenuation of the atmosphere on the thermal infrared radiation passing through it towards an airborne scanner have been used. These all depend on the use of varying the path length through the atmosphere. A split channel method cannot be used in the case of the Daedalus type scanner since the second thermal channel is not present.

4.1 Varying the Height and View Angle over a Water Surface

When the area of interest is an extended water surface there are two techniques which have been applied to determine the true surface temperature. The first relies on the surface being constant in temperature during the time taken to overfly the same area which is recorded at several different heights and view angles (Schott). It can be easily shown that if the radiance received at the scanner from the same surface area is determined as a function of the aircraft height then the transmittance and the atmospheric radiance can be calculated by applying a least squares analysis to equation 1. In order that the sky radiation be measured however it is necessary to use the fact that the emissivity of the water surface varies with view angle. Such variation is tabulated in the thermal band for water but the method could not be applied for other surfaces. It would nevertheless be possible to use an estimate of sky radiation as indicated earlier.

4.2 Varying the View Angle over a Water Surface

An alternative similar technique has been used by Callison. In this case the scene is examined for a region where the sea surface is constant in temperature. It is found that if video data from a

single logical data record is plotted as a function of view angle then
because of the variation of path length with view angle the apparent
surface temperature is less towards the edge of the scene.   In fact
averaged data from as many as 150 logical data records were used and
by a similar manipulation of equation 1 the effective atmospheric
transmittance determined.   This technique has the advantage that only
one flight is required and that even if the sea surface temperature is
changing the measurements are being taken at video rates and no such
change will affect the results.   It has the disadvantage however that
a large area of constant temperature is required for the calibration
to be effective.   This precludes its use in many applications.

4.3 *Varying the Height over Various Surfaces*

The dangers inherent in this type of technique have been discussed by
Djavadi.   In this experiment an aircraft carrying a Daedalus AADS
1230 thermal infrared scanner was flown at three different heights
over an area of the Tay Estuary and the adjoining areas of the city of
Dundee.   It was shown that during the time in which the aircraft
changed altitude considerable changes in temperature occurred – even
in the waters of the estuary where warmer fresh water masses replaced
the colder sea water during this time period.   The changes in
temperature were even more marked on the land surfaces and the
situation was complicated by the changing emissivity of the sandbanks
as they dried out. It was concluded that the use of this technique
for atmospheric corrections was unlikely to be a cost-effective way of
using aircraft and scanner time, particularly for land-based use where
changing atmospheric temperature and non equilibrium of surface
temperatures is not unusual.

5.   *CONCLUSION*

A brief review of the methods of extracting and handling Airborne
Thematic Mapper data has been given and the techniques of radiometric
correction outlined.   The allowance for atmospheric effects has been
discussed with particular reference to the thermal infrared band using
two methods.   Either method has its use and its limitations.   Where
circumstances permit the two methods can be shown to give similar
results (Callison, Blake and Anderson) but on the whole the direct
calculation of atmospheric correction is probably to be preferred on
the grounds of cost-effectiveness where the weather profile is
reasonably well known.

REFERENCES

Anderson J M and Wilson S B, The physical basis of current infrared
remote-sensing techniques and the interpretation of data from aerial
surveys, Int J Remote Sensing, Vol 5, 1-19, 1984.

Callison R D, Sea surface temperatures from Daedalus ATM data-preliminary results, Int J Remote Sensing, Vol 6, 1671, 1985; Ibid, Vol 7, 288 (erratum), 1986.

Callison R D, Blake P and Anderson J M, The quantitative use of airborne thematic mapper thermal infrared data, Int J Remote Sensing, Vol 8, 113-126, 1987.

Djavadi D and Anderson J M, Atmospheric correction of thermal infrared data using multi-height data acquisition, Int J Remote Sensing, Vol 8, 1879-1884, 1987.

Global Earth Sciences Limited, Airborne Thematic Mapper Computer Compatible Tape, 1985.

Goldstein R J, Application of aerial infrared thermography to the measurement of building heat loss, ASHRAE, Trans, Vol 84, 207, 1978.

Kneizys F X, Shettle E P, Gallery W O, Chetwynd J H, Jr, Abrev L W, Selby, J E A, Fenn R W and McLatchey R A, Atmospheric transmittance/radiance: computer code LOWTRAN 5. AFGL-TR-80-0067, Environmental Research Paper No 697, Optical Physics Division, Air Force Geophysics Laboratory, Hanscom, Massachusetts, 1980.

Ramsay, J W, Chiang H D, and Goldstein R J, A study of the incoming radiation from a clear sky, J Appl Met Vol 21, 566, 1982.

Roberts R E, Selby J E A and Biberman L M, Infrared continuum absorption by water vapour in the 8 μm - 12 μm window, App Optics, Vol 14, 2085, 1976.

Schott J R and Tourin R H, A completely airborne calibration of aerial infrared water temperature measurements, Proceedings of the 10th International Symposium on Remote Sensing of the Environment, University of Michigan, Ann Arbor, p 477, 1975.

Wilson A K, Calibration of ATM Data, Proceedings of the NERC 1985 Airborne Campaign Workshop, Monk's Wood, November 1986, along with supplement of February, 1988.

Wilson S B, Aerial thermographic measurement of surface temperature, MSc Thesis, University of Dundee, 1985.

MICROWAVE RADIOMETERS

L.T. PEDERSEN
Electromagnetics Institute
Technical University of Denmark
DK-2800 Lyngby
Denmark

ABSTRACT. The purpose of this paper is to give a general introduction to microwave radiometry, the physics, the instruments and the data processing and the parameter retrieval.

1. Introduction

All materials that have a temperature different from absolute 0, (0K) emit (and absorb) small amounts of microwave radiation. The purpose of microwave radiometry is to measure this radiation and from its spectral composition to derive information about the material.

2. Emission and transmission of microwaves

The thermal emission of electromagnetic energy from a medium is governed by Planck's blackbody-radiation law.

At microwave wavelengths (1m to 1 mm) and longer the Rayleigh-Jeans approximation applies

$$B = \frac{2fkT}{c^2} \Delta f \qquad (2.1)$$

where
B = Blackbody brightness ($Wm^{-2} sr^{-1}$)
f = frequency (Hz)
k = Boltzmann's constant
T = absolute temperature (K)
c = velocity of light (m/s)

A black-body is an idealized body which, when in thermodynamic equilibrium at a temperature T absorbs all incident radiation. Real materials absorb (and emit) less radiation and therefore are usually referred to as grey bodies.

The radiation of a grey body (at a particular wave-length) is equivalent to the radiation from a blackbody at a lower temperature. This temperature is referred to as the brightness temperature (T_B) of the grey body.

The ratio of the brightness temperature to the physical ("real") temperature of the grey body is called its emissivity (ε).

R. A. Vaughan (ed.), *Microwave Remote Sensing for Oceanographic and Marine Weather-Forecase Models*, 177–190.
© 1990 Kluwer Academic Publishers.

$$\varepsilon = \frac{T_B}{T} \qquad\qquad\qquad (2.2)$$

The emissivity therefore is a number between 0 and 1. A simple relationship exists between emission and absorption in local thermodynamic equilibrium, they must be equal, and since the radiation not absorbed by the medium must be transmitted through it, the following relationship between emission and transmission emerges

$$\varepsilon = (1-\tau) \qquad\qquad\qquad (2.3)$$

2.1. EMISSION FROM OCEAN AND ICE SURFACES

The emissivity of a specular sea surface is fairly low at most microwave wave-lengths. This is due to the fact that the permittivity of sea water is high at these frequencies. The permittivity is a well known function of salinity and temperature. The temperature dependence is stronger which allows measurements of the sea surface temperature, whereas the dependence on salinity is weak and salinity therefore cannot be measured with great accuracy.

The emissivity of the specular surface is different seen from different angles according to Snell's law. Therefore, a wind roughened surface will have different emissivity than the specular surface. The wind effect is also fairly well described and microwave radiometry is used to measure wind speeds (the wind direction effect is very small). Also foam on the sea surface increases the emissivity helping us to measure higher wind speeds.

First-year (FY) sea ice is a much stronger radiator of microwave energy than the sea surface, i.e. the emissivity and therefore the brightness temperature is higher. This can be attributed to the low permittivity and high absorption of the ice.

Old or multiyear (MY) ice that has survived a summer melt period will have a different signature. Melt-water from the ice surface will percolate through the ice causing a desalination. This low-salinity MY-ice will have a lower emissivity at the shorter wavelengths. Figure 1 shows a schematic presentation of the emissivity as a function of microwave frequency for sea water, FY-ice, and MY-ice at approximately 50° incidence angle and horizontal polarization.

2.2. ABSORPTION AND EMISSION BY THE ATMOSPHERE

The absorption, emission and transmission properties of the atmosphere are mainly due to the following effects

- Absorption by oxygen molecules
- Absorption by water molecules (vapour)
- Absorption and scattering by water particles (droplets)

The frequency dependence of these effects are known to a useful extent and microwave radiometry is used for quantitative measurements of water-vapour content and cloud liquid-water content of the atmosphere. The oxygen content of the atmosphere is well known and measurements of microwave radiation at different frequencies with different oxygen absorption can be used for atmospheric temperature retrievals as is done with the MSU-instrument of the TIROS Operational Vertical Sounder or TOVS.

Concluding this we may say that even though the amount is very low, the spectral composition of the microwave radiation from the surface of the Earth transmitted through the atmosphere contains valuable information about both the surface and the atmosphere.

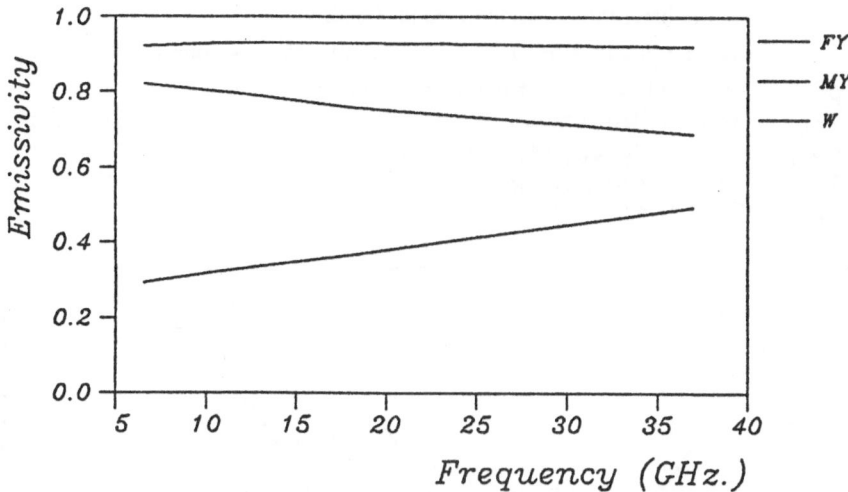

Figure 1. Emissivities of first-year ice (FY), (upper curve), multiyear ice (middle curve) and open water (lower curve) at horizontal polarization and 50° incidence angle.

2.3. RADIATIVE TRANSFER

Microwave radiation at frequencies below 50 GHz will penetrate most clouds, which means that it is possible to measure surface parameters under almost all weather conditions.

The atmosphere does contribute though, to the radiation received at a satellite platform. This contribution is caused by absorption of a certain (small) amount of the radiation originating from the surface and a corresponding emission from the atmosphere as a grey body.

The sum of these effects including radiation from the sky is illustrated in Figure 2. Considering the atmosphere as a single homogeneous layer, we can express the four terms in the following simple way.

1: Contribution from the sky

$$T_1 = T_{EXT} \cdot \tau \cdot \rho \cdot \tau$$

2: Contribution (reflected) from atmosphere

$$T_2 = T_{AIR} \cdot (1-\tau) \cdot \rho \cdot \tau$$

3: Contribution from the surface

$$T_3 = (W \cdot \varepsilon_W \cdot T_W + (FY \cdot \varepsilon_{FY} + MY \cdot \varepsilon_{MY}) \cdot T_{ICE}) \cdot \tau$$

4: Contribution from the atmosphere

$$T_4 = T_{AIR} (1-\tau)$$

where: ε_W, ε_{FY}, ε_{MY} are water, FY ice, and MY ice emissivities

      T_{EXT} is the brightness temperature of the sky

      τ is the transmission coefficient through the atmosphere

      ρ is the reflection coefficient at the surface

      T_{AIR} is the air temperature

      $(1-\tau)$ is the emissivity of the atmosphere

      T_W is water surface temperature

      T_{ICE} is ice temperature

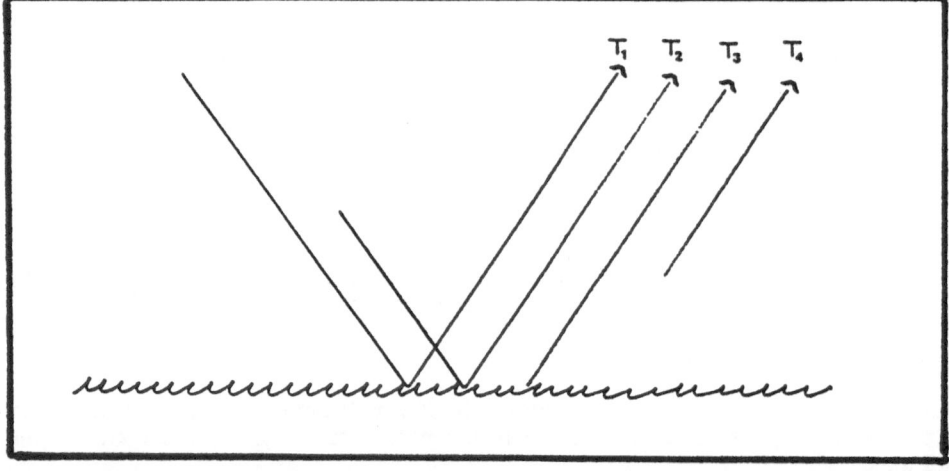

Figure 2. The four contributions to the signal received by an antenna at satellite altitude 1) reflected sky radiation 2) reflected atmospheric radiation (downwelling) 3) surface emission and 4) atmospheric radiation (upwelling).

The reflection coefficient at the surface ρ can be written as

$$\rho = W(1-\varepsilon_W)+FY(1-\varepsilon_{FY})+MY(1-\varepsilon_{MY})$$

where W, FY, and MY are the amounts of open water, FY-ice and MY-ice within the field of view of the instrument. The temperature apparent to an antenna at satellite altitude will be the sum of the four contributions

$$T_A = T_1+T_2+T_3+T_4$$

and will therefore contain information about both surface and atmospheric conditions.

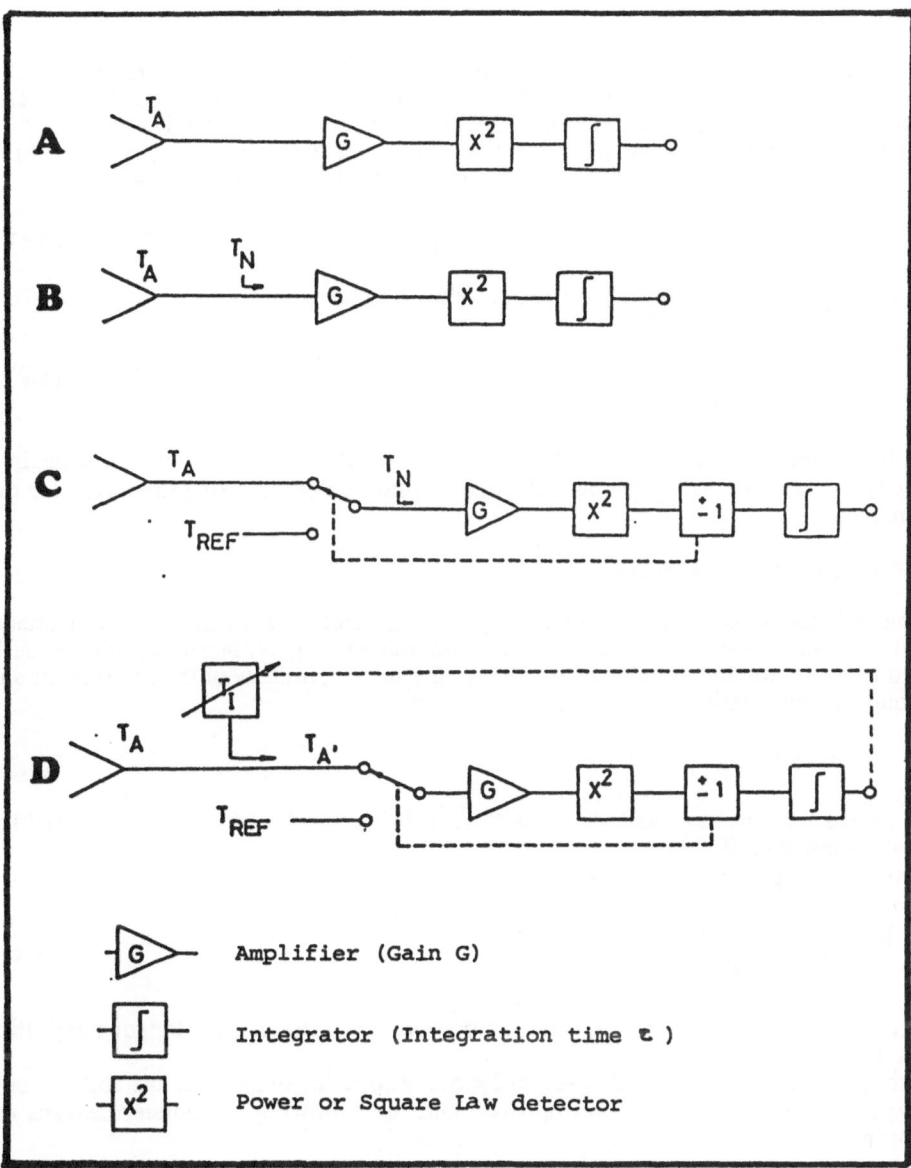

Figure 3. Block diagrams showing the basic principles of radiometer designs. a) The ideal receiver. b) Total power radiometer. c) Dicke-radiometer. d) Noise-injection radiometer. Figure adapted from Skou (1978).

3. The radiometer receiver

A microwave radiometer is simply a calibrated very sensitive receiver to measure T_A at a specific wave-lenght or rather wave-lenght interval called the receiver bandwidth B. The measurement situation is outlined in Figure 3 where also an ideal receiver is shown.

G is the amplification of the signal by the radiometer receiver, and the power at the output terminals now is proportional to input , bandwidth B and receiver gain G, i.e.

$$P = kBT_A \cdot G \tag{3.1}$$

The sensitivity of such an ideal receiver can be shown to be

$$\Delta T = \frac{T_A}{\sqrt{B\,\tau}} \tag{3.2}$$

where B is bandwidth and τ is the integration time. It is now evident that a desired sensitivity can be obtained by proper selection of integration time (the time used for the measurement).

3.1. THE TOTAL POWER RADIOMETER

Unfortunately life is not quite that simple with real radiometers. The electronic components constituting the radiometer will add noise to the desired signal. Normally this noise is referred to as an equivalent noise source T_N at the input terminals of the radiometer and equation (3.1) now reads

$$P = k \cdot B(T_A + T_N)\,G \tag{3.3}$$

which means that the received signal will vary if T_N or G varies with time, making absolute calibration difficult.

Also equation (3.2) is modified to

$$\Delta T = \frac{T_A + T_N}{\sqrt{B\,\tau}} \tag{3.4}$$

It is seen that a desired sensitivity still can be obtained, only using a longer integration time now.

With typical values of T_A of 300 K and T_N of 5-800 K and quite large unstabilities in G and T_N this situation needs to be improved and the following radiometer designs are attempts to do this.

3.2. THE DICKE RADIOMETER

In order to reduce the effect of variations in T_N a swithcing between antenna input and the reference temperature T_{REF} takes place at a frequency much higher than the fastest variations in G and T_N. Before integration the signal from the power detector is multiplied by +1 if the switch is in the T_A position and by -1 if in the T_{REF} position, resulting in a subtraction of the signals by the integrator

$$P = kB(T_A + T_N) \cdot G - kB(T_{REF} + T_N) \cdot G = kB(T_A - T_{REF}) \cdot G \qquad (3.5)$$

from which it is seen that T_N has been eliminated, while G is still present, with less weight though if T_{REF} is suitably selected ($T_{REF} \approx T_A$).

The sensitivity has decreased, however, by a factor of 2 since we are now only measuring the desired signal for half of the time

$$\Delta T = 2 \cdot \frac{T_A + T_N}{\sqrt{B\tau}} \qquad (3.6)$$

3.3. THE NOISE INJECTION RADIOMETER

The purpose of the noise injection principle is to utilize the fact that by setting $T_A = T_{REF}$ we can totally eliminate the effect of G on the measurement.

Reference temperature sources that can be controlled accurately over a temperature range of 100 to 300 K are impractical. Therefore T_{REF} is normally kept at a constant quite high level (ex. 300 K) and noise is added to T_A in order to reach the same temperature i.e.

$$T_A' = T_A + T_I = T_{REF} \qquad (3.7)$$

This leads to the power at the output

$$P = kB(T_A' - T_{REF}) \cdot G = 0 \qquad (3.8)$$

which is independent of G and T_N. The injected noise is controlled by a feed-back loop to keep P=0, and the absolute accuracy is now only dependent on the accuracy with which T_I is known.

3.4. CALIBRATION OF A MICROWAVE RADIOMETER

The purpose of the calibration process is to establish the relation between input antenna temperature and the output quantity (volts, counts etc.) of the radiometer.

All the above described radiometer designs lead to a linear relationship between the input and the output quantities, therefore the calibration procedure normally consists of measuring the output quantity for a known hot (~300 k) and a known cold (~100k) target. These targets may be

- a microwave load in liquid (nitrogen)
  $T_A = T_{PHYS}$

- a microwave absorber wall or room
  $T_A = T_{PHYS}$

- the sky in clear sky conditions with a low water vapour content
  $T_A = T_{SKY}$

Problems with calibration of spaceborne radiometers are changes in performance during launch or at very low/high temperatures in the satellite. These changes may result in a new relationship between radiometer input and output. For this reason spaceborne radiometers are often in-flight calibrated by pointing the antenna at the sky, or connecting the radiometer to a microwave load (black-body) instead of the antenna.

3.5. MICROWAVE RADIOMETER ANTENNAS

The choise of antenna for a microwave radiometer decides the spatial resolution of the output data via the so called antenna foot-print, or the field of view of the antenna.

Hence this choise must be given serious consideration and the satellite constructors often put restrictions on the performance of radiometers in this respect by limiting the size of the antenna.

Two useful quantities need to be known when talking about microwave antennas:

1) The antenna gain $G(\theta,\phi)$ which is the ratio of power density radiated by the subject antenna to the power density radiated by a lossless isotropic antenna $((\theta,\phi)$ indicates direction in space).

2) The antenna half power beamwidth $\beta_{1/2}$ which is the angle within which the power density is larger than or equal to half the maximum power density (wats/steradian) radiated by the antenna. Notice that the antenna will also receive signals from angles outside $\beta_{1/2}$, however, these signals will in most cases be small.

The purpose of the antenna on the spacecraft (or airplane) is to direct the radiometer towards a small area on the earth. Therefore we want a large gain in the direction towards the ground and a small half power beamwidth.

For any antenna the beamwidth is to first order equal to the wavelength divided by the antenna dimensions

$$\beta_{1/2} = \frac{\lambda}{D} \tag{3.9}$$

so a larger antenna gives a better spatial resolution since the antenna footprint size can be calculated as

$$F = \beta_{1/2} \cdot R \tag{3.10}$$

where R is satellite altitude.

This explains why we need a short wavelength to obtain fine spatial resolution with a given antenna size, and why f.ex. the SMMR instrument has different spatial resolution at the different wavelengths (they all use the same antenna).

3.5.1. *Antennas*. The antennas used by past and present radiometer systems normally come from one of the following three groups

Horn antennas: have a low gain and large half-power beamwidth, and have been used with a number of airborne and ground based systems.

Reflector antennas: where a parabolic shaped reflector is used to focus the signal received by a horn antenna. This results in high gain and small beamwith. Also using a multifrequency horn, the reflector antenna has multi-frequency capabilities.

Phased array antennas: The phase of the signal received by a number of antennas is controlled in order to obtain that only signals from one particular direction add up constructively when the differently shifted contributions are combined. This design was used by the NIMBUS-5 and 6 ESMR (Electronically scanned microwave radiometer) but has the disadvantage of being only capable of receiveing one frequency and that the phase shifters introduce quite high losses in the signal path.

All three types of antennas may be used for imaging systems. The first two by mechanically pointing the antenna towards different points on the ground (scanning) and the latter by electronically controlling the phase shift at the individual antennas and thereby the direction. The first two thus require moving parts on the satellite platform resulting in stability problems for larger antennas.

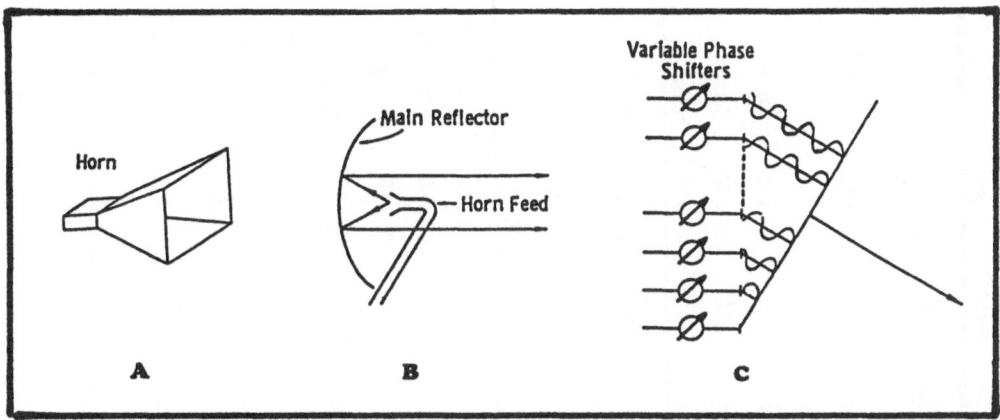

Figure 4. Different antenna principles, a) the horn antenna b) reflector antenna and c) phased array antenna. Figures from Ulaby et al. (1981).

3.5.2. *Scanning Considerations.* If we want a scanning or imaging radiometer system we have to consider the relationship between swath width, foot-print size, and integration time. In a scanning system the time between two scanlines must satisfy

$$t \leq \frac{F}{V} \tag{3.11}$$

where F is foot-print size (ex. 30 km) and V is sub-satellite velocity (ex. 7.5 km/s) in order to have two consecutive foot-print lines touching each other.

Also in the cross-track direction the similar constraint is that the time for one foot-print (i.e. the integration time) must satisfy

$$\tau \leq t / \frac{Sw}{F} \leq \frac{F^2}{Sw \cdot V} \tag{3.12}$$

where Sw is swath width (ex. 750 km). This upper limit to the integration time also sets a lower limit on the radiometer sensitivity according to (3.4) and (3.6).

The examples given are the values for the 37 GHz channel of the NIMBUS-7 SMMR and lead to

$$\tau \leq 160 \text{ msec.}$$

Spacecraft	Instrument	Operation	Frequency [GHz]	Polariz.	Antenna	Sw [km]	F [km]	Radiometer	B [MHz]	τ [ms]	ΔT [k]
NIMBUS-5	ESMR	1972-76	19.35	H	Phas.Array	3000	~25	Dicke	110	47	1.5
NIMBUS-6	ESMR	1975	37	H+V	Phas.array	1272	35 x 20	Dicke	230		
NIMBUS-7 (SEASAT)	SMMR	1978-87 1978	6.6	H+V	reflector	822	148 x 95	Dicke	250	126	0.9
			10.69	H+V	reflector	822	91 x 59	Dicke	250	62	0.9
			18	H+V	reflector	822	55 x 41	Dicke	250	62	1.2
			21	H+V	reflector	822	44 x 30	Dicke	250	62	1.5
			37	H+V	reflector	822	27 x 18	Dicke	250	30	1.5
NOAA-6-	MSU	1978-	50.31			2300	100-200	Dicke	200		0.3
			53.73			2300	100-200	Dicke	200		0.3
			54.96			2300	100-200	Dicke	200		0.3
			57.95			2300	100-200	Dicke	200		0.3
DMSP	SMM/I	1987-	19.35	H+V	reflector	1390	70 x 45	total power	250	8	0.8
			22.235	V	reflector	1390	60 x 40	total power	250	8	0.8
			37	H+V	reflector	1390	38 x 30	total power	900	8	0.6
			85.5	H+V	reflector	1390	16 x 14	total power	1400	4	1.1
MOS-1	MSR	1987-	23.8	H	reflector	317	~29	Dicke	400	10/47	≤1
			31.4	V	reflector	317	~22	Dicke	500	10/47	≤1
ERS-1	ATSR/M	1990	23.8	Nadir	reflector	22	22				
			23.8	Nadir	reflector	22	22				

Table 1. Instrument characteristics of some past and present spaceborne microwave radiometers. (Sw=swath width, F=foot print size, B=bandwidth, τ=integration time, and ΔT=sensitivity.

4. Data processing and parameter retrieval

The basis for most parameter retrieval methods is models of the relationship between brightness temperature at a particular set of frequencies and polarizations, and the corresponding physical parameters for the surface and/or atmosphere that emitted or affected the radiation.

A block diagram of such a model system is shown in figure 5.

A set of physical parameters that account for the main part of the microwave emission is fed into an ocean model (wind speed (ws) and sea surface temperature SST), an ice model (amounts of water (w), first-year (FY) and multiyear (MY) ice temperature) and an atmosphere model (profiles of cloud liquid water content, water-vapour content and atmospheric temperature profile).

The models produce emissivities, absorption coefficients, and transmission coefficients, and a radiative transfer model (RT) then combines these into a resulting antenna temperature (T_A).

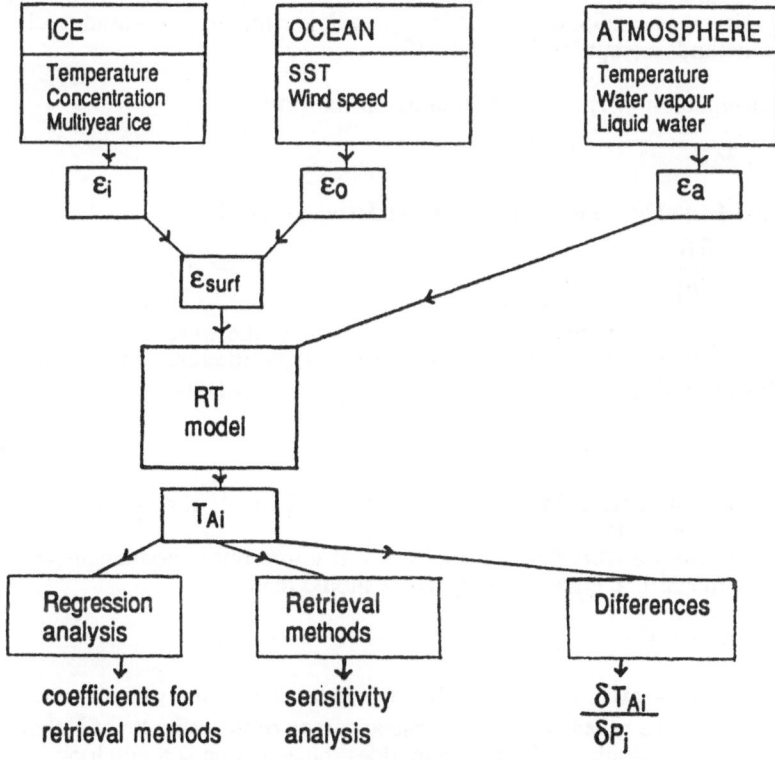

Figure 5.   Block diagram showing the major paths and uses of a model complex. Surface models produce emissivities (e) and transmission coefficients. The radiative transfer model (RT) takes care of the atmospheric transmission (in number of layers) and a set of antenna temperatures emerges.

The models are all relationships of the type

$$T_A = f(P) \tag{4.1}$$

where T_A is a vector containing apparent antenna temperatures at a number of frequencies and polarizations and P is a vector of physical parameters influencing the measurement. A retrieval method should read

$$P = g(T_A) \tag{4.2}$$

In order to find the function g (or f^{-1}) a common approach is to let the model produce a large set of T_B's corresponding to a large variety of surface and atmospheric conditions P. These two sets of data are then used in a regression analysis to produce simple equations relating a certain physical parameter to T_A.

Another approach is to relate measured T_A's to measured physical parameters also using regression techniques. However a number of the parameters such as ice extent and type and cloud liquid content are very difficult to measure at the same accuracy as can be obtained using microwave radiometry, so this approach has mainly been used to check the results of other methods except for SST and WS.

In the linear case equation (4.1) can be written in matrix form

$$T_A = MP \tag{4.3}$$

where the matrix M actually consists of the partial derivatives of f

$$M_{ij} = \frac{\delta T_{Ai}}{\delta P_j} \tag{4.4}$$

The equation (4.4) may be inverted in the case where the number of independent measurements is larger than the number of physical parameters (ex. 10 brightness temperatures to 7 physical parameters). Which leads to the last squares solution

$$p = (M^T M)^{-1} M^T T_A \tag{4.5}$$

that can be obtained if $M^T M$ can be inverted i.e. if the measurements are sufficiently independent.

Now in real life the measurement of T_A is not absolute but is connected with some uncertainty which means that equation (4.3) really should read

$$T_A = MP + e \tag{4.6}$$

The questions that now arise are: given that we know the value of e or actually the covariance of e here to be called R_e^{-1} what is the accuracy of the estimate of P that we can obtain from the equation similar to (S) and what does this equation actually look like.

Both questions can be answered by linear estimation theory, the estimate of P is

$$p = (M^T R_e^{-1} M)^{-1} M^T R_e^{-1} T_A \tag{4.7}$$

with covariance

$$\hat{S} = (M^T R_e^{-1} M)^{-1} \qquad (4.8)$$

The variance of the estimate of each of the physical parameters can be found as the corresponding diagonal element of the covariance matrix which actually defines how the error bars of the measurements map onto the error bars of the solution.

Assuming that the covariance of P (S_p) and the mean (P_0) is known (from climatology) we can obtain a better estimate

$$\hat{P} = (S_p^{-1} + M^T S_e^{-1} M)^{-1} \cdot (S_p^{-1} P_0 + M^T S_e T_A) \qquad (4.9)$$

and covariance

$$\hat{S} = (S_p^{-1} + M^T S_e^{-1} M)^{-1} \qquad (4.10)$$

i.e. the variance of the estiamted parameters are reduced by using apriori knowledge of the parameters, if this knowledge is correct that is.

The described retrieval method contains a number of difficulties such as

- evaluating M requires a good model
- evaluating S_e requires knowledge about system parameters or measurements not always readily available
- evaluating S_e requires knowledge about the covariances of the physical parameters, seasonal and regional changes etc, very seldom available
- in order to use (4.9) to retrieve P we must apply a stepwise approach evaluating M at each step because of the nonlinearities.

A somewhat similar approach may be used to include information about the spatial correlation between the physical parameters int he retrieval method. More details can be found in Rodgers (1976) and Rosenkranz (1978).

5. Conclusions

Microwave radiometry provides a tool for measuring a number of physical parameters of the Earth surface and the atmosphere.

The desire for better spatial resolution leads to very large antennas in space; several meters in diameter is necessary in order to obtain a resolution of a few kilometers at frequencies below 100 GHz.

At higher frequencies the atmospheric effects increase, and the future will probably bring systems to measure several atmospheric constituents at frequencies above 100 GHz.

Apriori knowledge of the covariance of the physical parameters can improve retrieval methods considerably, and also knowledge about the spatial and temporal correlation of the parameters can be used.

6. References

Rodgers, C.D., *Retrieval of Atmospheric Temperature and Composition from Remote Measurements of Thermal Radiation*, Reviews of Geophysics and Space Physics, Vol. **14**, No. 4, pp. 609-624, 1976.

Rosenkranz, P.W., *Inversion of data from diffraction-limited multiwavelength remote sensors, Linear case*, Radio Science, Vol. 13, No. 6, pp. 1003-10, 1978.

Skou, N., *Development of an Airbone meltifrequency Radiometer System*, Report R 203, Electromagnetics Institute, Lyngby, Denmark, 1978.

Ulaby, F.T., Moore, R.K., Fung, A.K., *Microwave remote sensing, active and passive*, Volume 1-3, Addison-Wesley Publishing Company, Reading, Massachusetts, 1981.

MICROWAVE ATMOSPHERIC SOUNDING (WATER VAPOR AND LIQUID WATER)

René BERNARD
C.R.P.E. CNET/CNRS,
38,40 rue du General Leclerc,
92131 Issy les Moulineaux.

Abstract:

The ATSR microwave radiometer will be flown on-board the ERS-1 satellite, and wil be devoted to the measurement of atmospheric water vapour and liquid water content, with as main purpose the knowledge of atmospheric propagation conditions for the altimeter. Although it is not scanning, it will however give useful information about water vapour and cloudiness below the satellite, in relation with other measurements of the air-sea interface.
The lecture introduces the general principle of microwave atmospheric sounding, with application to temperature and water vapour profiling, and to measurement of integrated water content. Then the ATSR microwave radiometer algorithm is described, and used to analyze the radiometer performances, and the methods to calibrate and validate the radiometer products.

R. A. Vaughan (ed.), Microwave Remote Sensing for Oceanographic and Marine Weather-Forecase Models, 191–216.
© *1990 Kluwer Academic Publishers.*

1. Introduction:

1.1 APPLICATIONS OF ATMOSPHERIC SOUNDING:

1.1.1 Climatology - Meteorology.

Atmospheric temperature or water vapour profiles are of basic importance for weather modelling and for energy transfer studies. Satellite remote sensing is the only way to obtain regularly such data over seas, on a global basis. Furthermore, microwave remote sensing gives the opportunity to access atmospheric data even below clouds.

Thus TOVS (TIROS Operational Vertical Sounder) is able to give vertical temperature profiles, and then geopotential heights and the derived thermal wind (figure 1- Chedin and Scott, 1983), with an accuracy matching that of conventional data (as assimilated by the ECMWF numerical model) gathered over Europe, where the observational network is dense. The advantage of the satellite data is obviously that the same accuracy may be obtained over southern hemisphere oceans, where surface data is sparse. In the example shown here, microwave data has been used to improve the profile restitution over cloudy areas.

Even with a low vertical resolution (or even integrated data), the satellite information may be still useful for weather forecasting or climatology studies. For instance, the figure 2 shows annual mean atmospheric water vapor map obtained from SMMR microwave radiometers (Prabakhara et al. 1982, Chang et al. 1984). The water vapor structure which characterize the southern tropical Pacific is clearly resolved, much better than on a map derived from conventional surface data. Actually, monthly or even weekly evolution of that structure, characteristic of the atmospheric circulation, can be resolved. From such data, it is then possible to study water vapor transport (Cadet 1986, Cadet and Greco, 1987a, 1987b), or energy budget over the global ocean. At shorter time scales, it is also possible to study the structure of atmospheric water content associated to fronts or cyclone (Figure 3, McMurdie et al., 1985).

1.1.2 Propagation Correction.

Another interest of microwave atmospheric sounding, more related to remote sensing techniques, is its ability to provide correction information for other remote sensing instruments, for instance propagation delay correction for altimetry, or atmospheric attenuation for microwave scatterometers or infra-Red radiometers.

The tropospheric propagation delay and attenuation are related to the complex refractive index, the imaginary part corresponding to the absorption, and the real part to a phase delay equivalent to an apparent path increase. The real part can be written:

$$Re(N) = N_0 + D(f) = 77.67 \frac{p}{T} + 71.7 \frac{e_v}{T} + 3.74 \, 10^{-5} \frac{e_v}{T^2} \qquad (1)$$

where p is the dry air pressure (mb), T the temperature (K) and e_v the water partial pressure (mb).

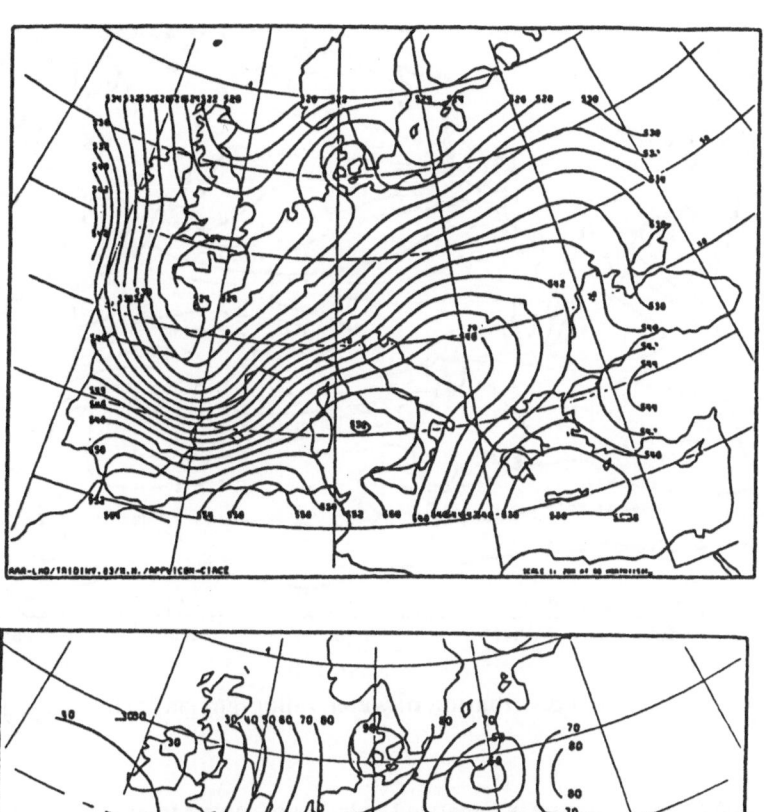

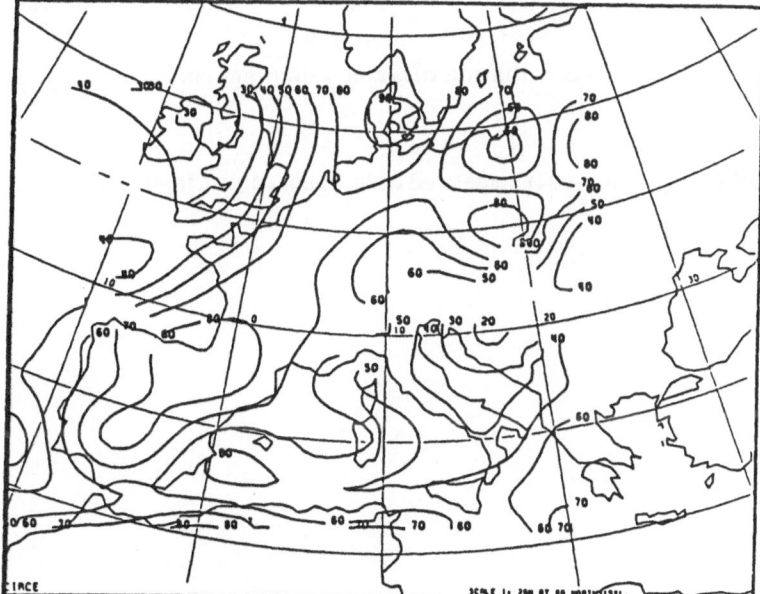

Figure 1: (from Chedin and Scott, 1983)
a)- Geopotential height 1000-500 mb from 3I inversion algorithm and TOVS data. Agreement with ECMWF model is very good (bias 1.5 dam, rms 2.2 dam).
b)- Relative humidity in the 1000-800mb layer from the same algorithm.

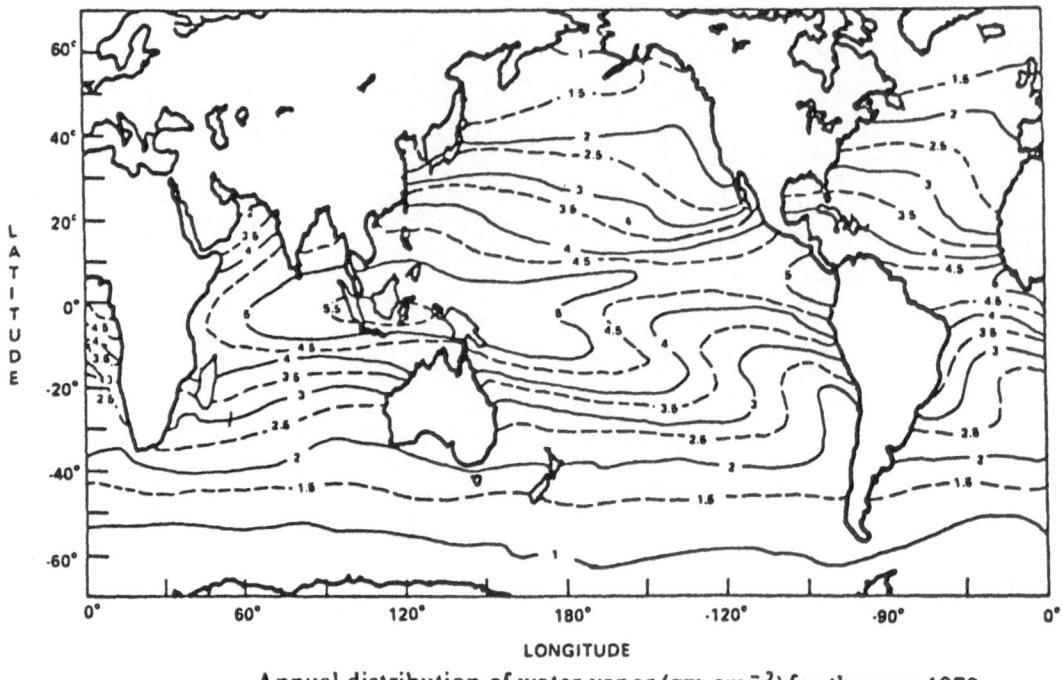

Annual distribution of water vapor (gm cm⁻²) for the year 1979.

Figure 2: (from Chang et al., 1984). Integrated water vapor content from SMMR radiometer data.

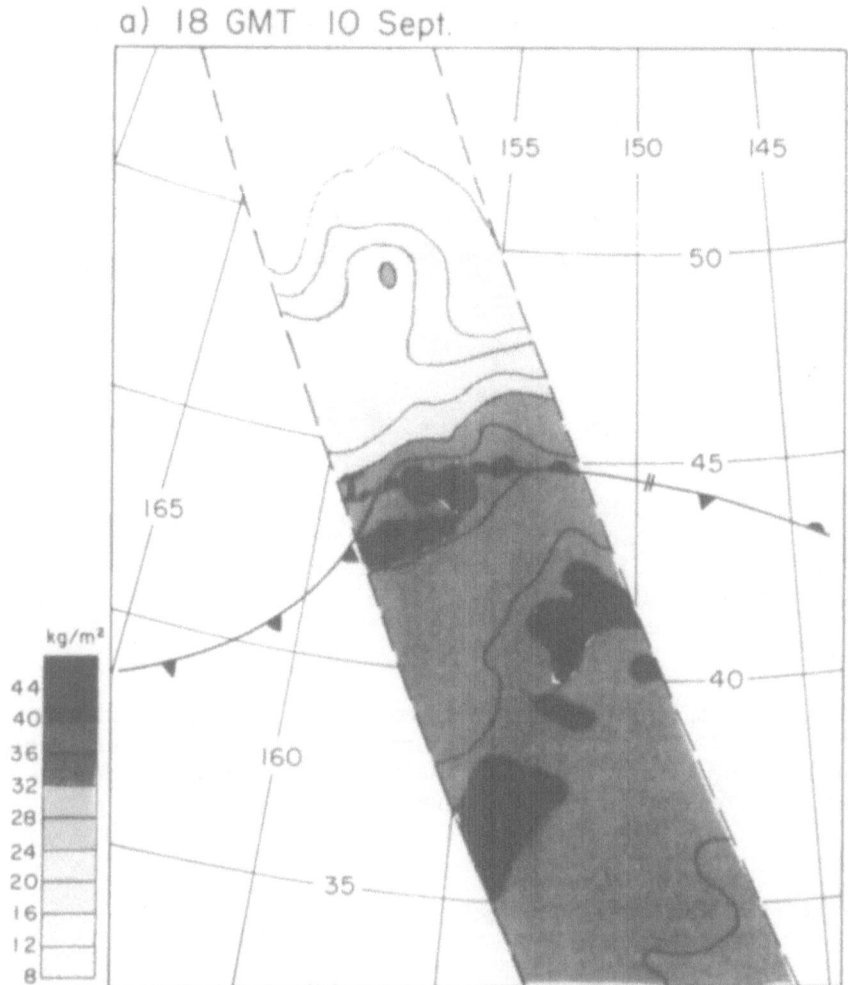

FIG. 3. Horizontal distribution of integrated water vapor in kg m⁻² with frontal analysis superimposed. The data are contoured every 4 kg m⁻² and shaded every 8 kg m⁻² with darker shading indicating higher water vapor content. (a) Overpass number 1084, 1912 GMT 10 September 1978. (b) Overpass number 1092, 0900 GMT 11 September 1978. (c) Overpass number 1098, 1840 GMT 11 September 1978. (d) Overpass number 1106, 0830 GMT 12 September 1978. (from McMurdie and Katsaros, 1985).

Figure 3: (from McMurdie and Katsaros, 1985). Water vapor content distribution across a front.

The propagation path increase can be then separated into a "dry" contribution:

$$dh_d = 77.67 \; 10^{-5} \int_0^\infty \frac{P}{T} = 7.767 \; 10^{-5} \frac{RP_0}{g} = 0.227 \; P_0 \qquad (2)$$

with dh in cm and P_0 being the surface pressure in mb,

and a "wet" contribution:

$$dh_w = 0.3744 \int_0^\infty \frac{e_v}{T^2} dz \; - \; 6 \; 10^{-6} \int_0^\infty \frac{e_v}{T} dz \qquad (3)$$

If the atmosphere is nearly isothermal (at least where there is a significant amount of water vapor), the "wet" path increase may be related to the integrated water vapor content W (g/cm^2):

$$dh_w = 6.18 \; W$$

This direct relationship between path correction and integrated water vapor content is indeed verified with a good accuracy over most of the atmospheric conditions, and integrated water vapor content measurements can be used directly to correct altimeter data for wet tropospheric delay.

2. Principle of microwave radiometry:

2.1 INTRODUCTION: THERMAL RADIATION.

Microwave radiometry is the detection of natural thermal radiation. For a perfect blackbody, the energy B (m^{-2} sr^{-1} Hz^{-1}) radiated by a surface of unit area in the unit solid angle is given by the Planck law, function of the frequency f (Hz) and the temperature T (K):

$$B \; df = \frac{2hf^3}{c^2} \frac{1}{\exp(\frac{hf}{kT}) - 1} \; df \qquad (4)$$

c (ms^{-2}) is the velocity of the light and h (J) the Planck constant.

In the microwave range, the radiated power is much lower than in the Infra-Red range, but can still be detected (even at very low temperatures such as the cosmic background radiation at 2.7 K). In that frequency range, hf/kT is small, the Raleigh-Jeans approximation applies:

$$B \, df = 2 \frac{f^2 \, kT}{c^2} \, df$$

and there is a linear relationship between the radiated power in a narrow frequency bandwidth and the temperature, which is then commonly used instead of energy.

Actual bodies (including gases) are not perfect blackbodies, and the actual radiated power is lower than the power radiated by a blackbody at the same temperature. It is however still proportional to the temperature, and the ratio of the actual radiated power to the blackbody radiated power is the emissivity, which can be function of the frequency, of the polarisation and of the direction of propagation.

The emissivity characterizes the radiating properties of a body, but also its absorption properties. A blackbody absorbs all the incident radiation: this is actually the definition of the blackbody. Thermodynamic equilibrium considerations show that for any body,'the absorption coefficient (fraction of the incident radiation which is absorbed) is equal to the emissivity.

2.2 RADIATIVE TRANSFER:

2.2.1 Absorption/emission by gases.
When propagating within a gas, an electromagnetic radiation may be partly absorbed (molecular or atomic absorption). Conversely, any gas molecule is able to radiate energy, following the Planck law, according to its temperature and to its absorption characteristics. The absorption properties of a gas depend of its nature, of its temperature and of the radiation frequency, and are often associated to narrow lines. In the microwave range (1 GHz to 200 GHz), the dominant absorption mechanism in the atmosphere is associated to vibration levels of oxygen and water vapour molecules (Fig 4), and the absorption coefficient is proportional to the number of molecules involved..

2.2.2 Scattering by particles.
In addition to molecular or atomic interactions, electromagnetic radiation may interact with small particles present in the atmosphere. Part of this interaction corresponds to absorption mechanisms, characterized by an absorption coefficient, with the associated emissivity, but another part may be the scattering of the radiation through reflection on the particles. This mechanism is function of the size of the particles relative to the radiation wavelength, and of their shape. One can then define a scattering function, giving the fraction $P(\mathbf{r},\mathbf{r}')$ of incident energy (propagating in the direction $\mathbf{r}$) scattered in the direction $\mathbf{r}'$, for an interaction within a volume unity:

$$dB(\mathbf{r}') = B(\mathbf{r}) \, P(\mathbf{r},\mathbf{r}') \, d\Omega$$

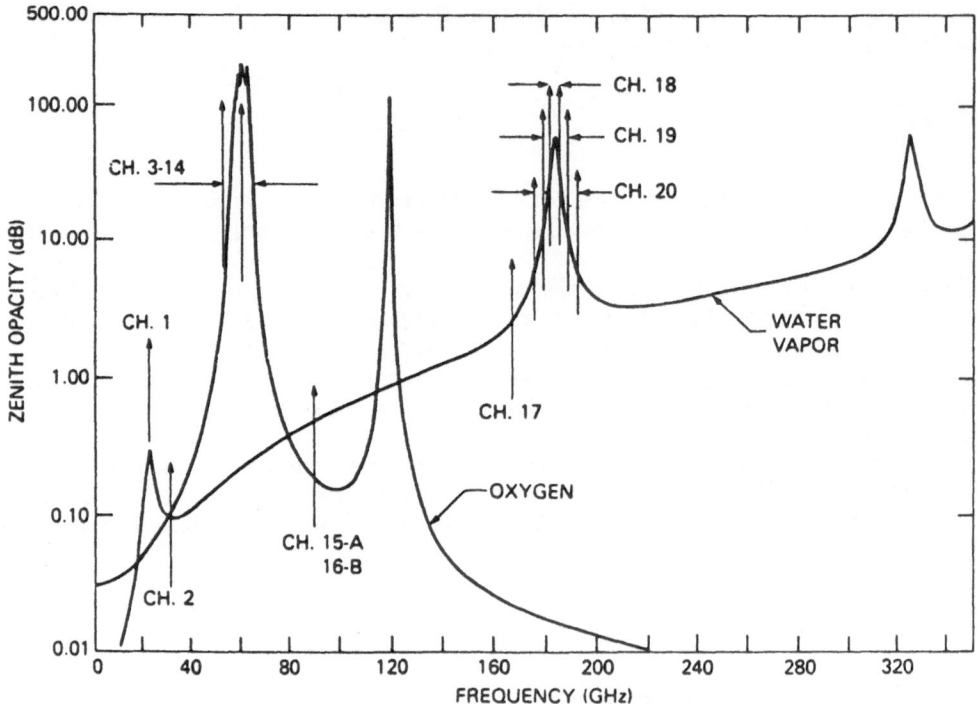

Figure 4: (from HMMR EOS Report, 1987). Atmospheric absorption lines and AMSU channels selection.

and a scattering coefficient k_d by integration over the sphere:

$$k_d(\mathbf{r}) = \int \int P(\mathbf{r},\mathbf{r}')\, d\Omega'$$

Both absorption and scattering coefficients are function of the particles density, size and shape, and of their dielectric properties. They can be modelized following the Mie theory. In the case of microwave propagation in the atmosphere, the most efficient scatterers are the liquid water droplets. For small drops (relative to the wavelength), the Rayleigh approximation may be applied, the scattering coefficient being negligible, and the absorption coefficient is only function of the liquid water density (whatever the size and shape of the drops). Ice particles due to their dielectric properties, are efficient only at higher frequencies (above 40 GHz).

2.2.3 Radiative transfer equation.

2.2.3.1 Non-scattering medium.
For an element of volume along the propagation path, the radiative transfer equation at a frequency f is written:

$$dB(r) = - k_a\ (B(r) + B_p(T_a))dr \qquad\qquad (5)$$

where

$$B_p(T_a) = k_a\left(2kT_a\frac{\delta f}{\lambda^2}\right)$$

is given by the Planck law, and k_a is the gas absorption coefficient (per unit volume) at the frequency f.
An equivalent form is:

$$dT_B(r) = - k_a\ (T_B - T_a)dr \qquad\qquad (6)$$

If the air temperature and the absorption coefficient are known along the radiation path, this equation is easily integrated between position r_1 and r_2:

$$T_B(r_2) = T_B(r_1)\exp\left(-\int_{r_1}^{r_2} k_a(r)\,dr\right) + \int_{r_1}^{r_2} k_a(r)\,T_a(r)\exp\left(-\int_{r}^{r_2} k_a(x)\,dx\right) \quad (7)$$

and one can define the transmission function:

$$\Gamma(r,r') = \exp\left[-\int_{r}^{r'} k_a(u)\,du\right] = \exp\left(-(\tau'-\tau)\right) \qquad\qquad (8)$$

where τ is the optical depth: $d\tau = k_a(z)\,dz$.

It is also convenient to define an "equivalent temperature", which is the temperature of an isothermal atmosphere having the same properties:

$$T_e(r,r') = \frac{\int_r^{r'} T_a(u)\frac{d\Gamma(u,r')}{du}\,du}{1 - \Gamma(r,r')} \qquad (9)$$

Equation (7) becomes:

$$T_b(r') = T_b(r)\,\Gamma(r,r') + T_e(r,r')\,[1-\Gamma(r,r')] \qquad (10)$$

One can see from this that the radiative transfer computation is easily done by dividing the atmosphere into layers, each being characterized by its transmissivity and its equivalent temperature. If the layer is small enough so that its temperature and absorption properties can be considered as constant, the integration is straightforward.

2.2.3.2 Scattering processes: Mie/Rayleigh.

For microwave propagation in the atmosphere, the scattering is mainly due to liquid water or to ice. The scattering mechanism is described by the Mie theory, and depends on the water dielectric constant and on the drop size and shape. The theory describes how incident radiation from one direction is scattered in any direction. However, according the drop size relative to the wavelength, several approximations are possible:

- When the drop size is small relative to the wavelength, the Rayleigh approximation is valid, and the only mechanism is the absorption. One can then apply the non-scattering medium equation, with an absorption coefficient function of the volumic water content.

- For larger drop size, the full Mie theory has to be used, and the radiative budget for one direction of propagation depends on the radiation propagating in all directions.

The radiative transfer equation is then written:

$$dT_B(r) = [k_a(T_a-T_B) + k_d(T_d-T_B)] \qquad (11)$$

where k_a includes molecular and particles absorption, and T_d is a diffusion source function:

$$T_d = \frac{1}{4\pi k_d} \int\int_{4\pi} T_B(\mathbf{r}')\,P(\mathbf{r},\mathbf{r}')\,d\Omega' \qquad (12)$$

which depends of the radiation propagating in all other directions. Then it is not possible to have a simple radiative transfer model, except if radiation is nearly isotropic. This is unlikely to happen at the upper cloud boundary, with a cold downward radiation. If the phase function is nearly isotropic, the effect is then equivalent to the reflexion of that downward radiation on the interface.

2.3 GENERAL EQUATION - WEIGHTING FUNCTIONS.

For a downward looking radiometer, the received brightness temperature may be written:

$$T_B = T_u (1-\Gamma_\infty) + e\ \Gamma_\infty T_m + r\ \Gamma_\infty (T_d (1-\Gamma_\infty) + T_c\ \Gamma_\infty) \qquad (13)$$

where one can distinguish the upward radiation from the atmosphere, the surface contribution attenuated on its way upward, the reflected downward atmospheric radiation and the downward sky radiation, reflected and attenuated twice across the atmosphere.

Here Γ_∞ is the total transmissivity of the atmosphere:

$$\Gamma_\infty = \exp(- \int_0^\infty k_a(r)\ dr)$$

T_u and T_d are equivalent upward and downward atmospheric temperatures

$$T_u = \frac{1}{(1-\Gamma_\infty)} \int_0^\infty T_a(u) \frac{d\Gamma(u,\infty)}{du} du$$

$$T_d = \frac{1}{(1-\Gamma_\infty)} \int_\infty^0 T_a(u) \frac{d\Gamma(u,0)}{du} du$$

The function $\dfrac{d\Gamma(u,\infty)}{du}$

is the weighting function which defines which part of the atmosphere contributes most to the observed radiation: if this weighting function is narrow enough, the equivalent temperature will be close to the temperature near the maximum of the weighting function. The figure 5 shows some examples of weighting functions for the AMSU microwave sounder.

202

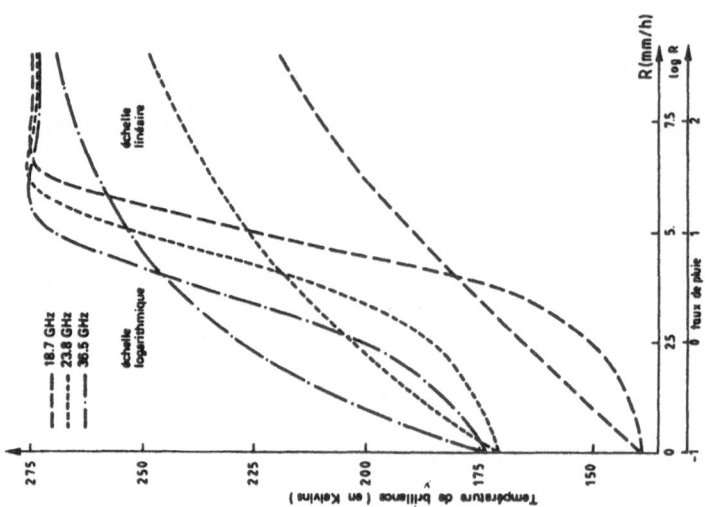

Figure 6: (from Millet, 1985). Brightness temperature sensitivity to rain rate at 18.7, 23.8 and 36.5 GHz. Cloud liquid content is 0.41gm^{-3} and the freezing level at 3000 m. The linear scale corresponds to small rain rate.

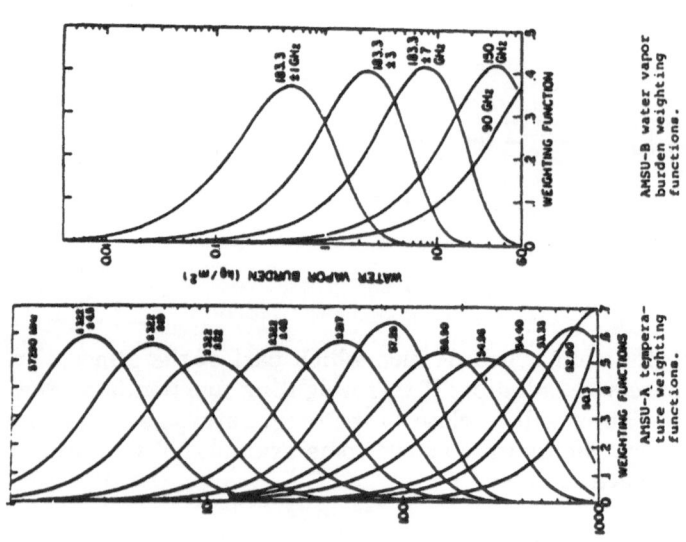

Figure 5: (from Stealin, 1987). AMSU weighting functions.

2.4 SURFACE CONTRIBUTION.

When observing downward from a satellite, as seen in the equation 13, the surface contributes to the received signal, if the total atmospheric optical depth is not too large. Actually microwave radiometry may be used to obtain surface characteristics, but even when interested in atmospheric sounding, it is necessary to deal with the surface contribution. The surface contributes by its own radiation, function of its emissivity and temperature, and by reflecting the downward atmospheric radiation. Most generally, emissivity and reflectivity are characterized by their directionality, being function of the incidence angle, and their dependance on polarisation. Thermodynamic considerations, and the symmetry of electromagnetic propagation allows to establish a relationship between the directional emissivity and the integrated bidirectional reflectivity:

Incident radiation impinging on a surface is partly scattered, and partly absorbed. If one defines a bidirectional reflexion coefficient $r(i,s) = 4\pi\, Bs/Bi$ where Bi is the incident radiation from direction i, and Bs the scattered radiation in the direction s, the absorption coefficient is:

$$a_i = 1 - \frac{1}{4\pi \cos(\theta_i)} \int\int_{2\pi} r(i,s) \cos(\theta_s)\, d\Omega_s \qquad (14)$$

and the emissivity in the direction i is equal to the absorption coefficient.

So, in the general case, the reflected part of equation 13 should be the integral over all the directions for the downward radiation. Two extreme cases can be considered:

One is specular reflexion: only the symmetrical direction contributes significantly to the radiation, and the radiative transfer has to be computed only for that direction.

The other corresponds to a lambertian surface which reflects isotropically the incident radiation. Emissivity is then independant of the direction, and the reflectivity varies as the cosine of the incident angle. It can then be shown that one can consider only the incident radiation at some mean incident angle close to 45°.

Which case has to be considered in the case of microwave radiation depends of the roughness of the surface, relative to the wavelength, and of the observation angle. The surface emissivity depends of the surface properties (dielectric properties and surface roughness) and of the incidence angle.

3. Principle of radiometer sounding: profiling / integrated content.

According to the transmissivity of the atmosphere at the selected frequency, radiometry can be used either to observe the surface or to probe the atmosphere. When Γ_∞ is close to 1, the equation 13 is reduced to the surface contribution, and the brightness temperature is then a function of the surface temperature, the surface roughness and the surface dielectric properties (such as for instance the soil moisture). When Γ_∞ is not equal to 1 one has to take into account the atmospheric contribution, either to correct the surface data, or as a tool to study atmospheric properties if the surface contribution can be estimated. Finally, if Γ_∞ is equal to 0, the measured brightness temperature is the equivalent temperature of equation 9, and corresponds to the atmospheric temperature weighted by the weighting function.

The absorption coefficient being a function of the frequency and proportional to the amount of absorbing material, it can be chosen so that different atmospheric levels are probed. Observation at different frequencies allows then to get atmospheric profiles. One can for instance choose different levels on the side of an absorption line, as shown on the figure 4. Which atmospheric parameter will be profiled depends on the nature of the absorbing constituent: if one choose an oxygen line such as 57 GHz or 118 GHz, the weighting function will be mainly function of the pressure, and we will then get directly the temperature function of the pressure level. On the contrary, if one select a water vapor line such as 183 GHz, the weighting function will be function of the integrated amount of water vapor (figure 5), and the measured temperature will correspond to the temperature at the level where a certain integrated amount of water is reached. One then needs to have also the temperature profile in order to retreive the water vapor profile.

Finally, if the absorption coefficient is small enough so that the maximum of the weighting function is never reached (such as with the 22.235 GHz water vapor line), the atmospheric contribution to the observed temperature will be mainly function of the total atmospheric water content.

3.1 INVERSION METHODS AND ALGORITHMS DEVELOPMENT.

It is clear from the integral form of the radiative transfer equation that there never is a direct relationship between any geophysical parameter (such as the temperature or the water content at some level) and the observation in one channel, and that the observations at different channels are not independent. It is then necessary to resort to inversion methods. It is not the aim of this lecture to expand on the various inversion methods. Let us just say that direct mathematical inversion is generally not possible, and statistical inversion is used most of the time, at least in order to get a first guess solution from which iterative inversion can be applied.

The general principle of statistical inversion is to generate some a-priori database, associating a set of observed data (brightness temperatures for the different channels), to the pertinent atmospheric or surface geophysical parameters. This database is established by applying the radiative transfer equation to a selected set of atmospheric and surface situations. The choice of that set is critical for the quality of the inversion. It can be either statistically representative of the actual situations (with possible subsets corresponding to different climatological situations), or else try to sample equally the extreme and average situations (such as the 3I initialisation scheme developed by Chedin and Scott, 1983, for the TOVS data inversion). An intermediate approach could be to use an uniform sampling for the geophysical parameters which are to be retrieved, and a statistically representative sampling of the parameters which may be influent, but are not resolved by the inversion. The database could also be obtained directly from simultaneous measurement of the brightness temperatures and of the geophysical parameters. However this method is generally not feasible, owing to the huge number of observations needed.

Once this database is established, the correspondance between geophysical parameters and radiometer observation is established through classical statistical inversion (least square or maximum likelihood methods). The inversion result may then be improved by mathematical inversion of the linearized radiative transfer equation, when possible.

One major problem with the development of profiling methods is that the observations at different frequencies are usually not independant, and the inversion matrix is not well conditioned. Simulations using the radiative transfer model and the inversion model are then necessary to select the optimum choice of channels (number and frequencies).

4. Integrated water vapour / liquid water sounding: ATSR/MW application.

When one looks for integrated water vapor or liquid water, ie when using weakly absorbed frequencies, the inversion may be simpler, as the number of parameters is weak. However, one has to deal with atmospheric and surface parameters, and their contribution to the brightness temperatures may not vary strongly with the frequency. Channel selection is then necessary, in order to have the highest sensitivity to the looked-for parameters (water content), and the weakest dependance to the ignored parameters, such as surface parameters or atmospheric profile structure. The approach which has been used for the development of the algorithms for the ATSR Microwave Sounder is described below, and it is shown how it can be used to evaluate the performances of that radiometer.

4.1 CHANNEL SELECTION AND ALGORITHM DEVELOPEMENT.

As one looks for integrated water content, liquid and vapor, and wants to be operational under most of the atmospheric situations, it is clear that we are limited to frequencies for which the Rayleigh approximation for scattering applies whatever the cloud water content, ie below 40 GHz. Water vapor sensitivity will be then through the 22.235 absorption line. Some other constraints have to be considered for channel selection, such as the necessity of interference-free channels (avoiding interferences from ground based or on-board transmitters).

Another remark is that integrated water content measurement requires that the transmissivity remains close to 1. In that case, a simplified radiative transfer equation (ignoring the reflected radiation) may be written:

$$T_B = T_u + \Gamma_\infty (eT_m - T_u) \qquad (15)$$

As we want to measure Γ_∞ , it is clear that one needs to have eT_m different of T_u, ie. a surface emissivity as low as possible. This limits the use of the method to the observations over the sea, as the emissivity of the land surface is close to one.

As stated above, channel selection and performance analysis are made through the developement of an inversion algorithm, based on the direct radiative transfer model. In the case of the ATSR Microwave Radiometer, the algorithm development had also to take into account the possible use of the altimeter wind measurement, obtained on the same spot, to get information on the surface roughness.

4.1.1 Radiative transfer model.
A radiative transfer model has to perform the integration of the radiative transfer equation, given atmospheric profiles parameters (P, T, q, rl) and spectroscopic parameters (absorption coefficients function of pressure, temperature and frequency). A comprehensive model should take into account the contribution of every absorption line in the vicinity of the observed frequency bandwidth. In the microwave range, the number of absorption lines is small, and the line width is large compared to the channel bandwidth. It is then possible to use a simplified model, computing the radiative transfer at one single frequency. The model used here has been implemented from Ulaby et al.(1981). It gives results close to those of a more comprehensive model, and is well adapted for sensitivity studies.

4.1.2 Atmospheric profiles models.
In order to simulate all the range of possible atmospheric situations, it is necessary to apply the radiative tranfer model to the largest possible set of atmospheric profile situations. It could be done through the use of a large set of actual radiosounding profiles. however, as the first aim is to determine the sensivity of the algorithm, it is preferable to synthetize atmospheric profiles such that the range of W, L, T(z)... is almost uniformly covered. The resulting algorithm may not be the most adequate for the actual situations, where the statistical distribution is not uniform, but is well adapted for this study.

4.1.3 Surface modelisation.
As the algorithm is not supposed to determine surface parameters, it is only necessary to simulate the range of possible variations of the surface emissivity, and of the relationship between the emissivity at different frequencies. However, as it is intended to simulate the use of the altimeter derived surface wind, a parametrisation of the wind/emissivity relationship for nadir observation has been included, adapted from Thomas and Francis, 1978.

A random variation of the emissivity at each frequency has been added, as the emissivity model cannot be considered as an exact model, and others parameters than surface wind may be significant (sea state, salinity...).

In order to decorrelate surface contribution and atmospheric contribution, the whole range of surface conditions has been aplied to each atmospheric profile, and that is finally a set of 561600 brightness temperatures which is generated in that example, associated to various surface and atmospheric situations, on which one can establish statistical relationships and possible algorithms.

4.1.4 Inversion method.

The resulting set of data includes brightness temperatures at 18, 21, 24 and 36 Ghz (reserved channels for radiometry), the integrated water vapour and liquid water contents, the altimeter path correction (as the ATSR-MS will be used to correct the altimeter measurement for tropospheric propagation delay), the surface wind and temperature. Relationships between W, L, dh and a combination of brightness temperatures and surface wind are then obtained from multilinear regression analysis. Actually, rather than T_b, it can be shown that it is better to use a log function of Tb :

$$f(T_b) = Log(T_0 - T_b), \qquad (16)$$

T_0 being taken equal to 290 K for 18-24 Ghz, and 280 K at 36 Ghz. It is possible to optimize that value, but the optimisation depends on actual statistical distribution of atmospheric profiles.

Then, the general algorithm form is:

$$\begin{pmatrix} W \\ L \\ dh \end{pmatrix} = \sum_i a_i \, Log(T_{0i} - T_{bi}) + a_0 + b\,U \qquad (17)$$

The residual variance for the variable X of the least square analysis is then:

$$<\sigma^2> = \left< \left\{ X - \sum_i a_i \, Log(T_{0i} - T_{bi}) - a_0 - bU \right\}^2 \right> \quad (18)$$

and gives an indication on the quality of the model, and then on the optimal performance of the algorithm (ie. on the statistical error which can be expected on that variable, due to the limitations of the model).

From the coefficients of the algorithm, it is also possible to determine its sensitivity to instrumental errors on the determination of the brigthness temperatures:

$$\delta X / \delta T_{bi} = a_{ix}/(T_0 - T_{bi}) \qquad (19)$$

A summary of this analysis is presented on the tables 1 and 2 for the altimeter correction (proportional to the water vapor content).

4.2 PERFORMANCES ANALYSIS.

The algorithm performance is directly derived from the residual of the inversion. The actual ATSR algorithm (using two channels at 24 and 36 GHz) will not use this functiorïal relationship, but rather a direct relationship between brightness temperatures and geophysical parameters (look-up tables), taking into account the non-linearity of the relationship. Surface wind correction from the altimeter can be added as a linear correction around the climatological mean wind (7 ms^{-1}). However the overall performances will not be very different.

It is clear from such an analysis that the best performance is obtained with a 3-channel radiometer, where the combination of two close frequencies, one in the water vapor line, and the other just below, allows to limits the effect of the surface variability (which will be the same at both frequencies), and the channel at 36 GHz being used for liquid water discrimination. However the two channel configuration of the ATSR, with correction from the altimeter wind, is nearly as performant (table 3).

Those performances are those of the algorithm itself, for a perfect radiometer, the limitation being due to natural variability (of atmospheric profiles and surface conditions) which cannot be taken into account with only two or three channels.

4.3 SENSITIVITY TO RECEIVER PERFORMANCES.

The actual brightness temperature measurement is not perfect, due to instrumental errors or limitations (thermal noise fluctuations). The sensitivity tables allow to estimate the effect of the measurement errors on the geophysical parameters determination. It can be seen that a very good absolute accuracy (better than 1 K) is required on the brightness temperature measurement in order to match the algorithm performance.

Different sources of errors can be identified and estimated in the radiometer. An error budget (K) is given below as it is estimated or measured for each ATSR channel:

Frequency stability:	0.20
Radiometric resolution:	0.33
Gain stability:	0.60
Calibration:	
Resolution	0.50
Coefficients	0.50
Sky temperature estimation:	
losses	1.20
secondary lobes	1.00
Antenna secondary lobes:	
earth view	1.00
structure	2.00
Antenna losses:	1.00
RMS total:	3.00K

The resulting error on the water vapor would then be 0.45 g/cm^2, and 0.08 Kg/m^2 for liquid water (compared to 0.2 g/cm and 0.11 Kg/m^2 due to the algorithm).

One remark has to be added about these errors. All the sources are considered to be independent, which is not necessarily the case. Part of the radiometer error budget is linked to the physical temperature of the radiometer or the surrounding structures. These temperatures will have an harmonic variation with the period of the orbit, and are not independent. Further averaging will not reduce their contribution. The same remark holds for the algorithm errors, as they are due to surface or atmospheric profile variations which can be correlated, and may have a strong latitudinal variation.

4.4 LIQUID WATER EFFECT.

4.4.1 Cloud water.

All the algorithms have been obtained from simulation over a large range of possible cloud liquid water content, and are able to separate the effect of liquid water and water vapour, the best performance for liquid water measurement being obtained when the 36 Ghz channel is used. However, the cloud effect on the brightness temperature may be very variable, for the same water content, as the liquid water absorption coefficient depends strongly of the temperature,and the cloud temperature depends on the cloud level. Then, the presence of clouds degrades the performance of the water vapour retrieval. The effect is largest for the 24/36 configuration (from 0.84 to 1.5 cm) than for the 18/21 one (from 0.88 to 1.3 cm), the effect being small with a 3 channel combination (from 0.73 to 0.83).

4.4.2 Rain effect.

The rain increases strongly the scattering, and tends then to make the atmosphere opaque to the microwave radiation, beginning at the higher frequencies, as shown on the figure 6. The rain effect depends strongly on the drop size distribution and on the rain thickness (ie on the freezing level). The brightness temperature tends then toward the rain physical temperature, and any information about water vapour content is lost.

The figure 7 shows the range of brightness temperatures which can be observed on our simulation set. The observed brightness temperatures as a function of rain rate are also shown for one atmospheric situation. It can be seen that for the 24/36 configuration, the information on water vapour is lost for rain rates higher than 4 mm/h, but that rain can be detected from thresholding the brightness temperatures. For rain rates smaller than 4 mm/h, rain cannot be distinguished from cloud, and introduces error on the water vapour determination which can reach 0.5 g/cm2 for rain rates smaller than 1 mm/h. An algorithm using the 18/21 configuration is less sensitive to rain, and can be used up to 4 mm/h. Again, a 3 channel configuration could be used with rain rates up to 10 mm/h.

All those estimations have been done assuming that the rain cell is completely filling the radiometer footprint. The effect of partial filling will be to reduce the increase of the brightness temperature due to the rain, and then prevent the possibility of rain detection even for high rain rate. However, the corresponding error on the water vapour estimation will also be reduced.

4.5 IMPROVEMENTS.

There may be improvements in the water vapor and liquid water determination, mainly related to the simultaneous operation of the Infra-Red Radiometer, which

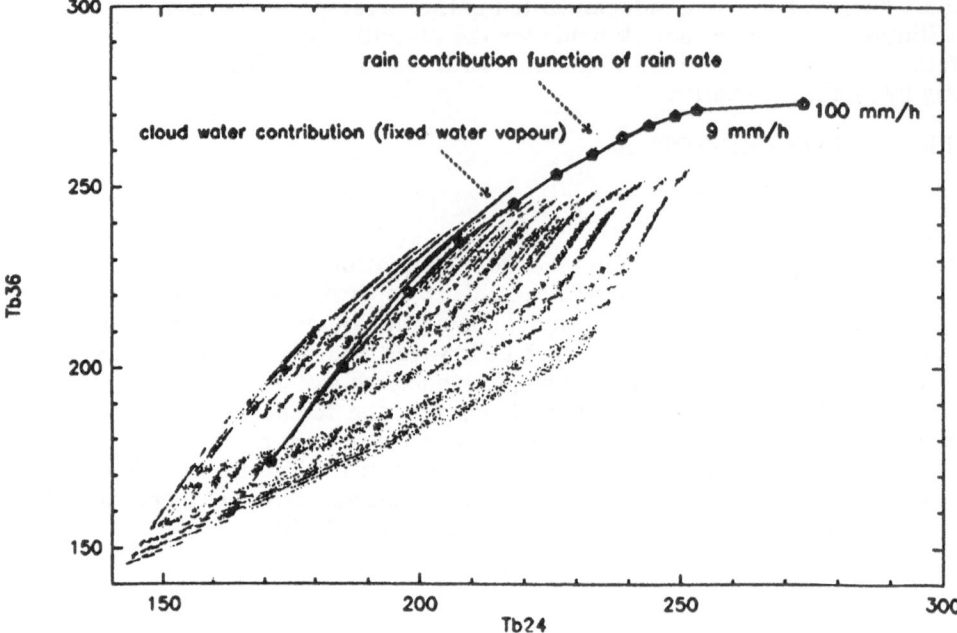

Figure 7: Brightness temperatures distribution at 23.8 and 36.5 GHz for the synthetic data set. Also shown is the distribution for a constant water vapor content and variable cloud liquid water content (thin line), and the distribution for constant liquid water and water vapor, and variable rain rate.

will gives a map of the surface temperatures within the Microwave Radiometer. It can be used to improve the estimation of the surface contribution, although the expected improvement will be low. A more interesting contribution will be through the cloud analysis, as it will allow first to estimate the cloud cover, and then to adapt the algorithm either to clear sky or to cloudy conditions. Furthermore we have seen that the indetermination of the cloud level was a source of error on the liquid water content determination. The measurement of the cloud top temperature will then give a valuable information for the liquid water algorithm.

5. ATSR/M radiometer operation.

5.1 CALIBRATION

The performances of the ATSR/M Microwave Radiometer have been estimated previously. They have to be estimated and controlled through several steps.

The first calibration is done before the launch, and measures most of the radiometer parameters, such as the antenna gain and antenna pattern (in order to be able to estimate the sidelobe contribution), the radiometer internal gain, and its variation with temperature. This last measurement is critical, as a 1K absolute accuracy is equivalent to a 0.015 dB accuracy on the gain determination. This is only possible by an end to end calibration, with a well known blackbody in front of the input antenna.

However, this gain calibration will be repeated regularly in orbit, using the cosmic noise input through a dedicated sky horn. It is then only necessary to know accurately the difference of gain beween the main antenna channel and the sky horn channel. As the difference correspond to passive component, it can be done at a single temperature. The pre-launch calibration will then have mainly to determine that correction, and the sensitivity of the gain to internal temperature variations. The amplitude of that variation will be small anyway, as the thermal design shows that temperature variations along the orbit are only a few degrees.

5.2 VALIDATION.

The instrument calibration concerns the brightness temperature measurement. However, the instrument performance is evaluated through its ability to measure water content, and validation is necessary, which includes the instrument and the algorithm performances. This validation will have two aspects. One will use point comparison to all available surface data (radiosounding). However, the amount of data thus available over open ocean is small. Data from islands can be used, but cannot be compared directly to radiometer data, as the radiometer does not operate well close to land masses, due to sidelobes contribution. The currently measured antenna pattern shows however that the radiometer data could be used reliably at distances not greater than 100 km.

Another continuous validation will be possible, using data from Numerical Weather Forecast Models. Those models include radiosonde data, and extrapolate

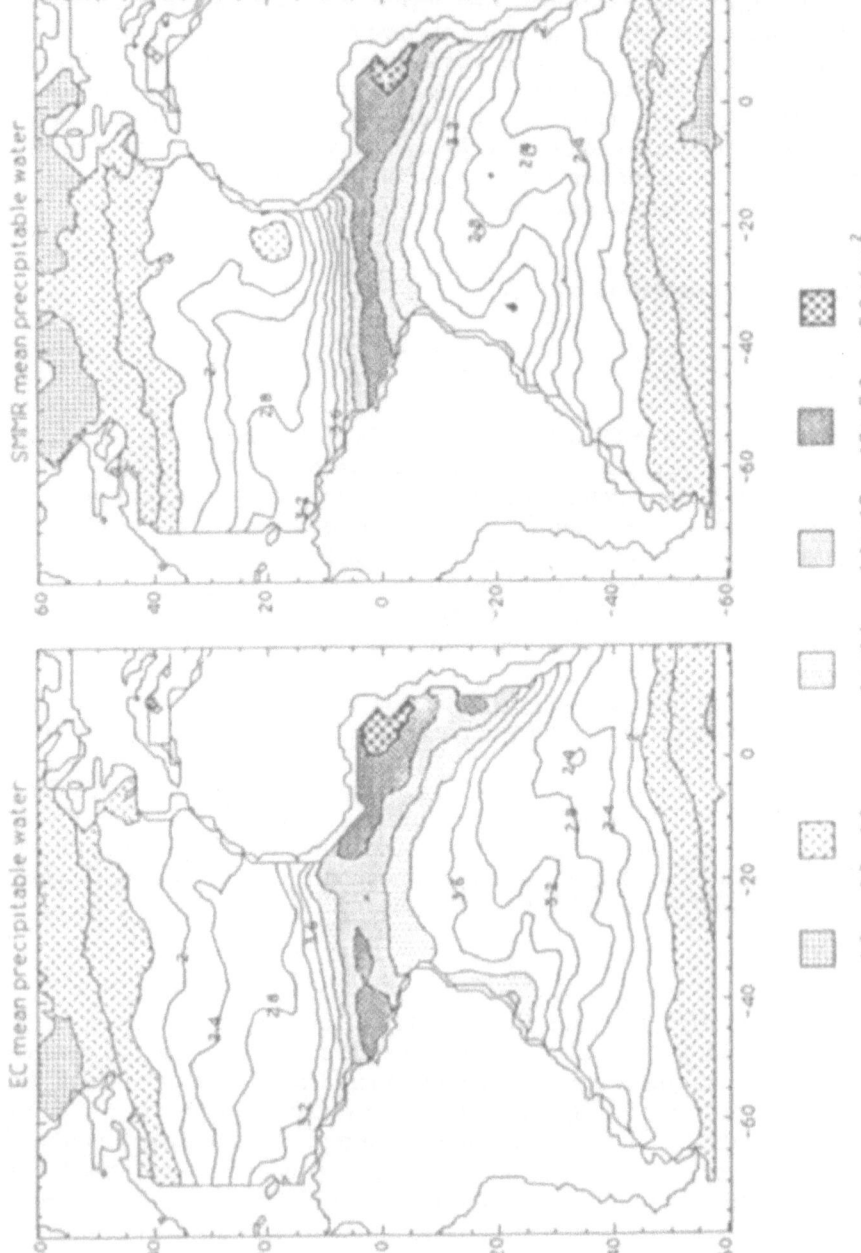

Figure 8: (from Eymard et al., 1989). comparison of integrated water vapor content from Nimbusb-7 SMMR radiometer(right), and from the ECMWF model (left).

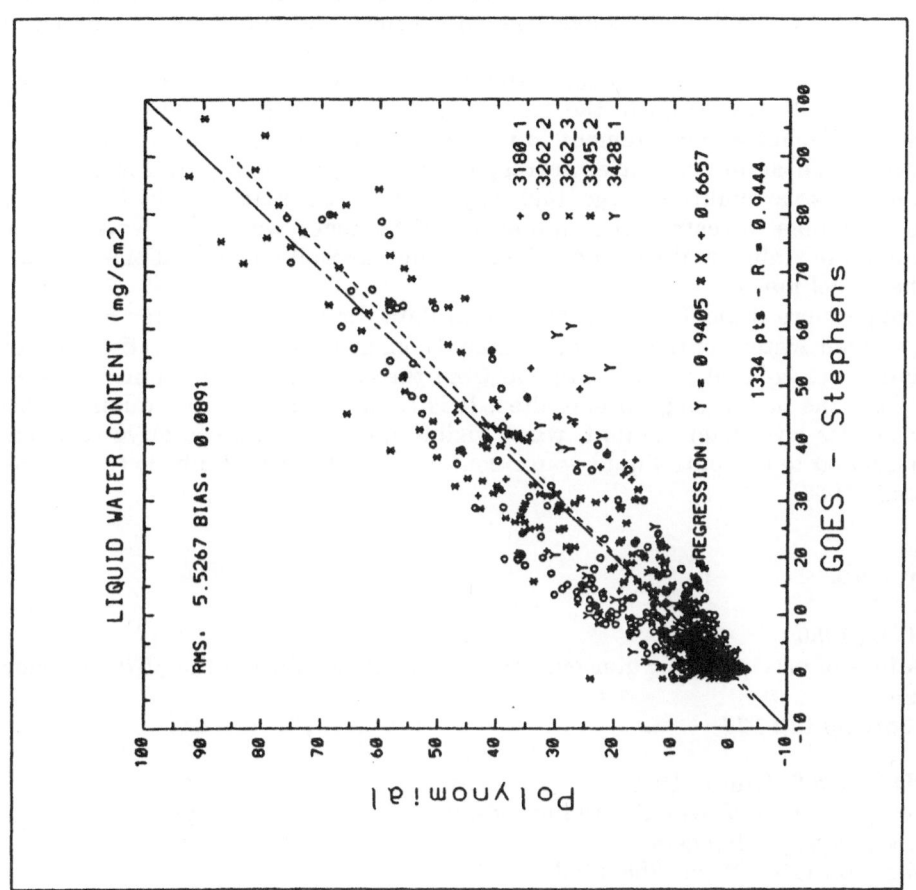

Figure 9: (from Lojou et al. 1989). Cloud liquid water content derived from Nimbus-7 SMMR radiometer and from GOES visible albedo.

them on the whole ocean surface through objective analysis. The comparison can then be done continously. Comparison of Nimbus-7 SMMR measurements and output of the ECMWF model have been done and are shown on the figure 8 (Eymard et al., 1988).

The general agreement is good, although some discrepancies can be seen, which could be due to errors in the model prediction (this is actually the goal of microwave radiometer measurement to improve the initialisation of weather forecast models). Those discrepancies correspond mainly to geographic zones where the water vapor variability is large, and may be poorly accounted for by the model. It is then possible to restrict the validation to the zones where the water vapor content is quite stable, and which can be identified as well on the radiometer data as on the model maps.

It is much more difficult to find validation data for the liquid water measurement, as cloud water content is not commonly measured. A rough estimation of the liquid water content can be obtained from the combined visible and infra-red images from the meteorological satellites. It is even possible to find quantitative comparison for low water content, when visible and infra-red radiative transfer may be applied to compute liquid water content from the cloud albedo (Figure 9-Lojou et al., 1988).

6. References.

Cadet, D.L., 1986
Fluctuations of precipitable water over the Indian Ocean during the 1979 summer monsoon.
Tellus. 38A, pp:**170-177**

Cadet, D.L. and S. Greco., 1987a
Water vapor transport over the Indian ocean during the 1979 Summer monsoon. Part II: Water vapor budgets.
Mon. Wea. Rev. 115 (10), pp:**2358-2366**

Cadet, D.L. and S. Greco., 1987b
Water vapor transport over the Indian ocean during the 1979 summer monsoon. Part I: water vapor fluxes.
Mon. Wea. rev. 115, pp:**653-663**

Chang, H.D., P.H. Hwang, T.T. Wilheit, A.T.C. Chang, D.H. Stealin and P.W. Rosenkranz., 1984
Monthly distributions of precipitable water from the Nimbus SMMR data.
J. Geophys. Res. 89-D4, pp:**5328-5334**

Chedin, A. and N.A. Scott., 1983
The Improved Initialisation Inversion procedure "3I".
1st International TOVS study conference. COSPAR-IAMAP-WMO LMD 117.

Eymard, L., C. Klapisz and R. Bernard, 1988
Comparison between Nimbus-7 SMMR and ECMWF model analyses: The problem of the surface latent heat flux.
J. Atmos. Ocean. Technol. to be published.

Instrument Panel Report, 1987
HMMR High resolution multifrequency microwave radiometer.
NASA Earth Observing System Report. Vol IIe

Lojou, J.Y., R. Frouin and R. Bernard., 1989
Comparison between SMMR and GOES liquid water content.
in preparation.

McMurdie, L.A. and K.B. Katsaros., 1985
Atmospheric water distribution in a midlatitude cyclone observed by the Seasat Scanning Multichannel Microwave Radiometer.
Mon. Wea. Rev. 113, pp:**584-598**

Millet, J.M., 1984
Determination du contenu en eau de l'atmosphere par radiometrie hyperfrequence.
These Universite Paris VII.

Prabhakara, C., G. Dalu, R.C. Lo and N.R. Nath., 1979
Remote sensing of seasonal distribution of precipitable water vapor over the oceans and the influence of boundary-layer structure.
Mon. Wea. Rev. 107, pp:**1388-1401**

Prabhakara, C., H.D. Chang and A.T.C. Chang., 1982
Remote sensing of precepitable water over the oceans from Nimbus 7 Microwave measurements.
J. Appl. Meteor. 21, pp:**59-68**

Stealin,D.H. 1987
Capabilities of the Advanced Microwave Sounding Unit (AMSU).
in Report of the Workshop on Space Systems Possibilities for a Global Energy and Water Cycle Experiment. WMO/TD-No 180

Thomas, D.P. and C.R. Francis, 1978
Development of an iterative algorithm for the extraction of sea and atmospheric parameters from Nimbus SMMR measurement.
British Aerospace Dynamics Group. ESS/SS/915

Ulaby, F.T., R.K. Moore and A.K. Fung., 1981
Microwave remote sensing: active and passive. Vol I: Microwave Remote sensing fundamentals and radiometry.
ARTECH House, Inc., Dedham, MA

Ulaby, F.T., R.K. Moore and A.K. Fung., 1986
Microwave remote sensing: active and passive. Vol III: from theory to applications.
ARTECH House, Inc. Dedham, MA

TABLE 1:	ALGORITHM PERFORMANCES: dh (cm)				
	no surface variations		e(u,f)	de = 0.02	de (df)
	no cloud	cloudy			
24/36	0.457	0.694	1.52	1.71	2.70
18/21	0.450	0.603	1.31	1.97	1.40
18/21/36	0.440	0.513	0.83	1.26	0.90
24/36/U			0.92	1.20	2.50
18/21/U			1.16	1.93	
18/21/36/U				0.74	1.01

TABLE 2	ALGORITHM PERFORMANCES (ATSR)					
RMS residual (for several data set):						
	(1)	(2)	(3)	(4)	(5)	(6)
W (g/cm2)						
24/36	0.068	0.099	0.132	0.252	0.286	0.469
24/36/U			0.115	0.144	0.196	0.428
L (Kg/m2)						
24/36		0.109		0.155	0.158	0.227
24/36/U				0.112	0.115	0.195

(1) Fixed ground condition, clear sky.
(2) "   cloudy.
(3) reference data base, clear sky.
(4) "   cloudy.
(5) surface emissivity +/-0.02
(6) surface emissivity variability +/- 0.04

TABLE 3: ALGORITHM SENSITIVITY TO RADIOMETER PERFORMANCES.				
	T24	T36		U
dW/dT (g/cm2/K)	0.128	0.075	dW/dU	0.035 (g/cm2/ms-1)
dL/dT (Kg/m2/K)	0.064	0.035	dL/dU	0.018 (Kg/m2/ms-1)

IN-SITU MEASUREMENTS FOR VALIDATION OF MICROWAVE DATA: PROBLEMS, ACCURACIES

KRISTINA B. KATSAROS
Dept. of Atmospheric Sciences
University of Washington
Seattle, WA 98195 USA

ABSTRACT. This review covers the relationship between remotely sensed microwave data and data obtained in-situ or from earth based remote sensing systems both for validation of algorithms and as supplemental information for deriving a geophysical quantity. Parameters obtained from the satellite borne microwave radiometers, the Scanning Multichannel Microwave Radiometer (SMMR) and the Special Sensor Microwave Imager (SSM/I) are emphasized. These include sea surface temperature and wind speed, integrated atmospheric water vapor, cloud liquid water and rain rate.

1. Introduction

In this chapter I will concentrate on just a few aspects of the vast problem of comparing remotely sensed data to measurements obtained *in-situ*. Since my experience is mainly with microwave radiometry, I will concentrate on the considerations peculiar to this field of remote sensing. Microwave radiometry has been mostly applied to observing the sea surface and the atmosphere over the oceans, so we will also restrict ourselves to remote sensing over the oceans. In general, the data obtained *in-situ* or remotely by earth-based systems should be considered an integral part of the procedure for obtaining information about the atmosphere and the earth's surface from space.

First of all, since the physical relationships between the signal received at the satellite and the geophysical parameter of interest is often only partially understood and in many cases is simply an empirical fit, the *in-situ* data are necessary for calibration of the algorithms. Secondly, there are temporal and spatial variations in this relationship which are not always theoretically included in the algorithm, and there are temporal drifts in the satellite signal due to deterioration of the hardware over time (e.g., Francis, 1987). For these reasons the satellite data need to be continuously correlated with anchor points in the form of *in-situ* measurements. A third aspect is that for certain objectives the satellite data only provide one factor of a desired quantity and supplemental earth based observations are needed to fulfill the objective.

One example of this type of objective is the calculation of atmospheric wind speed at one specified height from microwave radiometer signals or from the backscatter cross-section measured by a satellite scatterometer. Information on atmospheric stratification, which is not obtainable from space at present must then be added from supplemental

217

R. A. Vaughan (ed.), Microwave Remote Sensing for Oceanographic and Marine Weather-Forecase Models, 217–238.
© *1990 Kluwer Academic Publishers.*

sources in order to apply the appropriate corrections, such as those according to Businger and Dyer, for converting the surface wind speed to the speed at 10 m or some other specified height (e.g., Fleagle and Businger, 1980).

Another objective which requires mixed data is turbulent flux calculations of heat or moisture from the sea surface. Air temperature and humidity in the surface layer must be provided to supplement the estimates of surface wind speed and sea surface temperature obtained by remote sensing in order to use currently available parameterization schemes of the bulk aerodynamic type (e.g., Liu, et al., 1979). For long term averages an alternate method is available for obtaining the evaporation rate. Mean monthly microwave radiometer information on integrated atmospheric water vapor obtained from space by measurements in the 22 GHz water vapor line is correlated with mean monthly surface layer humidity and used as a stand-in for the surface layer humidity in the bulk parameterization schemes (Liu, 1984; Liu and Niiler, 1984), but on shorter time scales *in-situ* data are needed.

The subject of intercomparison between satellite data, surface data and sometimes other satellite data is a very broad one, and a large literature is available. Therefore I will refrain from giving details here, but will instead refer the reader to these publications where applicable.

Operational numerical models produce fields of various quantities based on *in-situ* observations and initialization schemes. These fields can also be used as intercomparison data for the satellite measurements. Vice versa, the satellite data can be used as input to the models. The passive microwave data on atmospheric water, in particular, provide coverage and resolution of several important parameters never before available, against which the water vapor, cloud and rain fields of the latest generation of numerical models from meso- to global scale can be tested. The interplay between numerical models and directly sensed satellite data is definitely an opportunity for the future and closely related to the topic of this article.

2. Microwave Instruments Under Consideration

The use of microwave radiances of millimeter to centimeter wave lengths from space to infer properties of both the ocean surface and the intervening atmospheric column were already being discussed in the early 1960's (e.g., Buettner, 1963). Several instruments were developed and flown both by the USSR and the USA space agencies (see review by Njoku, 1982). Some aspects of microwave radiometry have been covered elsewhere in this volume, so here I'll only describe the two most recent types of microwave radiometers to be flown in space. Scanning Multichannel Microwave Radiometers (SMMR's) were flown on two satellites, Seasat and Nimbus 7. They collected data from July through October 1978, and October 1978 through summer 1987, respectively. The SMMR's had ten channels employing five frequencies: 6.6, 10.7, 18, 21 and 37 GHz, all operating in both horizontal and vertical polarization.

The SSM/I (Special Sensor Microwave/Imager), launched in June 1987, is on a satellite in the Defense Meteorological Satellite Program of the USA. It has channels at 19.35, 22.235, 37 and 85.5 GHz; all except the 22.235 GHz channel operate in both horizontal and vertical polarization, the latter having only horizontal polarization. The SSM/I has similar abilities to the three higher frequencies of the SMMR's and in addition is sensitive to large ice particles in clouds. The characteristics of the two radiometers are summarized in Table 1. The sampling swaths and surface resolutions are also found in this table.

Table I. *Characteristics of the Two Polar Orbiting Radiometers - SMMR on Seasat and on Nimbus 7 and the SSM/I on the F8 DMSP Satellite*

	SMMR		SSM/I
FREQ. GHz	APPROX. RESOLUTION (km)	FREQ. GHz	APPROX. RESOLUTION (km)
6.6	150	--	--
10.7	100	--	--
18	65	19.35	55
21	60	22.235	50
37	35	37	35
--	--	85.5	15

Swath Width:
Seasat: 650 km
Nimbus 7:   780 km

Swath Width: 1400 km

Period of Operation:
Seasat: July-October 1978
Nimbus 7:   October 1978-
            Fall, 1987

Period of Operation:
July, 1987 - Present

3. Geophysical Parameters Derivable from the SMMR and SSM/I Measurements

Over the open ocean one can obtain sea surface temperature, wind speed, columnar atmospheric water vapor, cloud liquid water and rain rate or rain area coverage from the SMMR channels.

For SMMR data the sea surface temperature algorithm is mainly dependent on the 6.6 GHz brightness temperature (radiance), while surface wind speed is sensed especially by the 10.7 GHz channels, since surface roughness changes particularly affect the emittance at this frequency. The SSM/I does not have the two lower frequencies of the SMMR's, which are particularly sensitive to surface properties. Wentz, et al. (1986) suggest, however, that SSM/I can nonetheless be used to obtain surface wind speed in rain free areas. The 21 and 22 GHz channels of the two instruments sense the columnar atmospheric water vapor in a weak atmospheric emission band, while cloud droplets and rain size particles emit and cause multiple scattering of the weak surface emission from the ocean at all three of the higher frequencies of the SMMR's (Gloersen and Barath, 1977). The effects at the three lowest frequencies of the SSM/I are equivalent to those at the three highest frequencies of the SMMR's. Algorithms based on physical considerations as well as algorithms based on multiple regression equations have been used to solve for a desired parameter, while eliminating influences from others (e.g., Staelin, et al., 1976; Wilheit, et al., 1980; Bernstein and Morris, 1983; Pandey, 1983; Wentz, 1983; Gloersen, et al., 1984; Wilheit, et al., 1984; Spencer, 1986; Wentz, et al., 1986; Hollinger, et al., 1987; Olson, 1987).

Interpretation of SMMR or SSM/I signals in terms of rain rate can never be exact, since even exact measurements of the total water in rain size droplets in the column do not reflect details of dropsize distributions. The true fall velocities of the drops depend not only on their size, but also on the dynamics of the synoptic weather system in which the rain is occurring. Areal coverage of the rain, i.e., the fractional footprint filling appears to be a better parameter to derive from these low-resolution radiometers (Katsaros, et al., 1988). This latter parameter could possibly be empirically correlated to rain rate, but this idea has not yet been tested.

The 85 GHz radiation sensed only by the SSM/I is particularly sensitive to scattering by large ice crystals in clouds. Few empirical results have been presented to date (Katsaros, et al., 1988), but these channels show promise to improve our ability to diagnose cloud structure due to this unique property.

4. Sources of *In-Situ* Data

The Seasat satellite, on which the first SMMR was launched, was a proof-of-concept mission to demonstrate that valuable data could be obtained from space over oceanic regions (Born, et al., 1979). The need for such data is obvious, since large portions of the earth covered by ocean (70%) are very poorly sampled by conventional *in-situ* devices (Figure 1). Because Seasat's instruments primarily sample over the oceans, it was, therefore, also rather difficult to verify the algorithms for calculating the various parameters derived from SMMR signals against *in-situ* data. However, since the problem was recognized, a special intensive surface sampling experiment was planned during September 1978, the Gulf of Alaska Seasat Experiment (GOASEX - Lipes, et al., 1979). It was carried out around Weather Ship Papa at 50°N in the Pacific Ocean. A special effort was also made to draw on the intensive data collection in the JASIN (Joint Air Sea Interaction) Experiment, July-August 1978, which took place West of Scotland in the North Atlantic. Such programs must be an integral part of any future satellite instrument validation. However, most intercomparison data for validation of the algorithms used with remotely sensed data come from conventional sources. They are different for the different types of geophysical variables, i.e., for the surface parameters and for the atmospheric water parameters. These two groups will, therefore, be dealt with separately.

4.1. SEA SURFACE TEMPERATURE AND WIND SPEED

4.1.1. *Point Observations*. Ship and Buoy Data. Ship and buoy data provide true *in-situ* information for comparison to satellite derived surface parameters. The ship data come in two types: the synoptic reports of the Voluntary Observing Fleet (VOF), and reports from weather ships. The VOF consists of merchant ships, which carry an anemometer, dry and wet bulb thermometers, sea water thermometer either in the ship's cooling water intake or for immersion in a bucket sample. Bridge personnel make qualitative observation of the cloudiness, precipitation occurrence and the wave field. These data are typically recorded at 3-hourly intervals and still constitute a very important data source for the diagnosis and forecasting of the weather and serve as input to large-scale numerical models.

The Voluntary Observing Fleet. There are two or three major problems associated with surface ship observation of this kind. The accuracy of individual ship measurements are low, primarily because the ship represents a large obstacle to the air flow and, therefore, influences both wind speed and direction and air temperature and humidity measurements. Instruments are not calibrated as often or as carefully as for research work. However, because there are many ships and their errors mostly random, the errors in most cases can be averaged out. An additional problem is that shipping lanes are confined to very narrow paths over the ocean (Figure 1). Fortunately, they lead to population centers, so that the observations are useful for local short range forecasts. For verification of the global coverage from satellites they provide somewhat limited coverage, but in the regions where they are found they have provided a valuable data source for satellite verification. (Discussions of accuracy can be found in articles by Blanc, 1986; Dobson, et al., 1982; Tabata, 1978a; and Liu, 1988).

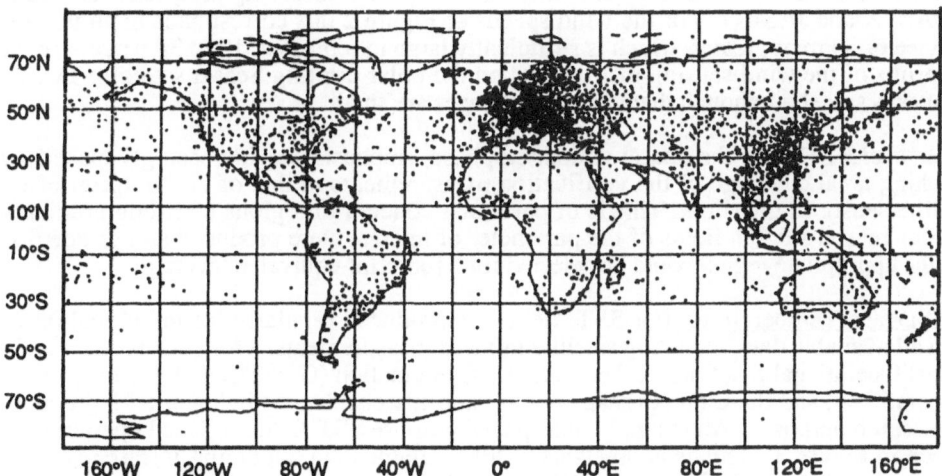

Figure 1: *Example of synoptic surface observation over the earth in a 6-hour period, around 00:00 UTC on March 23, 1983. This figure illustrates the limited amount of data available from large expanses of the oceans, especially in the southern hemisphere. Source: ECMWF, 1983.*

Weather Ships. In addition to the the Voluntary Observing Fleet there were until recently numerous weather ships in operation in the northern hemisphere but few remain in operation today. They are manned by various national weather services. Their data are of higher quality and more complete than those reported by the VOF's (Tabata, 1978b), especially for cloud and wave observations.

Buoys. Along the coastline of the USA and other continents, there are weather buoys, which report surface data to the international networks via satellite communication. Their wind and temperature data appear to be of higher quality than the typical data from the VOF's. Winds from these buoys have been used extensively in verification efforts (Wentz, et al., 1986). Sea surface temperature is measured on these buoys by a thermistor, sometimes mounted inside the hull of the buoy - for protection. The temperature data from such a thermistor is representative of the surrounding ocean except under extreme conditions of solar radiation. The main limitation is that NDBO (National Data Buoy Office) buoys cover only a small geographical area near continents, and, therefore, a somewhat narrow range of meteorological conditions.

Wind is one of the more difficult parameters to measure and intercompare, since the relationship to the satellite signal is rather indirect and changes in the driving force - the pressure gradient - occur rapidly due to moving atmospheric waves. Much consideration was, therefore, given to the difficulty of sampling the surface wind speed at a point and comparing it to areal averages from a satellite footprint (e.g., Pierson, 1983). Figure 2 from Pierson (1983) based on data obtained by M.A. Donelan illustrates that in addition to the dominance of the spectrum by long period fluctuations, eddies with periods shorter than 20 minutes can also have large amplitudes, but a slight dip in the spectra seems to be present at this period. Because of this fact, 20 minute averages of point measurements of wind-speed were collected from the buoys in the special observing programs for Seasat (GOASEX and JASIN). For the wind speeds of Figure 2 this corresponds to an upwind distance of 8 km or greater, which is sufficiently large to represent 10 to 30 percent of the footprints of the wind sensing instruments on the polar orbiting Seasat satellite. Twenty minutes is still short enough that temporal changes are small.

4.1.2. *Fields.* During the Seasat validation phase, emphasis was increasingly placed on matching areal coverage of the verification measurements to that of the footprint of the satellite sensor in question. This is of particular concern in regions of strong gradients. For this reason smooth fields of the parameter of interest were produced and gridded, so that the appropriate region could be selected for a footprint-equivalent average.

Sea Surface Temperature. For SST, fields are produced regularly for the global ocean from all available data, including satellite infrared measurements. One such product is the Global Operational Sea Surface Temperature Computation (GOSTCOMP). This product is operational, and therefore, perhaps not as carefully checked as fields of SST produced for research purposes. An example of a special purpose SST-field produced operationally is seen in Figure 3. This is an analysis of all available data, including infrared satellite data produced by the U.S. National Weather Service, mainly for identification of fronts useful for fishermen. Tabata (1981) suggests that the accuracy of satellite estimates of SST for the region shown in this figure is of the order of $\pm 0.5°C$.

Significant biases between SST's obtained by infrared remote sensing and bulk measurements at 2 m depth were reported by Schluessel, et al. (1987). Figure 4 reproduced from that article shows cool skin temperature deviations in the open ocean with means of 0.15°C, 0.45°C and 0.8°C, respectively, on three different days. The authors explain the differences from day to day in terms of varying infrared and turbulent heat losses from the sea and weak mixing in the near surface layer of the ocean in case of the largest deviation (see further Section 5 below).

Wind. Since there were several instruments on the Seasat satellite from which surface wind speeds could be inferred, the production of surface wind fields for verification was given much attention (Brown, et al., 1982; Cardone, et al., 1983). These methods start

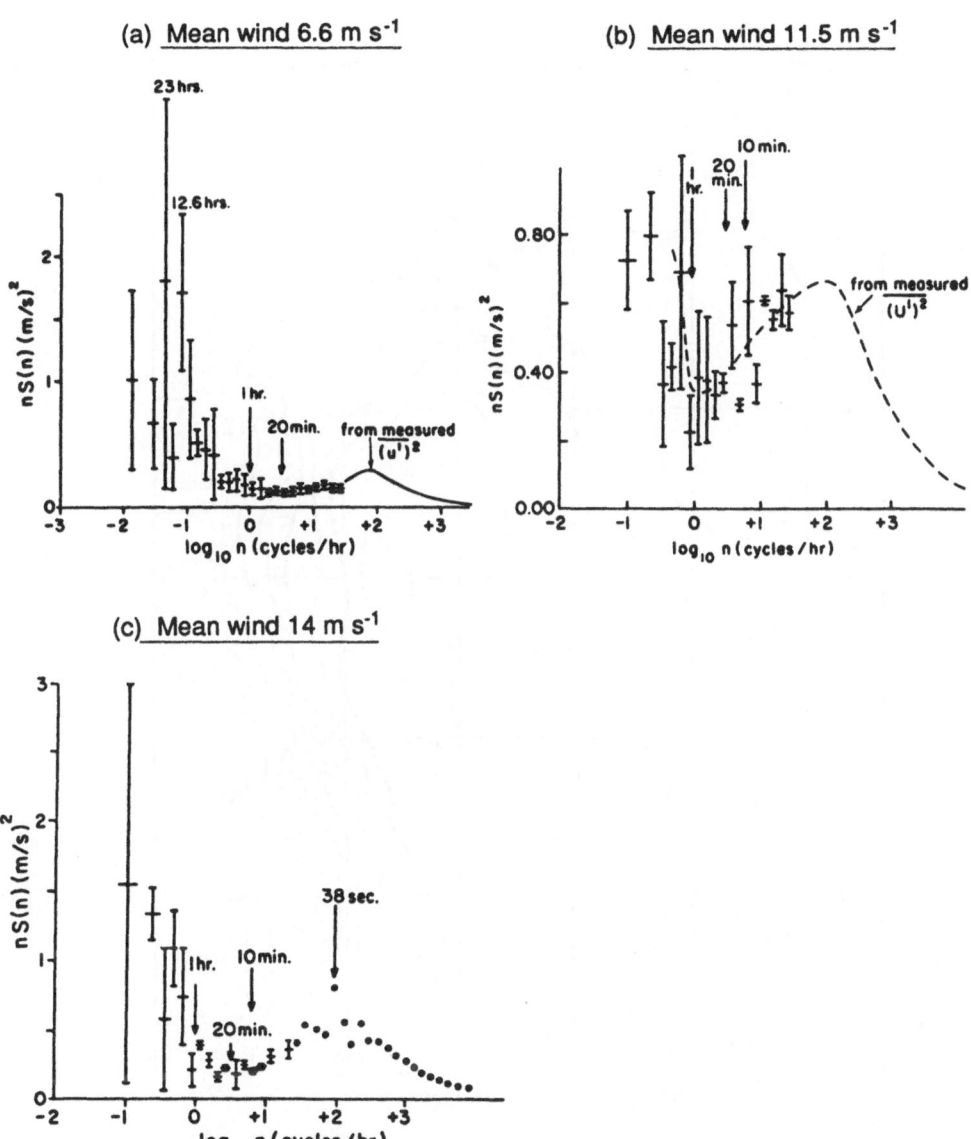

Figure 2: *Spectra of the wind for three average wind speeds calculated from data obtained with an anemometer at 11.5 m. height on the Canada Center for Inland Waters platform. n is frequency, S(n) spectral power density. (Courtesy M.A. Donelan. See Pierson, 1983, for details of data analysis).*

224

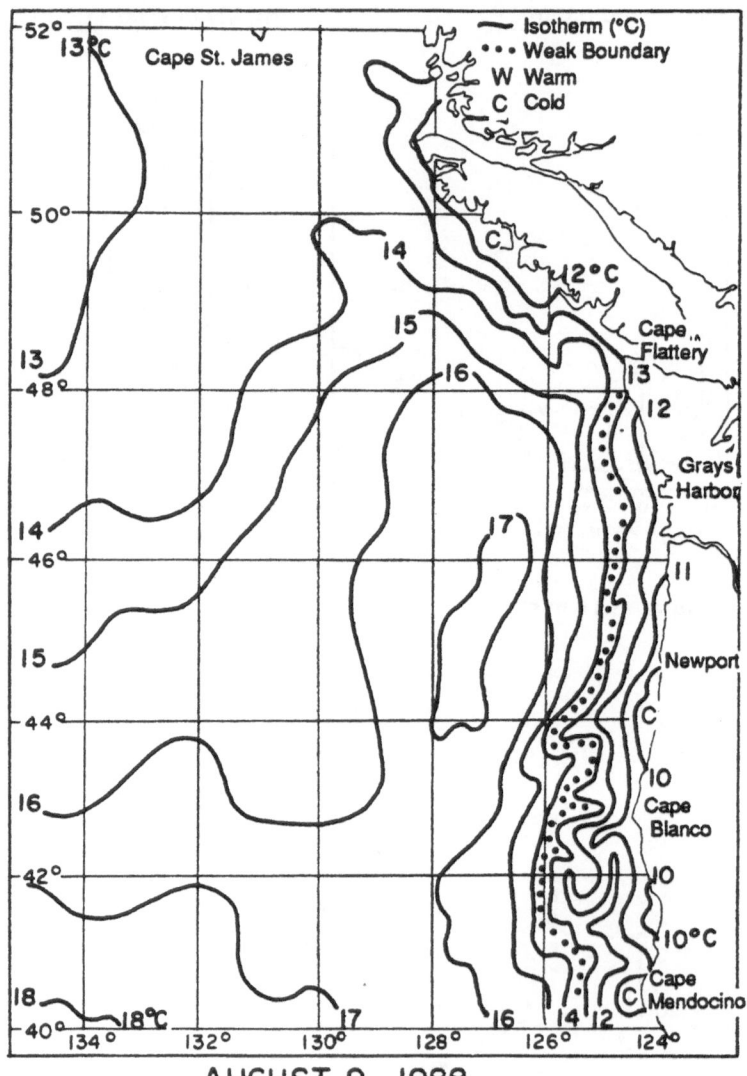

AUGUST 9, 1988

Figure 3: *Sea surface thermal analysis for August 9, 1988, produced routinely (about twice/week) for the Western Pacific Ocean by the U.S. National Weather Service, San Francisco, CA. Contours are labeled in °C. The dotted line indicates an oceanic surface front.*

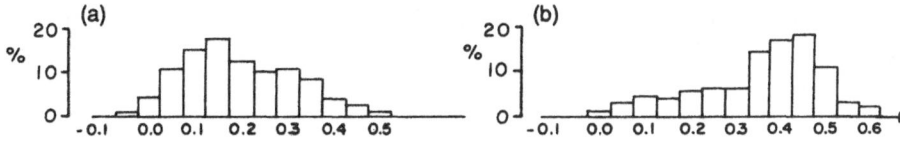

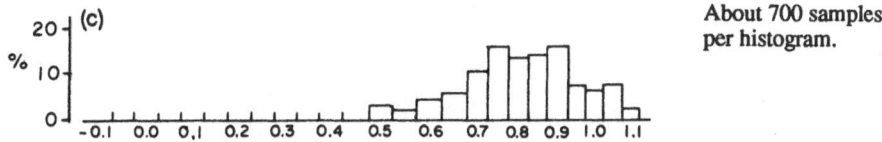

About 700 samples
per histogram.

Figure 4: *Deviation of the skin temperature obtained from an infrared radiometer on a ship (with correction for reflected emission from the sky) from SST measured at a depth of Z m for three days at sea in 1984 (after Schluessel, et al., 1987).*

a. *November 25, Z = 4m, $\overline{\Delta T}$ = 0.15°C, North Atlantic Ocean, strong downwelling radiative flux.*

b. *November 26, Z = 4m, $\overline{\Delta T}$ = 0.45°C, North Atlantic Ocean, clear sky.*

c. *July 14, Z = 2m, $\overline{\Delta T}$ = 0.8°C, Eastern North Pacific, strong evaporation, weak mixing.*

with maps of smoothed surface pressure, from which the geostrophic wind vector can be found. The change in both wind speed and direction between the free atmosphere and the surface due to surface friction and the stratification in the whole planetary boundary layer (PBL), as well as in the surface layer, are accounted for by application of a model (e.g., Brown and Liu, 1982). The results of comparing gridded fields of these surface winds to scatterometer winds and SMMR winds can be found in articles by Brown, et al. (1982) and Brown (1983). Comparisons of SSM/I winds, obtainable in areas with no rain and low cloud liquid water content, to buoy winds and wind speeds from fields are found in Figures 5 a,b (Bright and Brown, personal communication).

Another avenue for evaluating the accuracy of a surface wind field from a satellite sensor is to compare to winds produced with data from other satellite instruments. Wentz, et al. (1982) compared wind fields produced from the Seasat scatterometer, SMMR, and altimeter. They were found to have substantially different biases, which were not independent of locality. Through such intercomparisons, basic questions about the physical processes responsible for the signal received at the satellite, and questions concerning the calibration of the instruments were brought out. Concerns about the conversion from a signal reacting to sea surface roughness in the form of small scale waves (of the same order of magnitude as the electromagnetic wavelength of the scatterometer, the altimeter or the SMMR's 10.6 GHz radiometer) to wind speed at a certain height involve our understanding of wave generation and growth, small scale wave modulation by larger waves, and the atmospheric surface layer wind profile in relation to these (see NASA Satellite Surface Stress (S^3) Committee Report, O'Brien, et al., 1982). Some rather large discrepancies were found in extreme conditions between winds produced with the Seasat scatterometer, SASS, algorithm and surface winds obtained in-situ or from pressure fields (e.g., Woiceshyn, et. al., 1986).

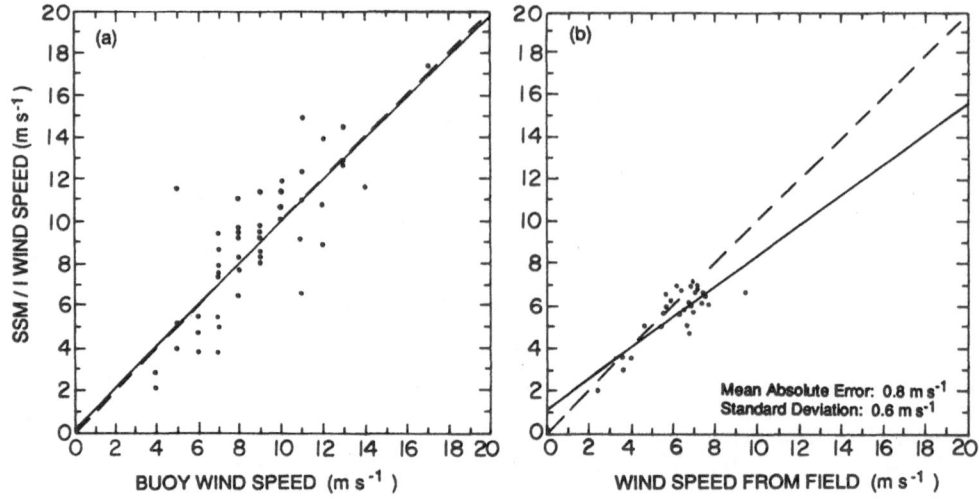

Figure 5: *Comparison of winds derived from SSM/I signals according to the algorithm of Wentz, et al. (1986).*

a. *Winds from NDBO buoys west of U.S. Pacific coast. Bias of 1.5 ms⁻¹ has been removed from the SSM/I wind speeds. All buoy values less than 4 ms⁻¹ were omitted from the plot.*
    _____ *regression, -------- perfect fit.*
b. *Winds derived from pressure fields employing model by Brown and Liu (1982).*
    _____ *regression, -------- perfect fit .*
(Courtesy of D. Bright & R.A. Brown).

4.2. ATMOSPHERIC WATER PARAMETERS

4.2.1. *Water Vapor.* The weather ships discussed above also launch radiosondes from which the temperature and humidity profiles in the troposphere can be determined. This information has been crucial in verifying the columnar water vapor (also called precipitable water), obtainable from both SMMR and SSM/I. The integrated radiosonde values of humidity and the values from SMMR or SSM/I show similar scatter and very little bias (Katsaros, et al., 1981; Prabhakara, et al., 1982; Alishouse, 1983; Chang, et al., 1984; McMurdie and Katsaros, 1985; McMurdie, 1988; Alishouse, 1988, personal communication concerning the SSM/I). Figure 6 shows one such intercomparisons for the Nimbus 7 SMMR which includes radiosonde data from tropical atolls.

Fields of water vapor are produced globally on a grid of 2.5° of latitude by the European Center for Medium Range Weather Forecasts (ECMWF). The numerical grid data from the First GARP (Global Atmospheric Research Program) Global Experiment (FGGE) in 1979 was integrated and compared to Nimbus 7 SMMR data (McMurdie, 1988). This intercomparison illustrated that the ECMWF and the SMMR maxima were in agreement (Figure 7). However, even though the maxima are of the same magnitude, the global model maxima ahead of cold fronts cover much larger areas, thereby implying that

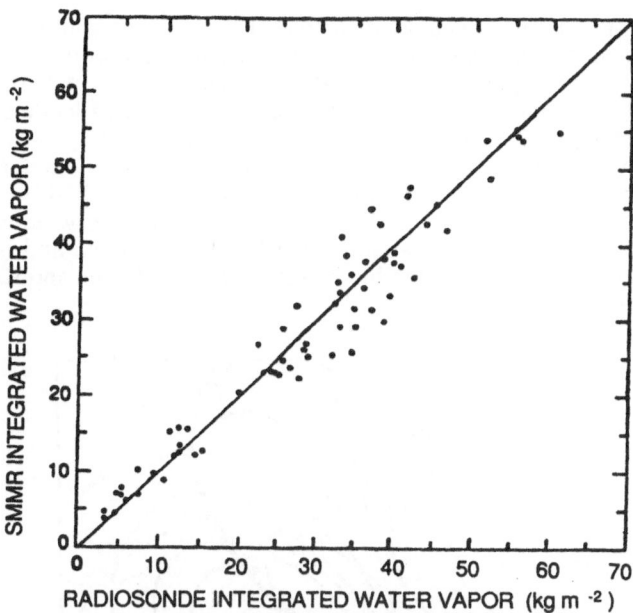

Figure 6: *Integrated water vapor calculated from radiosonde information from island and ship stations located in the tropics, midlatitudes and high latitudes vs. integrated water vapor derived from SMMR for the year 1979. Only SMMR passes within ±2 hrs. of the radiosonde launch and within a 1° latitude/longitude radius of the station were selected. 70 matchups were used for this figure. SMMR values have an r.m.s. difference of 2.7 kg m⁻² from the radiosonde values (after McMurdie, 1988).*

more water vapor is present in the atmosphere. Differences in the patterns occur at the strong gradients near midlatitude fronts and near low pressure centers and can be attributed largely to the lower resolution of the numerical model (250 km x 250 km compared to Nimbus 7 SMMR's 60 km x 60 km).

Limited area models have a typical grid resolution of 40 to 80 km which is comparable to that of SMMR geophysical data. Comparing the IWV-field from such a model, due to Sundqvist (1978) and Hammarstrand (1987), which is initialized with the ECMWF global model, to Nimbus 7 SMMR data over the North Atlantic, we found that the model carried too much water vapor in a frontal region (Katsaros, et al., 1988). In this case the maxima are too large in magnitude. Perhaps the limited area model causes additional moisture convergence in a field which starts out with too broad a maximum. As a consequence these fields are not very useful for testing the IWV from microwave radiometers. Rather, at present, the satellite derived IWV values, being well verified by radiosonde data, provide good tests for the models. The situation is different, however, with respect to integrated cloud liquid water and rain.

4.2.2. *Cloud Liquid Water.* No other direct method besides microwave radiometry exists for obtaining estimates of integrated cloud water content, ICLW, except possibly sampling within clouds by aircraft instrumented with dropsize counters or sensors for liquid water

228

content. Such sampling is costly and must be done at many levels within the cloud over a short time period to allow vertical integration. Few such data points exist. This author has obtained one by spiralling vertically for one hour inside a uniform warm frontal cloud band with the National Center for Atmospheric Research (NCAR) Electra aircraft, during the Storm Transfer and Response Experiment (STREX), in December, 1980. It is undoubtedly a fortuitous coincidence that in this instance the integrated cloud water content exactly matched the value, 0.4 kg/m^2, calculated from data obtained in the vicinity at about the same time by Nimbus 7 SMMR. From qualitative observations one can state that, in general, the derived cloud water content looks reasonable and matches cloud outlines seen in visible and infrared satellite images (e.g., McMurdie and Katsaros, 1985; Katsaros and Lewis, 1986).

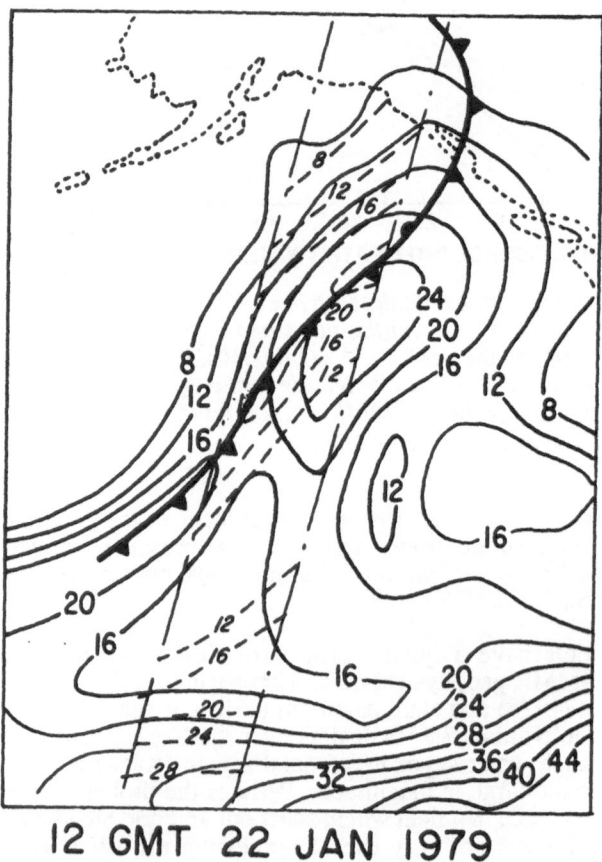

12 GMT 22 JAN 1979

Figure 7: *Comparison of SMMR and FGGE derived integrated water vapor (in kg m^{-2}) for 12 UTC 22 January 1979 in the North Pacific. The solid contours correspond to the FGGE derived integrated water vapor and the dashed contours correspond to SMMR's. The SMMR information lies between the dash-dot lines outlining the SMMR overpass. The contours are labeled every 4 kg m^{-2} (after McMurdie, 1988).*

Comparison to limited area numerical model results derived from parameterizations of cloud physics processes and based both on the dynamics of the situation and on complementary fields, such as fields of water vapor convergence, is another avenue for verifying this new meteorological observable. It is worthwhile to attempt even such indirect verification, since data on cloud liquid water may have important application to numerical weather forecasting, to atmospheric wet chemistry and probably to budget calculations of atmospheric radiation (Katsaros, 1989). As indicated by the discussion in the section above the model formulations are not yet definitive. Thus, comparison to numerical models will be an iterative process, in which model formulations are modified to compensate for obvious discrepancies and unrealistic results by comparison to the satellite signals, and the satellite algorithms are gradually corrected.

4.2.3. *Rain.* Problems with rain verification also stem from the fact that rain is not measured at all by any quantitative *in-situ* technique on the ocean. Only qualitative observations as to the nature of the rain are reported, i.e., whether it falls as drizzle or rain; light, moderate or heavy, or in the form of showers. Such observations could possibly be employed in a statistical sense, if some direct calibrations of their meaning were made. They appear to correctly reflect the relative intensity of the rain in midlatitude cyclones as sensed by SMMR (McMurdie and Katsaros, 1985).

Rain rate estimates from microwave satellite data can be compared to coastal rain gauge or radar measurements. For coastal rain gauge data, a rather long lag time occurs between the SMMR observation of a rain cell over the ocean (at least one footprint away from the coast) and the arrival of the same rain cell on the coast. This must be so, because the sidelobes of SMMR antennae must not have the "hot" land surface in their field of view. Evolution of the weather system over this lag time should be minimal for acceptable intercomparisons. Effects on the rain rate of the contrast between land and sea at a coastline and of orography must also be considered. Furthermore, the point measurement of a single surface sensor compared to the areal average of a satellite sensor is particularly troublesome for rain because of its areal variability. Nonetheless, Katsaros and Lewis (1986) found good correlation between measurements of convective rain cells west of the coastline of the states of Washington and Oregon by Nimbus 7 SMMR and subsequent onset of rain at several downstream rain gauges positioned along the coast. Using a portion of the rain rate algorithm by Wilheit and Chang (1980), which applies for low rain rates (< 5 mm/hr), they also found quantitative agreement within a factor of two or three. An interesting discussion of the importance of considering the resolution of satellite sensors when designing calibration procedures is given by Gabriel, et al. (1988). They discuss the multifractal nature of cloud and rain fields, suggesting that proper modeling of this aspect will aid development of future analysis methods of remotely sensed fields of these quantities.

Coastal radars which are co-located with the SMMR measurement can provide valuable intercomparison data. However, interpretation of the radar signals in terms of rain rate also involves an algorithm and certain assumptions. The areal coverage is much better than for rain gauges, but the resolution is still not well matched to the satellite data (Spencer, et al., 1983; Petty and Katsaros, 1988). In general, the incidence of rain and its areal coverage is accurately obtained by SMMR, but for heavy rain the existing rain algorithms saturate and information on the rain rate is lost (Katsaros and Petty, 1988). (A liquid water/rain algorithm employing the 6.6 and 10.6 GHz channels, which does not saturate, has been proposed by Prabakhara, et al., 1983, but it has not, to my knowledge, been tested against radar or other *in-situ* data.)

230

Comparison of microwave rain estimates to other remote sensing techniques (e.g., Martin and Scherer, 1973; Adler and Negri, 1988) have been suggested, but have not yet been carried out. Techniques based on the brightness in visible light and the coldness of cloud tops, sensed in the infrared, work best for convective storms (particularly over land). As discussed above, the heavy rain in such storms tend to saturate the emission based microwave algorithms employing 37 GHz. Scattering based algorithms also reach an upper limit for deep convective systems. Thus the visible/infrared and microwave techniques are best thought of as complementary rather than comparable.

Comparisons between rain produced from Nimbus 7 SMMR signals and in recent LAM's, such as the one by Sundqvist (1978) and Hammarstrand (1987), show that there is skill in both the remotely sensed data and their associated algorithms and the model formulation, Figure 8. Currently, however, only qualitative comparisons such as those seen in this figure have been attempted.

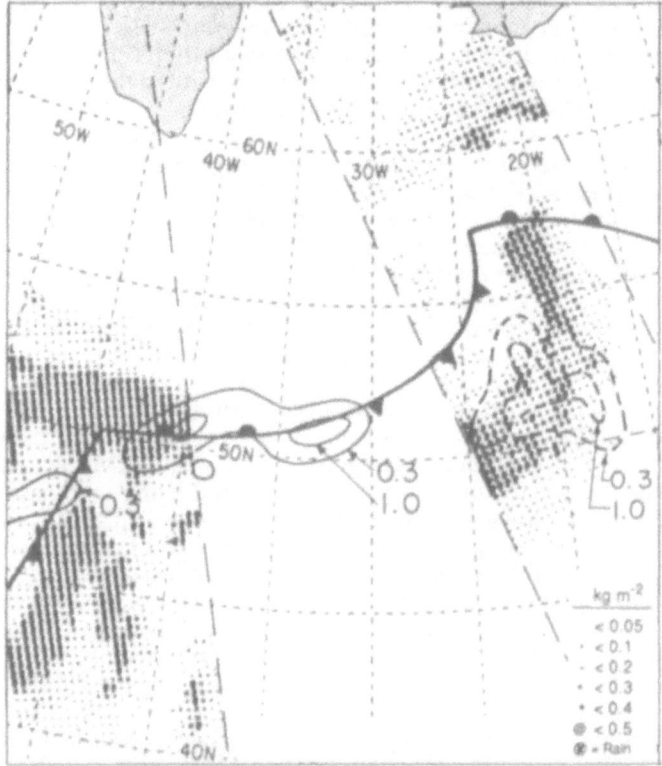

Figure 8: *SMMR-inferred integrated liquid water content (ICLW) observed over the North Atlantic at 1218 UTC (right swath) and 1358 UTC (left swath), 23 May 1983, compared with accumulated precipitation (mm) predicted by Sundqvist-Hammarstrand limited-area model for period 23 May, 00 UTC - 23 May, 18 UTC. Values of SMMR ICLW > 0.5 kg m⁻² are interpreted as rain of unknown intensity. Solid contours represent stratiform precipitation predicted by model; dashed contours represent convective precipitation. Frontal positions correspond to routine National Meteorological Center analysis, valid 12 UTC (after Katsaros, et al., 1988).*

5. Evaluation of Intercomparison Data

5.1. ACCURACY

Comparison data for the parameters obtained from space with microwave signals each have their own specific errors. Estimates of the currently obtainable accuracy of the intercomparison data for these parameters measured over the ocean are found in Table 2.

Table II. *Estimates of the Root-Mean-Square Errors in Comparison Data for SMMR and SSM/I Parameters*

PARAMETER	OPERATIONAL MEAS.	RESEARCH MEAS.
Sea Surface Temperature	0.7°-1.5° C	0.3° C
Wind speed, direction	2-4 m/sec, ± 30°	0.1 to 1 m/s, 5° to 20°
Integrated water vapor	3 kg/m^2	2 kg/m^2
Integrated cloud liquid water	Not applicable	Not yet well known
Rain rate* (midlatitudes)	factor of 2 or 3	50%
Rain area* (tropical cloud cluster)	50%	25%

* (very little experience with these)

For sea surface temperature and wind speeds we depend to a large extent on the VOF, whose accuracies of individual observations are low. They are point measurements and they are often measured instantaneously or averaged over a very short time interval (≤ 2 min.). Weather ship and buoy data are generally of higher quality, but very few, and they sample only a small portion of the global ocean.

Tabata (1978a) suggests that SST's obtained on weather ships can attain ± 0.1°C accuracy. Schluessel, et al. (1987) discuss the currently available accuracy of SST determinations from space by infrared techniques and conclude that under optimal conditions an accuracy of ± 0.3°C is possible, while ± 0.7°C is typical (see also Tabata, 1981; Strong, 1986). To obtain high accuracy the two channel procedure established by McClain, et al. (1983) and McClain, et al. (1985) should be used, whereby the variable attenuation in the infrared atmospheric window (8-12 μm) is deduced by sampling in two channels which have different absorption coefficients for the relevant atmospheric gases. Contamination by clouds in the field of view can be accounted for by a technique also involving multichannel sensing.

Even when the atmospheric attenuation has been properly accounted for, there remains the deviation of the surface temperature measured in the infrared from subsurface temperatures measured in conventional bucket samples, in the intake to the ship's engine or in the hull of buoys. The infrared radiation emanates from the top 50 μm to 0.5 mm of the water, depending on wavelength. This deviation of radiatively measured SST's is due to the "cool film", the aqueous thermal boundary layer, whose depth is of the order of a mm. Across the cool film, temperature drops of 0.2° to 0.5°C are common, the magnitude being a function of the net heat loss from the interface and intensity of the turbulence in the water (see review by Katsaros, 1980). Microwave signals at 6.6 GHz emanate from a depth of about 2 mm, well below the cool film. This should be considered when comparisons are made to SST's determined by infrared measurements. At the frequencies of SSM/I this so-called "penetration depth" is less than 0.5 mm.

Wind speed and direction have very large errors, if ship data are used, because of the flow distortion by the ships (± 4 m/sec, 30°) and are, therefore, almost useless unless many points are averaged. For wind comparison, buoy data and estimates from fields are preferable.

For validation of integrated water vapor, the situation is satisfactory because the *in situ* data are of the same accuracy as the satellite derived values. A few high quality radiosonde integrations from tropics to polar regions chosen to be away from sharp gradients are sufficient to establish the regression line between the satellite estimate and the *in-situ* data with the same amount of scatter along both axes. During the validation of the water vapor algorithm for the Seasat SMMR it was found that the statistics improved substantially, if only radiosonde data within 200 km of the footprint and within two hours of the overpass time were included (Katsaros, et al., 1981; Alishouse, 1983).

For both ICLW and rain the intercomparison data are at least as uncertain as the estimates produced by the microwave algorithms.

5.2. RESOLUTION, COVERAGE

The general difficulty of mismatch in areal sampling between the rather large footprints of microwave instruments and all surface based point sensors has been touched upon above. It is an inherent problem, which can be somewhat alleviated by taking time averages in cases, which have been selected for their steady state and lack of sharp gradients in the vicinity of the sample. If care is taken to select the intercomparison data on the basis of their representativeness for a footprint-sized area, improved calibrations can be obtained. When possible, smoothed fields of the variable provide the most appropriate intercomparison data. These can be gridded and the appropriate areal average used in the comparison.

6. Summary - Recommendations

Verification of geophysical variables over the oceans derived from satellite measurements is not an easy task, because of the difficulty of making direct measurements from platforms at sea, because of the inaccessibility of large expanses of the ocean and because of the great costs involved in oceanic operations. Sea surface temperature and wind vector verifications have been given much attention in the past. Sea surface temperature, being a rather slowly varying quantity and already available from space by infrared observations when clouds do not interfere, can nowadays be obtained to rather high accuracy (0.7°C) over the whole global ocean. Wind, because it has a much shorter time scale, is much more difficult to verify. Producing fields of surface winds from surface pressure maps with a surface layer-PBL model appears to be the most reasonable method of obtaining reliable surface wind estimates. However, this procedure requires information on stratification (air temperature in particular), which is not always available. Integrated water vapor has been verified to everyone's complete satisfaction, in spite of the limited supply of radiosonde stations. For the other two atmospheric water parameters, the situation is much less satisfactory, since no really quantitative intercomparison data exist over the ocean. Obviously, the integrated cloud liquid water and rain rate (or rain occurrence), being unique new parameters not readily available by other measuring techniques, should eventually be verified. In these cases, we have to employ specialized research tools or indirect methods, such as comparison to model output and extrapolation to coastal radars and rain gauges, in order to gradually gain confidence.

For remotely sensed data to reach their potential usefulness, they must also be quantitatively well known. *In-situ* measurements at so-called anchor points should, therefore, be an integral part of the remote-sensing system, and not a haphazard post-launch consideration. The expenditure for establishing the accuracies of the calculated geophysical parameters early in a satellite mission is a very good investment. The confusion that otherwise develops - as errors are gradually discovered - is very costly. During the course of the mission, the signals from the satellite sensor should be monitored for deterioration and calibration drifts. This is best done at the sensor itself but data of the geophysical parameters should also be used to monitor the validity of the algorithms, since seasonal and regional variations may not have been properly included in their development.

In addition to *in-situ* data being valuable for validation and monitoring, they also serve as complementary information for obtaining important derived quantities, such as the mean wind speed at a specified height (or the geostrophic wind and pressure gradient) or the surface heat and water vapor fluxes. In the future, for accomplishing these purposes to an optimal extent, the relevant *in-situ* data or complementary satellite data should be included in the planning of the complete remote sensing system.

7. Acknowledgments

I'm grateful to Ms. Janet Meadows who typed the manuscript, to Ms. Kay Dewar, who produced the figures, to Mark A. Donelan for permission to use his wind spectra in Figure 2, to Robert A. Brown and David Bright for unpublished material used in Figure 5, and to Lynn A. McMurdie for the use of her Ph.D. material (Figures 6 and 7). This article was written with support from NASA's Global Processes Program, Grant NAG 5-943.

8. References

Adler, R.F. and A.J. Negri, 1988: 'A satellite technique to estimate tropical convective and stratiform rainfall'. *J. Appl. Met.,* **27,** 30-51.

Alishouse, J.C., 1983: 'Total precipitable water and rainfall determination from the Seasat Scanning Multichannel Microwave Radiometer (SMMR)'. *J. Geophys. Res.,* **88,** 1929-1935.

Bernstein, R.L. and J.H. Morris, 1983: 'Tropical and midlatitude North Pacific sea surface temperature variability from the Seasat SMMR'. *J. Geophys. Res.,* **88,** 1877-1891.

Blanc, Theodore V., 1985: 'The effect of inaccuracies in weather-ship data on bulk-derived estimates of flux, stability and sea-surface roughness'. *J. Atmos. and Oceanic Techn.,* **3,** 12-26.

Born, G.H., J.A. Dunne and D.B. Lame, 1979: 'Seasat mission overview'. *Science,* **204,** 1405-1406.

Brown, R.A., 1983: 'On a satellite scatterometer as an anemometer'. *J. Geophys. Res.,* **88,** 1663-1673.

Brown, R.A., V.J. Cardone, T. Guymer, J. Hawkins, J.E. Overland, W.J. Pierson, S. Peteherych, J.C. Wilkerson, P.M. Woiceshyn and M. Wurtele, 1982: 'Surface wind analyses for Seasat'. *J. Geophys. Res.,* **87,** 3355-3364.

Brown, R.A. and W.T. Liu, 1982: 'An operational large scale marine planetary boundary layer model'. *J. Appl. Meteor.,* **21,** 1-22.

Buettner, K.J.K., 1963: 'Regenortung vom Wettersartelliten mit Hilfe von Zentimeterwellen'. *Die Naturwissenschaften,* **50,** 591-592.

Cardone, V., T. Chester and R. Lipes, 1983: 'Evaluation of Seasat SMMR wind speed measurements'. *J. Geophys. Res.,* **88,** 1709-1726.

Chang, H.D., P.H. Hwang, T.T. Wilheit, A.T.C. Chang, D.H. Staelin and P.W. Rosenkranz, 1984: 'Monthly distributions of precipitable water from the Nimbus 7 SMMR data'. *J. Geophys. Res.,* **89,** 5328-5334.

Dobson, F.W., F.P. Bretherton, D.M. Burridge, J. Crease, E.B. Kraus and T. H. vondar Haar, 1982: 'The CAGE (Can the Atlantic Gain Energy) Experiment: A feasibility study'. *World Climate Program,* **22,** WMO, Geneva, 95 pp.

Fleagle, R.G. and J.A. Businger, 1980: *An Introduction to Atmospheric Physics,* 2nd Edition. Academic Press, New York, 432 pp.

Francis, E.A., 1987: *Calibration of the Nimbus-7 Scanning Multichannel Microwave Radiometer (SMMR), 1979-1984.* M.Sci. Thesis, Oregon State University, 248 pp.

235

Gabriel, P., S. Lovejoy, G.L. Austin and D. Schertzer, 1988: Multifractal analysis of resolution dependence in satellite imagery. *Geophys. Res. Let.,* **15,** 1373-1376.

Gloersen, P. and F.T. Barath, 1977: 'A Scanning Multichannel Microwave Radiometer for Nimbus-G and Seasat-A'. *IEEE J. Ocean. Eng.,* **OE-2,** 172-178.

Gloersen, P.D., D.J. Cavalieri, A.T.C. Chang, T.T. Wilheit, W.J. Campbell, O.M. Johannessen, K.B. Katsaros, K.F. Kunzi, D.B. Ross, D. Staelin, E.P.L. Windsor, F.T. Barath, P. Gudmandsen, E. Langham and R.O. Ramseier, 1984: 'A summary of results from the first Nimbus 7 SMMR observations'. *J. Geophys. Res.,* **89,** 5335-5344.

Hammarstrand, U., 1987: Prediction of Cloudiness Using a Scheme for Consistent Treatment of Stratiform and Convective Condensation and Cloudiness in a Limited Area Model. *Special Volume of the J. Met. Soc. Japan. Short- and Medium-range Numerical Weather Prediction.* Collection of papers presented at the WMO/IUGG NWP Symposium, Tokyo, 4-8 August 1986, 187-197.

Hollinger, J., R. Lo., G. Pie, R. Savage and J. Pierce, 1987: *Special Sensor Microwave/Imager User's Guide.* Naval Research Laboratory, Washington, D.C.,177 pp.

Katsaros, K.B., 1980: 'The aqueous thermal boundary layer'. *Bound. Layer Meteor.,* **18,** 107-127.

Katsaros, K.B., 1989: Parameterization Schemes and Models for Estimating the Surface Radiation Budget. In *Surface Waves and Fluxes,* G.L. Geernaert and W.J. Plant, eds. D. Reidel Publishers, Dordrecht, Holland (in press).

Katsaros, K.B., P.K. Taylor, J.C. Alishouse and R.J. Lipes, 1981: Quality of Seasat SMMR (Scanning Multichannel Microwave Radiometer) atmospheric water determinations. In *Oceanography from Space,* Gower, ed. Plenum Publishing Corp., New York, pp. 691-706.

Katsaros, K.B. and R.M. Lewis, 1986: 'Mesoscale and synoptic scale features of North Pacific weather systems observed with the Scanning Multichannel Microwave Radiometer on Nimbus 7'. *J. Geophys. Res.,* **91,** 2321-2330.

Katsaros, K.B. and G.W. Petty, 1988: 'Precipitation observed over the South China Sea by the Scanning Multichannel Microwave Radiometer on Nimbus 7 during winter MONEX. *First Annual Report on Grant NAG5-943,* pp. 46.

Katsaros, K.B., G.W. Petty and U. Hammarstrand, 1988: 'Liquid water and water vapor in midlatitude cyclones observed by microwave radiometry and compared to model calculations'. *Proceedings of Palmèn Memorial Symposium,* August 29-Sept. 2, Helsinki, Finland.

Lipes, R.G., R.L. Bernstein, V.J. Cardone, K.B. Katsaros, E.J. Njoku, A.L. Riley, D.B. Ross, C.T. Swift and F.J. Wentz, 1979: 'Seasat Scanning Multichannel Microwave Radiometer: Results of the Gulf of Alaska workshop'. *Science,* **204,** 1415-1417.

Liu, W.T., 1984: *Estimation of latent heat flux with SEASAT-SMMR, a case study in N. Atlantic, Large-Scale Oceanographic Experiments with Satellites*, C. Gautier and M. Fieux, eds. D. Reidel Pubs.

Liu, W.T., 1988: 'Moisture and latent heat flux variabilities in the tropical Pacific derived from satellite data'. *J. Geophys. Res.*, **93**, 6749-6760.

Liu, W.T., K.B. Katsaros and J.A. Businger, 1979: 'Bulk parameterization of air-sea exchanges of heat and water vapor including the molecular constraints at the interface'. *J. Atmos. Sci.*, **36**, 1722-1735.

Liu, W.T. and P.P. Niiler, 1984: 'Determination of monthly mean humidity in the atmospheric surface layer over oceans from satellite data'. *J. Phys. Oceanogr.*, **14**, 1451-1457.

Martin, D.W. and W.D. Scherer, 1973: 'Review of satellite rainfall estimation methods'. *Bull. Amer. Meteor. Soc.*, **54**, 661-674.

McClain, E.P., W.G. Pichel, C.C. Walton, Z. Ahmad and J. Sutton, 1983: 'Multichannel improvements to satellite-derived global sea surface temperatures'. *Adv. Space Res.*, **2**, 43-47, Pergamon Press.

McClain, E.P., W.G. Pichel and C.C. Walton, 1985: 'Comparative performance of AVHRR-based multichannel sea surface temperatures'. *J. Geophys. Res.-Oceans*, **90**, 11587-11601.

McMurdie, L.A. and K.B. Katsaros, 1985: 'Atmospheric water distribution in a mid-latitude cyclone observed by the Seasat Scanning Multichannel Microwave Radiometer'. *Mon. Wea. Rev.*, **113**, 584-598.

McMurdie, L.A., 1988: *Interpretation of Integrated Water Vapor Patterns in Oceanic Midlatitude Cyclones Derived from the Scanning Multichannel Microwave Radiometer*. Ph.D. Thesis. Dept. of Atmospheric Sciences, University of Washington, Seattle, WA, 98195.

Njoku, E.G., 1982: 'Passive microwave remote sensing from space - a review'. *Proc. of the IEEE*, **70**, No. 7, 728-750.

Njoku, E.G. and L. Swanson, 1983: 'Global measurements of sea surface temperature, wind speed, and atmospheric water content from satellite microwave radiometry'. *Mon. Wea. Rev.*, **111**, 1977-1987.

O'Brien, J., R. Kirk, L. McGoldrick, J. Witte, R. Atlas, E. Bracalente, O. Brown, R. Haney, D.E. Harrison, D. Honhart, H. Hurlburt, R. Johnson, L. Jones, K. Katsaros, R. Lambertson, S. Peteherych, W. Pierson, J. Price, D. Ross, R. Stewart, and P. Woiceshyn, 1982: Scientific Opportunities Using Satellite Surface Wind Stress Measurements Over the Ocean. *Report of the Satellite Surface Stress Working Group*. N.Y.I.T. Press, Fort Lauderdale, FL, 153 pp.

Olson, W.S., 1987: *Estimation of Rainfall Rates in Tropical Cyclones by Passive Microwave Radiometry*. Ph.D. Dissertation, University of Wisconsin, Madison, WI, 282 pp.

Pandey, P.C., 1983: *Linear retrieval and global measurements of wind speed from the Seasat SMMR*. JPL Publication 83-5, NASA, Pasadena, CA., 26 pp.

Petty, G.W. and K.B. Katsaros, 1988: 'Precipitation observed over the South China Sea by the Nimbus 7 Scanning Multichannel Microwave Radiometer during winter MONEX'. *J. Appl. Meteor.*, (submitted).

Pierson, W.J., 1983: 'The measurement of the synoptic scale wind over the ocean'. *J. Geophys. Res.*, **88**, 1683-1708.

Prabhakara, C., H.D. Chang and A.T.C. Chang, 1982: 'Remote sensing of precipitable water over the oceans from Nimbus 7 microwave measurements'. *J. Appl. Meteor.*, **21**, 59-68.

Prabhakara, C., I. Wang, A.T.C. Chang and P. Gloersen, 1983: 'A statistical examination of Nimbus-7 SMMR data and remote sensing of sea surface temperature, liquid water content in the atmosphere and surface wind speed'. *Bull. Am. Meteor. Soc.*, **22**, 2023-2037.

Schluessel, P., H.-Y. Shin, W.J. Emery and H. Grassl, 1987: 'Comparison of satellite-derived sea surface temperatures with *in-situ* skin measurements'. *J. Geophys. Res.*, **92**, 2859-2874.

Spencer, R.W., B.B. Hinton and W.S. Olson, 1983: 'Nimbus-7 37 GHz radiances correlated with radar rain rates over the Gulf of Mexico'. *J. Clim. Appl. Meteor.*, **22**, 2095-2099.

Spencer, R.W., 1986: 'A satellite passive 37 GHz scattering-based method for measuring oceanic rain rates'. *J. Climate Appl. Meteor.*, **25**, 754-766.

Staelin, D.H., K.F. Kunzi, R.L. Pettyjohn, R.K.L. Poon, R.W. Wilcox and J.W. Waters, 1976: 'Remote sensing of atmospheric water vapor and liquid water with the Nimbus 5 microwave spectrometer'. *J. Appl. Meteor.*, **15**, 1204-1214.

Strong, A.E., 1986: 'Monitoring El Niño using satellite based sea surface temperatures'. *Ocean-Air Interactions*, **1**, 11-28.

Sundqvist, H., 1978: 'A parameterization scheme for non-convective condensation including prediction of cloud water content'. *Quart.J. Roy. Meteor. Soc.*, **104**, 677-690.

Tabata, S., 1978a: 'Comparison of observations of sea surface temperatures at Ocean Station P and NOAA buoy stations and those made by merchant ships traveling in their vicinities, in the Northeast Pacific Ocean'. *J. Appl. Meteor.*, **17**, 375-385.

Tabata, S., 1978b: 'An evaluation of the quality of sea surface temperatures and salinities measured at Station P and Line P in the Northeast Pacific Ocean'. *J. Phys. Ocean.*, **8**, 970-986.

238

Tabata, S., 1981: On the accuracy of satellite-observed sea surface temperatures. In *Oceanography from Space*, J.F.R. Gower, ed. Plenum Publishing Corp., pp. 145-157.

Wentz, F.J., 1983: 'A model function for ocean microwave brightness temperatures'. *J. Geophys. Res.*, **88**, 1892-1908.

Wentz, F.J., L.A. Mattox and S. Peteherych, 1986: 'New algorithms for microwave measurements of ocean winds: applications to Seasat and the Special Sensor Microwave/Imager'. *J. Geophys. Res.*, **91**, 2289-2307.

Wentz, F.J., V.J. Cardone and L.S. Fedor, 1982: 'Intercomparison of wind speeds inferred by the SASS, altimeter, and SMMR'. *J. Geophys. Res.*, **87**, 3378-3384.

Wilheit, T.T., J.R. Greaves, J.A. Gatlin, D. Han, B.M. Krupp, A.S. Milman and E.S. Chang, 1984: 'Retrieval of ocean surface parameters from the Scanning Multifrequency Microwave Radiometer (SMMR) on the Nimbus-7 satellite'. *IEEE Trans. Geosci. Rem. Sens.*, **GE-22**, 133-143.

Wilheit, T.T. and A.T.C. Chang, 1980: 'An algorithm for retrieval of ocean surface and atmospheric parameters from the observations of the Scanning Multichannel Microwave Radiometer (SMMR)'. *Radio Science*, **15**, 525-544.

Woiceshyn, P.M., M.G. Wurtele, D.H. Boggs, L.F. McGoldrick and S. Peteherych, 1986: 'The necessity for a new parameterization of an empirical model for wind/ocean scatterometry'. *J. Geophys. Res.*, **91**, 2273-2288.

PASSIVE MICROWAVE SATELLITE IMAGERY FOR IMPROVED RAINFALL MONITORING
AND FORECASTING OVER SEA AREAS

Eric C. Barrett
Remote Sensing Unit
Department of Geography
University of Bristol
Bristol BS8 1SS
UK

ABSTRACT. The use of passive microwave imagery from satellites has
been investigated much less than visible and infared imagery in
support of global rainfall monitoring and forecasting, but, being more
physically direct, has highter potential for such applications. This
paper reviews the principles and practices of passive microwave
rainfall evaluation over seas and oceans, illustrates the range of
results which have been obtained, and discusses the potential of this
approach for rainfall forecasting over oceans and windward coastal
zones.

1. INTRODUCTION

Rainfall is important for many reasons. Over land it is vital for
life. Over water it does not have this significance, but is the
primary input into the oceanic hydrological cycle, whose role in the
dynamics of the Earth's atmosphere is enormous. Unfortunately,
rainfall is particularly difficult to monitor satisfactorily by
conventional (in situ) means because of its highly variable space and
time distributions, plus the practical and economic difficulties of
maintaining raingauge and surface radar networks over many types of
surfaces. These include not only high altitude areas, high latitudes,
and deserts, but, even more extensively, the world's oceans and major
seas.
    It is generally agreed, therefore, that present supplies of
rainfall data are inadequate for most applications, ranging from
hydrological science and water management, through agrometeorology, to
weather forecasting and climate studies over most of the globe. These
data are inadequate particularly for near real-time operations, which
call for speedy and regular data flows, and for research, especially
when sets of relatively dense and homogeneous data are required.
Since the late 1960s it has been recognised that spaceborne remote
sensing can provide valuable information relating to precipitation,
and is destined to become a major component of any global-scale
precipitation monitoring system. (NASA, 1986).

(Paper presented by J Bailey)

239

*R. A. Vaughan (ed.), Microwave Remote Sensing for Oceanographic and
Marine Weather-Forecase Models, 239–252.*
© *1990 Kluwer Academic Publishers.*

Table 1. Satellite Microwave Imaging Instruments

Instrument.	ESMR-5	ESMR-6	SMMR	SSM/I
Spacecraft.	Nimbus-5	Nimbus-6	Nimbus-7/Seasat	DMSP 5D-2/SX
Launch date.	12-72	06-75	10-78 / 06-78	06-87
Frequencies/ footprint sizes.				
6.6GHz	_____	_____	VH 121 * 79 km	_____
10.7GHz	_____	_____	VH 74 * 49 km	_____
18.0GHz	_____	_____	VH 44 * 29 km	_____
19.35GHz	x 25 * 25 km	_____		VH 69 * 43 km
21.0GHz	_____	_____	VH 38 * 25 km	_____
22.235GHz	_____	_____		V 50 * 40 km
37.0GHz	_____	VH 25 * 45 km	VH 21 * 14 km	VH 37 * 28 km
85.5GHz	_____	_____		VH 15 * 13 km
Primary aims of mission;	Liquid water contents of clouds, sea ice and open sea coverage. Surface composition and soil type, with surface features and surface moisture.		Sea surface temperature, near sea surface winds. Water vapour, liquid water content and cloud droplet size; Rainfall rate.	Continual ocean wind speed and ice coverage, age and extent. Intensity of precipitation, cloud water content & land surface moisture.

Key: ESMR = Electrically Scanning Microwave Radiometer
      SMMR = Scanning Multispectral Microwave Radiometer
      SSM/I = Microwave Imager

      V = vertical polarization
      H = horizontal polarization
      x = non-polarized

Of the many different techniques which have been tested for satellite-assisted rainfall monitoring systems the great majority have been based on visible (VIS) and/or infrared (IR) imagery. Comprehensive reviews of these have been provided by Atlas & Thiele (1981), Barrett & Martin (1981), and more recently but more briefly by WMO (1986). The principal problems with VIS/IR techniques stem from their physical indirectness: the task of deducing rain - which falls from the bottoms of clouds, from images of the cloud tops is clearly a very difficult one. However, these techniques have the great advantage of being based on imagery currently available on a quite regular and dependable basis from operational families of civilian satellites, including geostationary satellites providing data as frequently as every 30 minutes.

The most widely tested VIS/IR techniques have produced their best results from areas dominated by convectional clouds, particularly overland in the tropics. They have been tested, less in middle latitudes where stratiform and layered clouds are most widespread. They have been tested least over mid-latitude oceans, partly because of the high incidences of stratiform and layered clouds in these regions. In all of these zones research has been hindered by the sparseness of conventional (in situ) rainfall data for calibration and verification, but also because most end users of precipitation data have been concerned with rainfall over land.

Passive microwave (PM) data have been researched and assessed for rainfall monitoring since as long ago as the early 1970s, mainly based on imagery from microwave radiometers on Nimbus satellites. (See Table 1). Although the Nimbus family is effectively dead, interest in PM data for rainfall monitoring has recently been revived by the new USAF Block 5D satellite series. Indeed, it is now widely recognised that microwave techniques hold special promise for improved satellite monitoring of rainfall, for the microwave signals are physically much more directly related than the VIS/IR signals to the hydrometeors themselves, being much less affected by the clouds in which the hydrometeors are embedded.

2. PHYSICAL PRINCIPLES

For most meteorological purposes the microwave region is defined as the interval from about 3 to 300 GHz (1 GHz = 10^9 cycles s^{-1}). In wavelength, this interval extends from 10 to 0.1 cm. It merges with IR radiation at higher frequencies, and radio waves at lower frequencies. The wavelengths exploited by most meteorological radar also lie within the microwave region.

Microwave radiation may be absorbed, reflected or scattered. What happens to it depends on elements of the Earth and its atmosphere. Their influence, in turn, depends on temperature, dielectric state, and microphysical properties such as roughness, size and shape of active constituents, which include water, ice, land, O_2 & H_2O gases, cloud droplets and hydrometeors. (See Barrett & Martin, 1981). Microwave radiation is characterised by intensity and polarization.

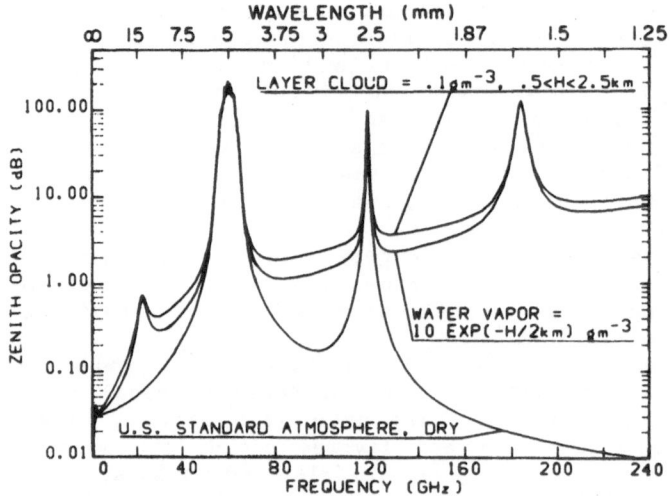

Figure 1. Zenith opacity of the atmosphere as a function of frequency. Opacity due to exygen (lower curve), opacity with 20 kg m⁻² water vapour added to the oxygen (middle curve), opacity with 0.2 kg m⁻² stratus cloud added to the oxygen and water vapour (upper curve). After Rosenkranz, 1978, from Barrett & Martin, 1981.

Since intensity is a linear function of temperature it is convenient to assess the former by a brightness temperature, $T_{b\lambda}$, where:

$$T_{b\lambda} = \varepsilon\lambda T,$$

and ε is the emissivity and T the blackbody function.

The emissivities of surface materials are known to be quite variable (Staelin, 1969); those of soils and vegetation are much higher than those of water. Dry land emissivities are generally between 0.85 to 0.95 at wavelengths of 1 to 1.5 cm (Wilheit, 1972). The presence of surface water lowers approximately in proportion to its fractional surface content so that puddled soil may have the emissivity of a sea or lake, which is ~ 0.4 at low microwave frequencies. At some wavelengths (including the important 19 GHz window), ε decreases with increasing thermal temperature, thereby compensating increasing blackbody emission so that brightness temperature is nearly constant over a wide range of surface temperatures. Effects of salinity on ε are significant only at very low microwave frequencies, but roughness, including foam, increase it throughout.

Polarization describes the eccentricity and orientation of the ellipse which the electric field vector of a simple electromagnetic wave will trace at a point in space. In microwave radiometry polarization is usually measured by the difference in brightness temperature between the horizontal and vertical components of intensity, i.e. $T_{bH} - T_{bV}$, with vertical defined as the local normal to the surface (Savage, 1976).

Polarization of reflected, transmitted and emitted PM radiation depends on molecular or crystalline properties and the surface roughness of a medium. Like emissivity, it is a function of angle of view. Water, having a high dielectric constant produces highly polarized radiation at oblique viewing angles; for example, at 37 GHz, calculated brightness temperature differences between horizontally and vertically polarized radiation emerging from a standard atmosphere exceeds 30 K (Rodgers et al., 1979). Radiation emitted by atmospheric gases is not polarized, and there is no significant polarization in either cloud drops or raindrops. Differences in polarization are therefore a primary basis for distinguishing between dry land, wet land, and rain over land in some PM rainfall estimation schemes.

Once PM radiation has been emitted, it may be affected by atmospheric absorption (by the gaseous atmosphere especially oxygen and water vapour, and cloud particles, and scattering (by cloud particles). Combining these effects, the atmosphere may be modelled for opacity, as illustrated by Figure 1. From such results, some of the best microwave window channels for rainfall evaluation can be identified: less than 20 GHz, around 30 GHz and between 80 - 90 GHz. Whilst the atmosphere is more transparent at the lower of these frequencies than the higher, it must be remembered that the signal strength increases with the frequency of the radiation: hence the higher spatial resolution at the higher frequencies (e.g. 85.5 GHz on SSM/I), as specified in Table 1.

The intention of PM rainfall monitoring methods is to try to relate the upwelling PM radiation to instantaneous rain rate. Fortunately for present purposes this is easier over water than over land. Present PM rainfall techniques exploit two two main physical regimes:

2.1 Absorption/emission

Here rainfall can be observed through the emission of thermal energy (e.g. Wilheit et al., 1982). Since a cold background is required in order to make such assessments, this is only practical over oceans, where, as we have seen, emissivities are both low and fairly constant, and so the rain areas are warmer than their environments. The liquid raindrops themselves are the dominant contributors to this absorption and emission, providing a direct physical relationship between the rainfall and the observed microwave radiances.

2.2 Scattering

The rainfall is evidenced via the scattering within the rain column, which reduces or even eliminates upwelling radiation, whilst reflecting the cold cosmic background towards the satellite (Spencer et al., 1983). Whilst this holds good over any surface, including the relationship between scattering and rain rate is less direct than in the absorption/emission regime because the observed scattered radiation may be dominantly from the frozen hydrometeors aloft rather than the rain itself.

Generally, at frequencies below the 22 GHz water vapour line (see Fig. 1), emission/absorption dominates; above the 60 GHz complex scattering dominates; between 22 & 60 GHz either may dominate according to the specifics of the situation. As frequencies increase towards and beyond the 85.5 GHz channel of the SSM/I, the effects of ice layers become very important. It has been predicted that, with ice layers as thin as 0.5 km, it is possible to obtain very low brightness temperatures, which may lead to rain rate over-estimation. However these low temperatures appear most likely to be associated with ice particles of the kinds generally associated with rain drop formation processes and should, therefore, be indicative of heavy convective precipitation (NASA, 1987).

3. PASSIVE MICROWAVE RAINFALL MONITORING ALGORITHS AND RESULTS

As a result of the above, it was both appropriate and ultimately profitable for several workers to evaluate 19.35 GHz brightness temperatures from the pioneer PM instrument ESMR on Nimbus 5 in terms of rain rates over the oceans, e.g. by Allison et al. (1974) for selected tropical cyclones. Figure 2 illustrates one model used for this, and presents results combining satellite, radar and raingauge data (Wilheit et al., 1977). Subsequent work by advocates of a multi-system (VIS, IR, PM) approach have shown that significant, sometimes even dramatic, improvements in the satellite estimates can be obtained if VIS & IR imagery is used to obviate beam-filling problems which are associated with the large fields-of- view of the PM sensors (Smith & Kidder, 1978).

Two applications of the Wilheit model may be summarised here. First, the effort by Kidder & Vonder Haar (1977) to infer seasonal frequency of rain over the world's tropical and subtropical oceans from the ESMR-5 data. Selecting 0.75 mm h^{-1} as a threshold for rain, they distinguished light, medium, heavy and very heavy rain rate classes. The brightness levels coresponding to each class varied by latitude according to observed zonal mean freezing levels. The total rain frequency map Kidder & Vonder Haar produced compared well with earlier climatic maps based on ship observations. Small differences, e.g. over the equatorial Pacific, probably stemmed from actual anomalies of precipitation in the ESMR-5 map period. Larger differences in higher latitudes, e.g. over the north-west Pacific, are less easily explained. It is possible that these may have been due to increases in surface emissivity resulting from foam produced by strong surface winds.

A second application of Wilheit's model was the monumental "Satellite-derived Global Oceanic Rainfall Atlas" of Rao et al. (1976). This was produced by averaging rain rates over grid cells of 4° latitude by 5° longitude, and which contained at least 25% water

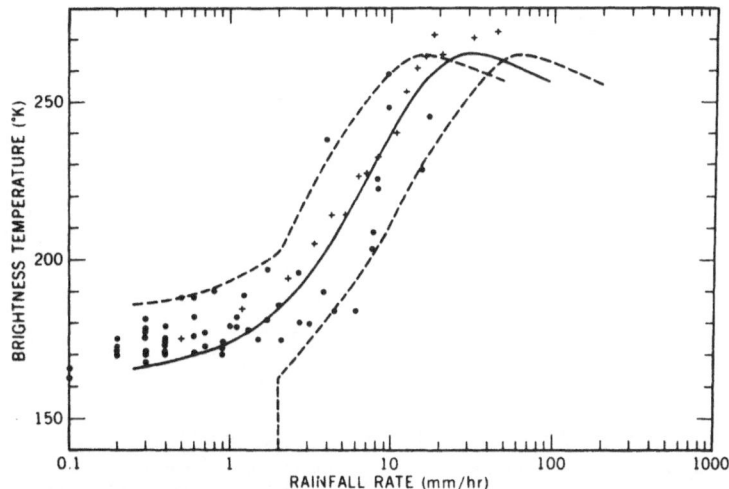

Figure 2. Brightness temperature as a function of rainrate: ESMR-5 versus WSR-57 radar (dots) and inferred from ground-based measurements of brightness temperature and direct measurements of rainrate (crosses). The solid line is the calculated brightness temperature of a 4 km freezing level. The dashed lines represent departure of 1 mm h^{-1} or a factor of 2 in rainrate (whichever is greater) from the calculated curve. From Wilheit et al., 1977.

surface. The Atlas includes:

(a) Maps of weekly, monthly and seasonal maps of rain rates over the world's oceans between 72°N & S between 11 December 1972 & 28 February 1975, and annual maps for the two full years, 1973 & 1974. (See Fig. 3).

(b) Graphs of zonally-averaged rainfall rates for the Pacific, Atlantic and Indian Oceans separately, and for the "Global Oceans" together, for the same periods (see Fig. 4); and

(c) Sectors selected from the above, printed in colour to illustrate more clearly the utility of the maps for studies of key phenomena, e.g. the Indian monsoon, and the El Nino effect.

Rao et al. recognized limitations in the Atlas, saying that "No claim is made for reliability in absolute values of rain rate better than a factor of two....". Comparisons of monthly and annual ESMR-5 rainfall estimates with existing climatological maps revealed differences within this range, from -30 to +80% only. This experience led to detailed modifications of the Wilheit model, and to suggestions for its further refinement.

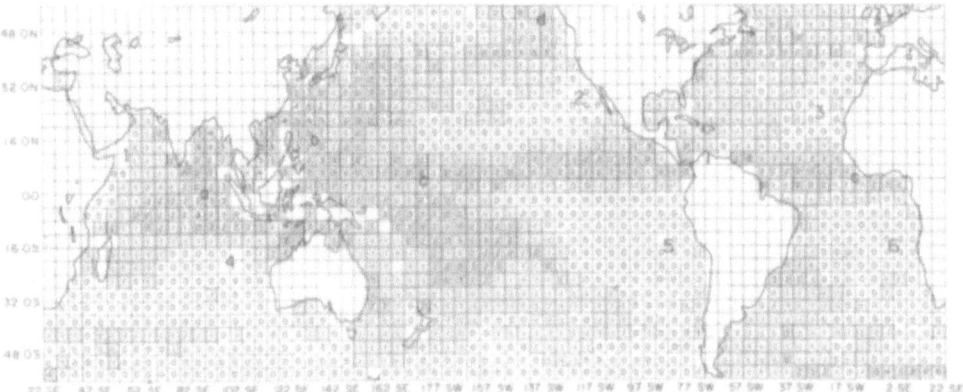

Figure 3. ESMR-derived global oceanic rainfall rate map for January 1974 - December 1974. Areas of heavy rain are labelled a through d, and areas of light rain 1 through 6. After Rao _et al_., 1976.

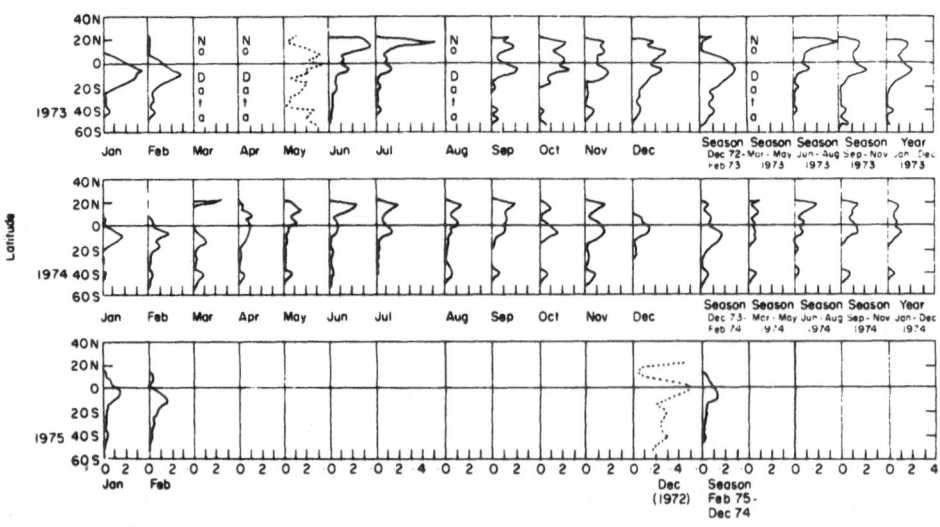

Figure 4. ESMR-derived graphs of zonally averaged rainrate versus latitude in the Indian Ocean. Broken lines represent curves based on inadequate data. After Rao _et al_., 1976.

Although the Atlas was criticised by some because of the methods which were used for processing and analysing the ESMR data, it remains an important pioneer document which threw important light not only on major features of the global rainfall distribution, but also the short-period fluctuations to which this may be subject. It is in this area of climate dynamics that satellites may have most to offer through passive microwave rainfall monitoring in the near-term future.

Absorption/emission methods such as the above were expected to benefit considerably from the advent of higher frequency data, beginning with the ESMR-6 instrument carried by Nimbus-6, for increasing the frequency from 19.35 to 37.0 GHz approximately doubles the sensitivity to water vapour and oxygen but triples sensitivity to liquid water droplets. Furthermore, on ESMR-6 both polarization components were measured. However, at 37 GHz the increase of T_b in relation to increasing rain rate reverses at only about $4mm\ h^{-1}$ compared with about $20\ mm\ h^{-1}$ at 19.35 GHz. This is the main reason why this instrument was used at first mainly over land. It is only more recently, as the science has advanced, that 37 GHz data have been used significantly for oceanic rain rate measurement, most notably by Spencer (1986), but using data from the Nimbus-7 SMMR instrument. (See Table 1).

Spencer's method requires an estimate of the effective radiating temperature of the cloudy portion of the atmosphere, plus a brightness temperature measurement of the cloud-free ocean surface. These two measurements bound all possible combinations of clear and cloudy conditions within a footprint in terms of bipolarized brightness temperatures. Any satellite-observed T_b lower than these values is assumed to evidence scattering, which is due only to precipitation-sized hydrometeors at 37 GHz. The technique involves a linear transformation between dual polarized brightness temperature and rain rate, so there is, therefore, no nonlinear "footprint filling" effect. Unique footprint-averaged rain rates result from the application of this algorithm, which may be represented as follows:

$$R = a(bT_b^{H37} - T_b^{V37} + C)$$

where (b) is the slope of the emission line in V,H space, and is generally about 0.5. This value is relatively unchanged with changing ocean surface and associated conditions. For R = 0, the value of (c) can be computed from the T_b of the cloud-free ocean surface, and (a) from observed T_b values at some rain-rates along the "observed" line. These principles are further elucidated by Figs. 5(a) & 5(b). They have been applied to a number of radar vs. SMMR case studies in tropical and sub-tropical regions, yielding correlation coefficients of around 0.90. (Spencer et al., 1983).

Unfortunately, however, the scattering signal at 37 GHz is less strong than the emission signal of light rainfall at 19 GHz; thus the advantages of the scattering method for heavy precipitation might best be augmented by the emission method to map light precipitation.

Most recently, some pointers to future techniques and associated operations have been identified since the launch of the DMSP Block 5D

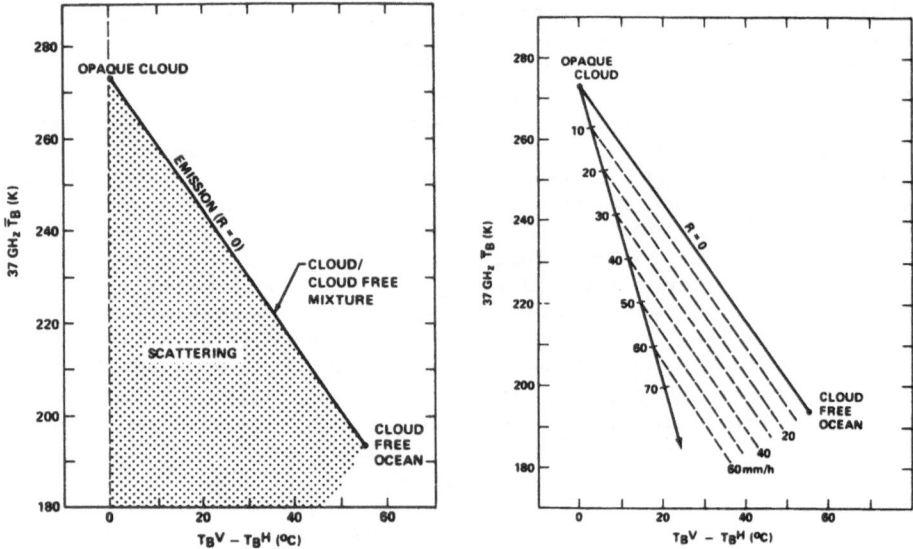

Figure 5. (a, left): Bipolarization of 37 GHz T_B vs rain rate showing the effect that emission by clouds and scattering by precipitation will have on the radiances upwelling from the highly polarized, radiometrically cold ocean surface. Increasing coverage of a footprint by clouds will result in a linear increase in T_B due to emission by their hydrometeors, matched by a linear decrease in polarization. We can extend this trend up to a hypothetical limit, that being the effective radiation temperature of a totally opaque cloud. Any scattering 37 GHz will be due only to precipitation and will result in T lower than what can be attributed to emission alone (represented here by a zero rain rate, or "emission" line). (b, right): Bipolarization plot of the zero rain rate (emission only) line from Fig. 5a, with the observed relationship between radar rain rates and SMMR T during the summer over the United States. From these two lines, a family of rain rate lines result from any combination of values along these two curves (i.e., any combination of cloudy, precipitating, and cloud-free conditions within a footprint). From Spencer, 1986.

military weather satellite in June 1987. This, for the first time, has extended satellite PM observational capabilities up to 85.5 GHz. (See Table 1). The first SSM/I-based algorithm testing and development has already been undertaken by Spencer et al. (in press), and tested in part over ocean areas. Theory tells us that, as we proceed from the lowest SSM/I frequency (19.35 GHz) to the highest (85.5 GHz) we see dramatic increases in volume scattering and

absorption coefficients as well as single scattering albedo. It is
this last effect, produced by ice, which is of greatest importance for
sharpening contrasts between the PM radiances of precipitation from
its surroundings: if ice is present above the freezing level, very low
brightness temperatures can result especially at 85.5 GHz, as
upwelling radiation fails to penetrate the ice layers. Radiative
transfer models, such as that of Wu & Weinman (1984), suggest that T_b
depressions of tens of degrees are commonplace for moderate to heavy
rain rates at low frequencies (18-19 GHz), increasing to almost 200
degrees at 85 GHz.

For these reasons Spencer et al. (in press) have been able to
remark that: "volume scattering by precipitation is probably the most
striking geophysical passive signature in the microwave spectrum." In
their recent work, distinctions are drawn between:

(a) The Effective Radiating Temperature (ERT) of cloud. This is a
    linear extrapolation from the cold ocean background of the net
    bipolarization warming or cooling produced by all hydrometeors to
    zero polarization difference. Of course, this reflects only the
    trend of oceanic T_B change by clouds and precipitation, hence
    another quantity is needed, viz:

(b) The Polarization Corrected Brightness Temperature (PCT). This
    takes into account the degree of the change caused by clouds and
    precipitation. Empirical indications are that 85.5 GHz cloud
    observations below about 250 to 260° K have a good probabality of
    being influenced by precipitation.

Analyses of early SSM/I images using the polarization correction
of brightness temperature have begun to yield exciting results, as
Fig. 6 illustrates. This shows that simultaneous WSR-57 radar and
SSM/I 85.5 GHz coverages reveal pleasing similarities of shape and
intensity of the rain areas, with the sole exception of the very light
rain area (level 1 radar reflectivity) on the southeast side of the
precipitation region. Other initial results from colder environments
raise the hope that a constant or nearly constant T_b threshold may be
used for delineating rain areas at 85.5 GHz in both warm and cold
environments, though much more work is needed to validate this.

4. THE FUTURE

Surprisingly little attention has been paid to the use of PM data for
rainfall monitoring over oceans and seas despite the dearth of
rainfall data obtained therefrom by conventional means. However,
excellent groundwork has been laid for very useful retrievals of maps
of instantaneous rain rates from selected areas and/or individual
storms, and for subsequent larger area and/or longer term inventories
of oceanic rainfall. Unfortunately, problems remain with the PM
algorithms, but the use of multiple-frequency methods should allow
information to be gathered on key rain-system structure

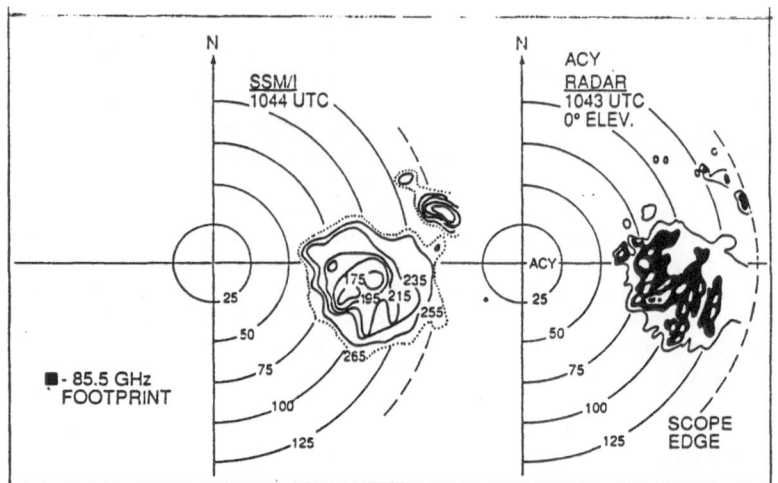

Figure 6. Radar/SSM/I rain signatures for rain systems observed by the
        Atlantic City, New Jersey (ACY) radar at 1043 UTC 31 July
        1987. From Spencer et al., in press.

characteristics (e.g. the relative quantities of liquid versus frozen
hydrometeors), and the use of VIS/IR data will be useful in helping to
resolve ambiguities and in identifying the types of rain systems
imaged by the PM data. Problems of data coverage may soon be the more
restrictive, at least in the short term: although SSM/I data coverage
is quasi-global on a daily basis (and thus much superior to SMMR,
which was only switched on half the time), the coverage is less than
complete in low and middle latitudes.

      Since rainfall varies so rapidly through time it would be much
more advantageous if PM data could be obtained from geostationary
platforms.   However, no firm plans for suitable geostationary
microwave sensors have yet been drawn up, not least because of
problems of low signal strength in the passive microwave region, and
related requirements for large, unwieldy antennae.   Therefore the
primary uses of PM image data in the foreseeable future, which are
most likely to include meteorological case studies and climatological
inventorying, will have to be based on DMSP SSM/I data, data from
planned microwave imagers on the Polar Platforms of the mid-1990s and
onwards, and from the US/Japanese TRMM satellite designed specifically
for tropical climate rainfall studies (Simpson et al., 1988).
However, it is already being proposed to use PM data to calibrate
geostationary VIS/IR rainfall monitoring techniques for numerical
weather forecast model initialisation over water, and associated
quantitative rainfall forecasting for coastal zones (Barrett et al.,
in press).

      Indeed, the greatest value of PM image data in the foreseeable
future seem likely to lie in combined VIS/IR/PM ("trispectral")

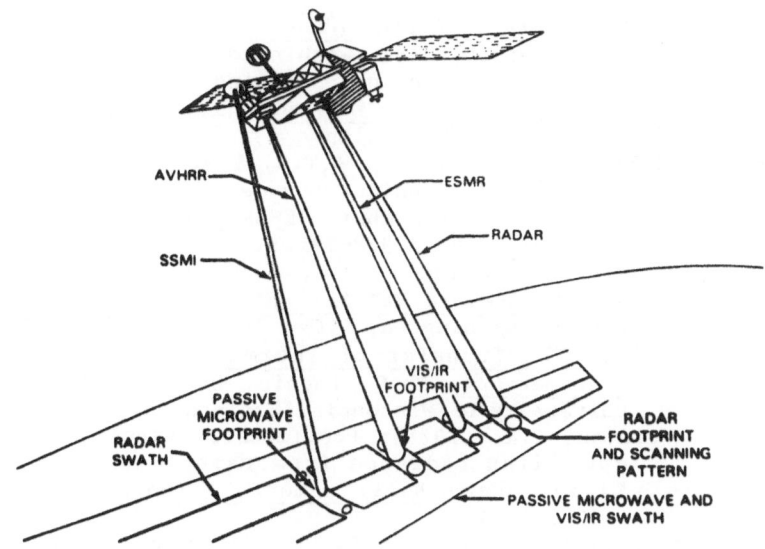

Figure 7. Artist's impression of the Tropical Rainfall Monitoring
    Mission Satellite. (Courtesy, NASA).

techniques for satellite-improved rainfall monitoring, over both
continents and oceans (Barrett _et al_., in press), and work has begun
towards that end.

References

Allison, L., Rodgers, E., Wilheit, T. & Fett, R. (1974): Tropical
        cyclone rainfall as measured by the Nimbus 5
        electrically scanning microwave radiometer. _Bull._
        Amer. Meteorol. Soc., 55, p.1074-1089.
Atlas, D.W. & Thiele, O. 1981: _Precipitation Measurements from Space._
        NASA, Goddard Space Flight Center, Greenbelt,
        Maryland, 246pp.
Barrett, E.C. & Martin, D.W. (1981): _The Use of Satellite Data in_
        Rainfall Monitoring. Academic Press, London, 340pp.
Barrett, E.C., D'Souza, G., Power, C.H. & Kidd, C. (in press): Towards
        trispectral satellite rainfall monitoring
        algorithms. _Proceedings, International Symposium on_
        Tropical Precipitation Measurements, Tokyo, Japan,
        28-30 Oct. 1987. NASA, Goddard Space Flight Center,
        Greenbelt, Maryland.
NASA 1986:   _Earth System Science: Overview._ National
        Aeronautics and Space Adminstration, Washington,
        D.C. 20546, USA.
NASA 1987:   _HMMR High-Resolution Multifrequency Microwave_

252

Radiometer, Earth Observing System. Instrument Panel Report, NASA, Washington, D.C., 59pp.

NASA 1988: Report of the Science Steering Group for a Tropical Rain Measuring Mission. Final Draft, NASA, Washington D.C., 131pp.

Rodgers, E., Siddalingaiah, H., Chang, A.T.C. & Wilheit, T. (1979): A statistical technique for determining rainfall over land employing Nimbus 6 ESMR measurements. J. Appl. Meteorol., 18, p.978-991.

Rao, M.S.V., Abbott, W.V. & Theon, J.S. (1976): Satellite-Derived Global Oceanic Rainfall Atlas (1973 & 1974). NASA SP-410, Washington, D.C., 31pp. plus appendices.

Savage, R.C. (1976): The Transfer of Thermal Microwaves through Hydrometeors. Ph.D. thesis, Dept. of Meteorology, University of Wisconsin, Madison, 147pp.

Savage, R.C. & Weinman, J.A. (1975): Preliminary calculations of the upwelling radiance from rainclouds at 37.0 and 19.35 GHz. Bull. Amer. Meteorol. Soc. 56, p.1272-1274.

Simpson, J., Adler, R.F. & North, G.R. (1988): A proposed Tropical Rainfall Measuring Mission (TRMM) Satellite. Bull. Amer. Meteorol. Soc. 69 (3), p.278-295.

Smith, E.A. & Kidder, S.Q. (1978): A Multispectral Satellite Approach to Rainfall Estimates. Unpublished manuscript, Colorado State University, Fort Collins, 26pp. plus tables and figures.

Spencer, R.W., Olson, W.S., Rongzhang, W., Martin, D.W., Weinman, J.A. & Santek, D.A. (1983): Heavy thunderstorms observed over land by the Nimbus 7 Scanning Multichannel Microwave Radiometer. J. Climate Appl. Meteorol., 22, p.1041-1046.

Spencer, R.W. (1986): A satellite passive 37 GHz scattering-based method for measuring oceanic rain rates. J. Clim. and Applied Meteorol. 25 (6), p.754-766.

Spencer, R.W., Goodman, H.M. & Hood, R.E. (in press): Precipitation retrieval over land and ocean with the SSM/I, Part 1: identification and characteristiscs of the scattering signal. J. of Atmos. & Ocean. Tech..

Wilheit, T. (1972): The Electrically Scanning Microwave Radiometer (ESMR) experiment. In The Nimbus 5 User's Guide (R.R.Sabatini, ed.), Goddard Space Flight Center, Greenbelt, Md., p.59-105.

Wilheit, T.T., Chang, A.T.C., Rao, M.S.V., Rodgers, E.B. & Theon, J.S. (1977): A satellite technique for quantitatively mapping rainfall rates over the oceans. J. Appl. Meteorol. 16, p.551-560.

World Meteorological Organisation (1986): Report of the Workshop on Global Large-scale Precipitation Data Sets for the World Climate Research Programme, Camp Springs 1985. WMO Geneva, WCRP-1182, 44pp.

THE POTENTIAL IMPACT OF ADVANCED SPACEBORNE MICROWAVE SENSORS ONTO MARINE WEATHER NOWCASTING AND VERY SHORT RANGE FORECAST

A.RATIER
On leave from Direction de la Météorologie
Centre National d'Etudes Spatiales
18 Avenue Edouard Belin
31055 Toulouse Cédex FRANCE

ABSTRACT. Very short range weather forecast/nowcasting is briefly presented as a specific challenging meteorological discipline with objectives, observational requirements and methods of its own. The potential contribution of advanced operational microwave sensors, active and passive, to be flown aboard ERS1 and other polar orbiting satellites is considered, essentially as a valuable and important addition to geostationary and ground-based systems, likely to impact onto the nowcasting of high latitude weather systems influenced by air/sea interactions and boundary layer convergence. Possible critical shortcomings of microwave observations are examined, and ways to address some scatterometer issues are discussed and illustrated. The final discussion of the respective nowcasting merits of SEASAT, NROSS and ERS1 instrumental payloads stresses the major interest of combining as far as feasible, sensors onboard the same polar platform.

1- INTRODUCTION TO MARINE WEATHER NOWCASTING

1.1- Objectives

Marine weather nowcasting attempts to ensure the continuous monitoring and short term pre-diction of the environmental conditions at sea, thereby contributing to the safety of marine users and systems but also increasing the feasibility, efficiency and economical return of marine or coastal activities. Along these lines a high priority is assigned to the timely warning of dangerous sudden events, whereas flexibility is needed to comply with requirements specific to various economical marine activities. The target product is a real time detailed description of weather and sea state to be extrapolated 3 to 12 hours ahead by techniques pertaining to very short range forecasting (hereafter vsrf), 'detailed' meaning as close as possible to the user need and perception of the weather. This objective implies that small scales down to local site-specific phenomena be tentatively resolved, such as squall lines, organized convection, sea breezes or deep convection possibly embedded into

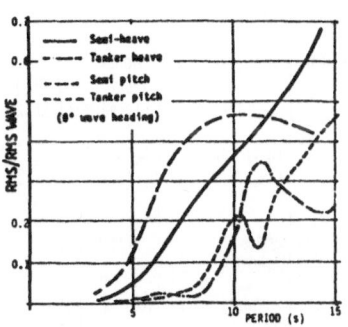

Fig.1: Response of a semisubmersible platform and a tanker to wave period. Note the resonance near and above 10 s.

R. A. Vaughan (ed.), Microwave Remote Sensing for Oceanographic and Marine Weather-Forecase Models, 253–271.
© *1990 Kluwer Academic Publishers.*

frontal systems, and that identified features be tracked and their time of arrival at a given site predicted. Moreover the wealth of significant weather elements and atmospheric parameters likely to endanger or affect marine activities must be encompassed, including cloudiness, precipitations, frost, lightening, fog and other hydrometeors, boundary layer wind, temperature and moisture. Ocean elements must be also considered like wind sea and swell, surface or subsurface currents, internal waves, storm surges, water turbidity, sea surface temperature and sea ice known to impact onto activities such as drilling, fishing, coastal engineering, ship routing. As illustrated by figure 1, the need of a given end-user can be more specific or focussed on subtle details like wave period and call for tailored products. A meaningful warning product therefore relies on the experience of the user need gained from mutual dialogue, but also on the anticipation of remote effects of local events, especially when resonant responses are expected, as in the case of strong marine frontal shears or pecular wind patterns forcing rapid shallow water response or swells likely to cause damages, when overflowing flat terrains after breaking at the coast.

1.2- Major challenges of nowcasting/vsrf

1.2.1- Seek for information: the integrated approach. Producing a three dimensional atmospheric snapshot extending down to the air/sea interface and resolving a wide spectrum of scales with a significant coverage is a real challenge. Table I confronts the horizontal resolution/coverage achieved by some observing systems with the typical scales of significant phenomena and highlights the satisfactory coverage of marine zones by remote sensing systems that still poorly resolve the complex three dimensional convective or mesoscale structures, yet known to produce severe weather. Moreover the required continuous and timely access to a tentatively comprehensive set of simultaneously observed parameters is again demanding from observing systems in terms of availability, operationality and measurement capabilities. None of the marine observing systems can

Phenomena	Scale	Observing system	horizontal resolution	revisiting rate	parameters sensed	coverage
Baroclinic waves	2000–10000km	Doppler/non coherent radars (ground-based)	0.1–5km 3D	15 mn	hydrometeor reflectivity	200 – 300 km
Cyclones	1000–3000km					
Frontal systems Hurricanes	200–2000km	satellite imagery	2D			
Polar lows	400 km	Meteosat NOAA	4 km (nadir) 1km (nadir)	30 mn 1,5–12 h	Cloudiness surface	< 55 deg latitude
Sea breeze	30 km	NOAA Sounding	20–110 km(nadir) (HIRS–MSU) 3D	1,5–12 h	Temperature moisture Soundings	Global Cloud free (HIRS) all weather (MSU: temp. only)
Squall line Thunderstorm	20km					
line convection	5x100 km	ligthning detection system	impact location 2D	1h	ligthning detection	800 km
Cumulus	1–5km	Sky wave radar	20 km 2D	3h	Surface wind direction/Sea state	2000x2000 km
		in situ (ship, buoy)	local (1D to 2D)	1–6h	Boundary layer/ Profiles	very sparse shipping lines

Table I: Horizontal scales of phenomena (left) versus capabilities of marine observing systems

individually meet those stringent multiform specifications. Ground-based radar or satellite imagery does sample the smaller scales and short term variations, but is mainly used as only detecting more or less integrated, poorly quantitative or implicit signatures of three dimensional phenomena. Shipborne radiosondes deliver detailed temperature moisture and wind profiles at synoptic times but are sparse in space and concentrated along the shipping lines or at weather stations. Ground-based systems adapted to coastal protection marginally cover the open sea, contrary to polar satellites that observe the globe at a revisiting rate limited by orbital constraints and instrument swath width. Thus the major issue of reconstructing a consistent picture of the atmosphere and sea state is underdetermined by any individual observing system and even by observations as such, since redundant and corroborative information is needed for quality control and to ensure the reliability of extreme event warning and hence the credibility of the nowcasting service. It should be further noted that independent observations are also needed to validate vsrf products beyond the user feed back. The path to success therefore implies that individual observing systems be striven to their limits, while merging and combining all informations available emerge as the cornerstone function of any nowcasting system, referred to as the horizontal integration process. Some of the relevant systems are described by plate 5 and further in the text.

The hierarchy of practical objectives then ranges from detection and pattern analysis, up to tendency analysis and very short term extrapolation or prediction, and is in practice fulfilled to a variable extent through the so-called vertical integration process, depending on the amount and relevance of the data available. Data processing and interpretation is the core task leading first to value-added composite intermediate products and ultimately to the issue of warning, diagnostic and forecast products. Fortunately subjective experience gained from years of data and weather monitoring can be fed therein, especially for the purpose of interpreting, cross-checking and relating observations in the light of the understanding of phenomena typical of a region, site or season or with pecular signatures or impact in their regional setting.

1.2.2- Timeliness. The timeliness constraint applies to nowcasting understood as an end-to-end service. This implies that horizontal and vertical integration processes as well as product dissemination be accomplished within a few hour time, which clearly impacts on the data, methods, tools used, that exclude lengthy or non flexible procedures and high bit rates of complex data, and leads to design a nowcasting system as an integrated system (fig 2) in the spirit of the Swedish PROMIS system.

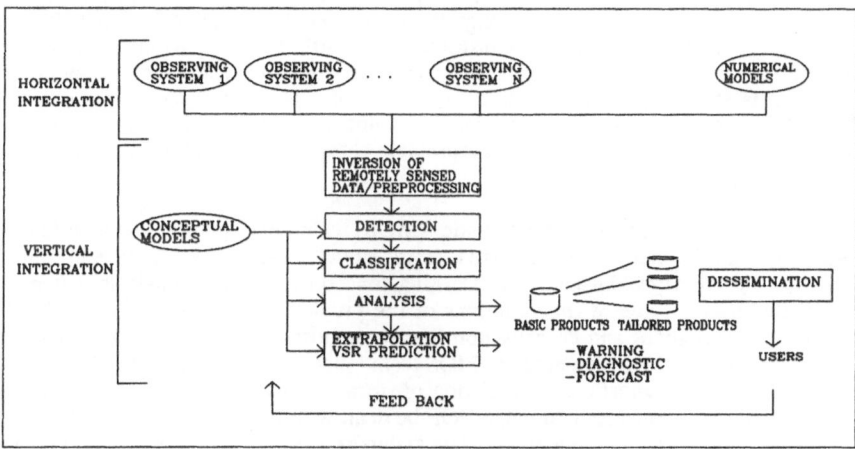

Fig.2: Nowcasting as an integrated system

256

1.3- Methods and tools

1.3.1- Numerical short term prediction. Recent advances in mesoscale modelling led to the operational implementation of high resolution (15 to 50 km) limited area numerical models for short range weather forecast, up to 36 hours ahead. Nested into coarser models providing the required forecast lateral boundary conditions, the French Peridot [ref 2] and the UKMO [ref 3] systems are examples of such models, that tend to predict with a reasonable skill synoptically forced mesoscale systems like frontal mesoscale circulation or topographically forced flows like sea breezes, provided such flows be well enough documented by lateral boundary forcing and to a lesser extent by observations. As an example, when fed in research mode with all observations available, the operational Peridot system captured the rapid development of 7 june 1987 [ref 4] (fig 3) missed by medium range forecast systems. Such a system should similarly help diagnose and qualitatively pre-

Fig.3: Peridot research mode 6 hour forecast for the 7 June squall line case (from Geleyn et al.,1988). θ_w^* isolines are plotted.

-dict cases of marine explosive cyclogenesis or polar lows, presumably triggered by baroclinic instability in their early development, as pointed out by Anthes et al., and Fedor et al. [refs 5,6], and confirmed by recent ECMWF coarser simulations accurately locating the QE2 storm low center [ref7]. Nevertheless, CISK driven organized convection fostered by mesoscale boundary convergence and local diabatic processes are also recognized to substantially contribute to the subsequent deepening of explosive cyclogenesis or polar lows [refs 6,8]. Locally forced phenomena are the most serious concern, as their predictability is not proven and may both depend on locally favourable initial conditions seldom resolved by observations and the explicit formulation or accurate parameterization of diabatic processes.The accurate location and intensity of extreme events is considered beyond the quantitative predicting capabilities of current mesoscale models, together with poorly handled significant weather elements like fog, precipitations, cloudiness or lightning. In addition duty cycle constraints leave little flexibility to take advantage of the continuous real time asynoptic data stream from imaging satellites and ground-based systems, and the spin-up problem is another strong limitation to the use of such models for very short range prediction. Nevertheless model outputs are considered as documenting mesoscale-alpha and beta flows and providing advection fields as well as interpretable clues of mesoscale developments when scale interactions are involved.

1.3.2- Model Output Statistics. Model output statistics (MOS) techniques assume that additional or detailed information can be statistically inferred from model outputs and past and present observations, using regressions established from learning time series. The rationale is that model outputs represent the driving force of the variations of the predictand parameters. Stratifying the learning file into weather regimes and deriving non linear composite physical predictors is crucial prior to the intrinsically linear regression analysis. MOS techniques are applicable at fixed location at sea, but also to the low cost diagnostic and prediction of storm surges along a shore line, provided that non local atmospheric forcing predictors like EOF be dealt with [ref 9], (fig 4). MOS belong to statistical methods also applied to remotely sensed data and indirect observations, in order to retrieve detailed informations or indexes useful to nowcasting.

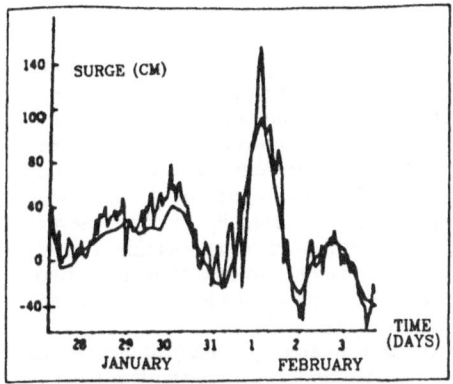

Fig.4: Storm surge statistical prediction at Dunkirk, using pseudo stress ($\|\nabla Z_{1000}\| k \wedge \nabla Z_{1000}$) EOF predictors derived from 1000 hpa geopotential (from [ref 9])

1.3.3- <u>Direct semi-subjective use of observations</u>. To compensate for the constraints and limited skill of numerical models, more flexible and direct approaches based on the direct combination of imagery and other remotely sensed data are needed as the only reliable way to locate, delineate, analyse and sometimes anticipate rapid convective or meso-beta developments, as well as significant weather elements. Geostationary and ground-based imagery afford the temporal repetitivity and redundancy needed to analyse and extrapolate tendencies and pattern displacements in the very short term. By then the first problem to tackle is data compression and classification, whereby large volumes of complex data are tentatively reduced to a more manageable but synthetic set of products. To come up with high data rates and the complexity of weather patterns or measurement physics, this task is necessarily computer-assisted, and the capabilities of advanced graphic devices are amply exploited for interactive contrast enhancement, thresholding, pattern recognition, feature or image correlation. Because the loss of information or clues present in a single data source is a constant threat, feedback and interactive procedures for cross-checking and verification dominate the process, and postprocessing data into more dynamically significant or conservative variables like vorticity is extremely useful when feasible. Once classified, compressed and postprocessed data remain representative of various scales, parameters or phenomena and vertical levels, and consistently coupling them together becomes the key to success. The detailed analysis or research mode numerical modelling of well-documented extreme events, as well as conceptual models of the unresolved phenomena provide guidance for this coupling effort and for subjectively reconstructing the missing information. It is for instance recognized that strong positive relative vorticity advection associated to a mid troposphere jet aloft favours strong organized convection along the zero relative vorticity isopleth [ref 10], and energetic alternate vorticity bands can therefore be considered as fostering explosive cyclogenesis. The NOAA 10 derived thermal vorticity map [ref 11] for the severe squall line case of 7 June 1987 over South western France is illustrative in this respect, when compared to vorticity patterns for the QE2 storm case. (plate 6). Conceptual models of precipitating mesoscale systems like those proposed by Browning [ref 12] provide a framework, whereby frontal systems can be classified from the interpretation of cloud or radar imagery, and significant weather can be diagnosed (fig 5).

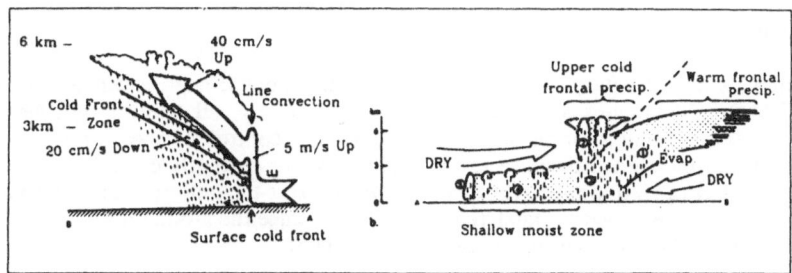

Fig.5: Conceptual models of precipitation systems: cross frontal sections (from Browning, 1985)

1.4- Data requirements

1.4.1- <u>Observations</u>. The hierarchy of objectives pursued and the variety of methods and tools applicable to weather nowcasting/vsrf involve a broad spectrum of data requirements. Timeliness, reliability and continuity constraints of the operational service demand availability and operationality from the observing systems, while internal consistency and redundancy is appreciated as contributing to the data quality control or relative calibration and to the reliability of extreme event warning or as giving access to gradients and dynamically significant fields. Reliability applies to the observing system as a whole, from the hardware to the output data that must keep marginally affected by severe environmental conditions associated to extreme events. Possibilities of accomodating relative accuracy and implicit or qualitative signatures within the somewhat subjective detection and pattern analysis relax the critical need for resolving the subtle quantitative variations of a lot of variables and preserve the relevance of the unique sampling performances of imagery.

A basic cut off exists between scales and variables resolved or treated by numerical models and finer scales and significant weather elements. Mesoscale numerical models can currently assimilate only measurements linked as linearly as possible to the variables they quantitatively resolve and predict, which excludes any intrinsic use of cloud, precipitation or subgrid scale data. The measurement transfer function must be quantitatively known within prescribed statistical errors close to white noise and its perturbating factors controlled by the model environment, so as to allow automated integrated quality control and assimilation. Regarding the finer scales the dominant use of computer-assisted subjective and interactive approaches tends to emphasize the high time/space resolution and relax the quantitative requirements on transfer function and absolute accuracy. Systematic or correlated errors due to perturbating effects, proximity of the coast or ill-calibration are tolerated to the extent that contrasts and structures are preserved at the scales of interest, but data must be easy to handle, process and interpret.

1.4.2- <u>Data retrieval</u>. When the warning of extreme, ill-predicted and poorly observed events is at stake, preserving the intrinsic information content of each observation becomes crucial, and turns to a challenge, when non linear and non univocal transfer functions like the radiative transfer equation or backscatter models must be inverted prior to the exploitation of remotely sensed data. The inversion procedures must be physically rather than statistically based and attempt to maintain the original resolution, reject spurious data, filter noise, correct for perturbating factors, and avoid involving incestuously model outputs or other external data of limited accuracy or reliability, used either as initial estimate to an iterative retrieval process, or as a dealiasing or quality control background. Using a model first guess field for those purposes can lead to discard the possibly unique information content of a given observation, or to misinterpret it, when multiple solutions exist to ill-posed retrieval problems. Exemplary in those respects is the Improved Initialization Inversion method developed by Chédin et al. [ref 13] to retrieve satellite soundings without external information originating from numerical models.

1.5- Contribution of spaceborne systems

Geostationary satellites permanently cover the latitudes under 55 degrees with a high temporal repetitivity of about 30 minutes and a fixed observing geometry, all features priceless to the continuous monitoring of short term variability and to the control of measurement artefacts. Unfortunately due to their great distance to the earth they can currently embark only optical or microwave radiometers operating at frequencies above 100 Ghz, as the only ones achieving mesoscale horizontal resolution compatible with moderate antenna size, since the ground resolution achievable from altitude H at a wave length λ using an antenna of size D is:

$$\triangle X = 1.2 \ (\lambda /D) \ H$$

Similarly budget link and resolution considerations preclude the use of active sensors from the geostationary orbit. Yet highly repetitive geostationary visible and infrared imagery at a nadir resolution of about 4 km is a major ingredient of mid latitude weather nowcasting ideally reinforced by polar orbiting platforms, which on the one hand optimally sample and cover the high latitudes, where a given point can be revisited every 1.5 to 2 hours depending on the instrument swath width and on the other hand accommodate from their lower orbits active and passive low frequency microwave sensors providing original parameters or images at a fairly useful resolution.

2- POTENTIAL OF MICROWAVE SENSORS

2.1- Scatterometers

2.1.1. <u>Scatterometers and nowcasting</u>. At 20 to 50 degree off nadir incidence scatterometers measure the normalized radar backscatter cross section (hereafter σ_0) proportional via the radar equation to the power backscattered towards the instrument by the tilted capillary-gravity waves according to the Bragg resonant scattering mechanism. The wind sensing capability implies that the surface roughness should solely respond to the surface wind as a result of various interaction/equilibrium processes. In the absence of fully consistent measurement theory, current semi-empirical backscatter models (1) established from circle flight airborne campaigns relate σ_0 to the neutral surface wind and observing geometry and are calibrated against reference wind data or statistical constraints [ref 14].

$$\widehat{\sigma_0}_{pp} = A(\theta, V) V^{\delta(\theta)} \left[1 + B1(\theta, V) \cos(\phi) + B2(\theta, V) \cos(2\phi) \right] \quad (1)$$

pp, V, ϕ, θ, respectively denote polarization mode, wind speed, aspect and incidence angles.
Due to the biharmonic nature of (1) multiantenna measurements are needed to retrieve the wind vector and averaging and co-locating individual footprint data yield a resolution of about 50 km for the final wind product in the case of ERS1. Kp measurement noise by-products (2) are supplied as the noise standard deviation normalized by the averaged sigma naught, which is the product of the speckle and a thermal contribution signal and sensor dependent in nature.

$$Kp^2 = (1/Neq)(1 + 2/SNR + 1/SNR^2(1 + Tsn/TN)) \quad (2)$$

SNR is the signal to noise ratio, Neq the number of independent measurement samples and Tsn and Tn instrumental integration times for the noisy signal and pure noise respectively.

The nowcasting potential of scatterometry comes from its giving access to the surface wind vector, which is an environmental parameter of major impact at sea either directly or through its action on wave and Ekmann drift current generation, fog dissipation, or ice pack drift. Whathever the shortcomings inherent to single level information, mapping the surface mesoscale wind is of considerable value to the detection and characterization of potentially active mesoscale patterns. Indeed low level southerly jets frequently observed ahead of shallow active cold fronts should be detected for they have scales resolvable by the instrument and are usually embedded into neutrally or unstably stratified air, and have therefore clear surface signatures. Vortical boundary layer winds associated to early development of tropical cyclones can also be imaged, as well as high latitude strong mesoscale winds enhancing boundary layer convergence or latent heat and humidity transfer from the sea, both suspected to trigger or bear signature of CISK-like mechanisms most important at some of the deepening stages of marine explosive cyclogenesis or polar lows. Despite the limited coverage due to the scatterometer single swath, ERS1 fast delivery wind products are expected to be sufficiently internally redundant and dynamically consistent to allow the reliable timely detection of extreme events at high latitudes, when coupled with external cloud imagery and upper level soundings.

2.1.2. <u>Critical issues</u>. Regional internal inconsistencies found in SEASAT scatterometer wind datasets shed light onto residual issues possibly detrimental to the nowcasting potential of scatterometry. The upwind directional trapping observed on ascending passes over the tropical Indian Ocean (fig 6)

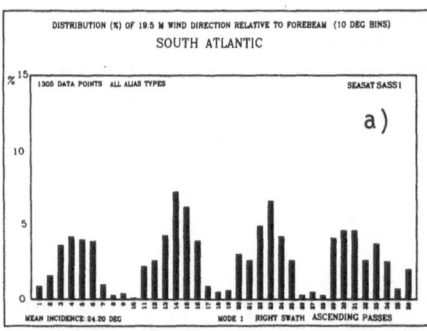

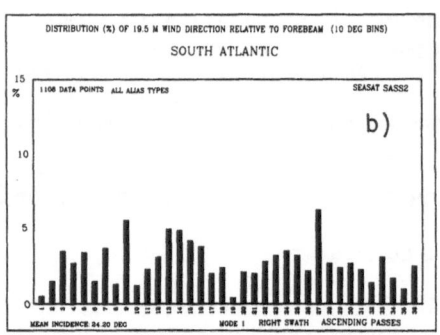

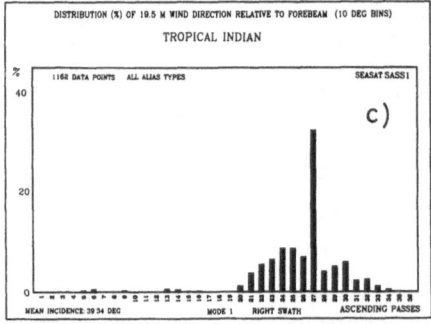

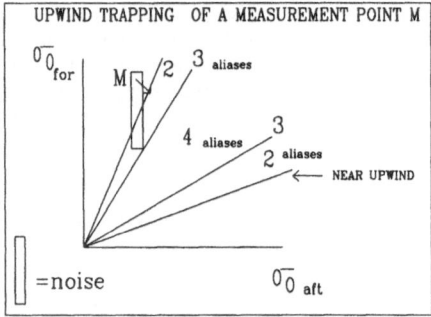

Fig.6: Directional distributions of SEASAT winds referred to the forebeam azimuth. a) and b): Ascending passes over the South Atlantic (SASS1 and SASS2 datasets) c): Ascending passes over the Indian Ocean (SASS1) d): Interpretation of the upwind trapping noticed in c): the non linear retrieval process 'projects' a measurement point onto the upwind edge of the backscatter model domain, when it falls outside due to noise or model underestimated wind sensitivity

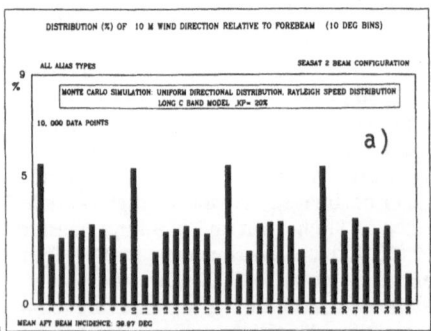

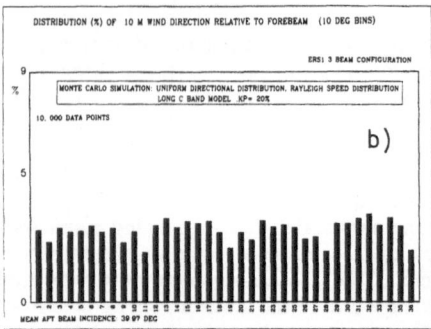

Fig.7: Monte Carlo simulations of noise induced wind trapping based on 20% Kp, the long C band model and a uniform wind distribution: a) SEASAT 2 beam configuration. b) ERS1 configuration.

contrasts with the absence of such upwind winds in the directional distribution over the South Atlantic and reflects the dependence of errors on wind/antennae relative geometry, as expected from the formulation of the backscatter model (1). Such errors are partially masked when ascending and descending passes, several incidence angles and regional data subsets are merged prior to the analysis. Monte carlo simulations assuming a given instrument noise level and various types of backscatter model errors demonstrate that upwind trapping can be induced either by noise or by underestimating the wind sensitivity of the backscatter model, as explained by fig. 6d, whereas the opposite type of error originates in the overestimation of wind sensitivity. Noise induced trapping effects are inherent to two beam scatterometry (fig 7). It should be noted that directional errors are necessarily correlated to speed errors via the backscatter model formulation (1), thus contributing to the distortion of space derivatives, as corroborated by artefact symetrical cross track variations of SEASAT derived slopes of kinetic energy spectra (fig 8). These results show how irrelevant, biases, RMS errors and specifications in those terms can be to scatterometry, and the more so because nowcasting requires that actual wind structures be preserved.

Nevertheless tracing back such complex systematic errors to their genuine cause is tricky, for a variety of factors can combine, propagate throughout the non linear processing chain and degrade the geophysical performances of the instrument. First and besides the previously illustrated instrumental noise effect, inaccurate satellite attitude control, antenna mispointing or heavy rain attenuation, all possibly unproperly corrected in real time can damage the basic sigma naught measurements. Then intrinsic wind sensitivity variations of a backscatter model as a function of wind and incidence can cause distortions in wind patterns throughout the swath. Interestingly the directional sensitivity of the ERS1 C band model [ref 15] expressed in terms of the excess to unity of the upwind to downwind backscatter ratio increases with wind speed and incidence (fig 9a) and is low at the inner edge of the swath under low wind, while speed sensitivity diminishes at high wind despite the noticeable absence of saturation (fig 9b). The very poor directional sensitivity found at most incidence angles for winds below 4 m/s should be aggravated by the increase of Kp governed by low signal to noise ratio according to (2).

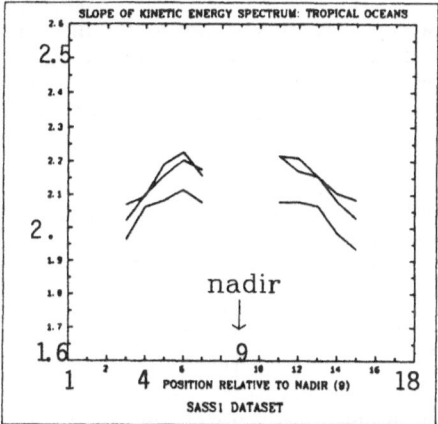

Fig.8: Irrealistic cross-track variations of SEASAT derived slopes of kinetic energy spectra for tropical oceans (Indian, Atlantic and Pacific). Incidence increases on both sides of nadir spot number 9 (Ratier and Roquet, 1987, unpublished).

Additional shortcomings in formulation, calibration and representativeness of the backscatter model result in serious complex systematic errors. The simplified assumption of $\overline{\sigma_0}$ being solely responsive to neutral surface wind is disputed by a number of theoretical and data analysis investigations, hinting at perturbating effects of viscosity, fetch, boundary layer stability, water depth, surface tension, sea state... Using dual polarized co-located SEASAT data Woiceshyn et al. [ref 16] rejected the power law formulation on internal consistency arguments, while Donelan and Pierson [ref 17] arrived at the same conclusion on theoretical arguments consistent with the results of figures 6a and 6c. A subsequent controversial matter is the interface variable that scatterometer measurements should be referred to, thought to be friction velocity, surface wind to wave velocity ratio, roughness length, drag coefficient, neutral surface wind depending on the authors. According to Danard [ref. 18] variations of near neutral surface stability over sharp SST gradients can induce significant boundary layer convergence, which points at the importance of knowing whether stress or wind is measured.

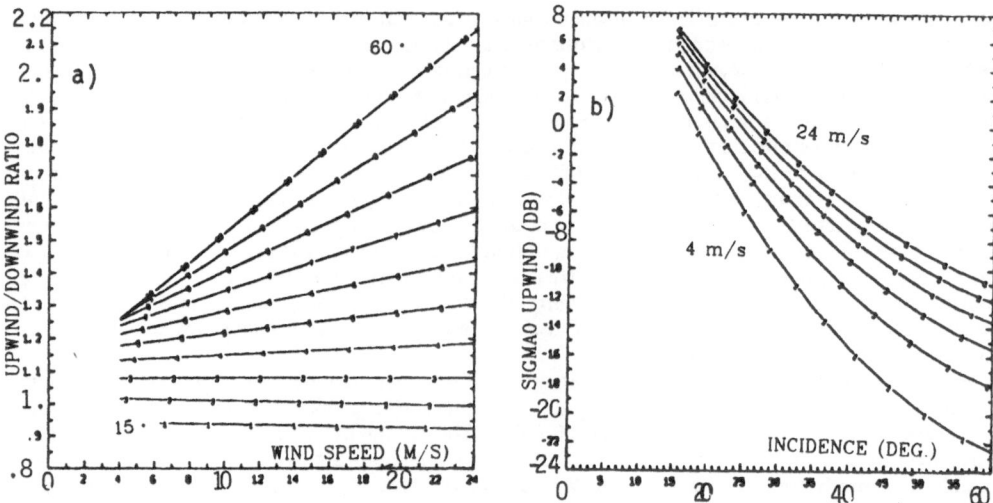

Fig.9: Sensitivity of the ERS1 C band backscatter model a) directional sensitivity (upwind to downwind ratio). Curves are plotted for increasing incidence from 15 to 60 degrees as a function of wind speed. b) speed sensitivity: Upwind sigma naught curves for increasing wind speed from 4 to 24 m/s as a function of incidence (adapted from Long, 1985)

Scatterometer data are calibrated against reference datasets of limited representativeness and the number of tuned coefficients supposed to reflect the sophistication of the backscatter model may yield excellent but paradoxically misleading fitting to site specific environmental conditions or biased reference datasets.

Though apparently irksome listing these possible error sources is worth, for they are undeniably likely to combine with maximum potential impact onto nowcasting, as a result of approximate or limited control affordable in real time and of emphasis on complex severe environmental conditions. In this respect the real time all-weather wind capability of scatterometers, cannot be taken for granted at this stage, and robust wind retrieval algorithms are also needed.

2.1.3. <u>Wind retrieval</u>. The wind retrieval has been so far a two step procedure (3). A local fitting of measurements to the backscatter model function first produces a set of alias wind vectors, including the true one to be selected by the subsequent dealiasing procedure.

$$
\begin{array}{ccccc}
\text{Backscatter} & & \text{Alias} & & \text{Unique} \\
\text{Observations} & & \text{Winds} & & \text{Wind} \\[4pt]
\begin{bmatrix} \sigma_i^0 \\[6pt] Kp_i \end{bmatrix} & \xrightarrow[\substack{\text{BACKSCATTER} \\ \text{MODEL}}]{\text{FITTING TO}} & \begin{bmatrix} V_l \end{bmatrix} & \xrightarrow{\text{DEALIASING}} & \begin{bmatrix} V \end{bmatrix} \quad (3) \\[6pt]
i=1,\ Nant & & l=1,\ Nalias & &
\end{array}
$$

For a 2 beam single polarized instrument the ambiguity problem arises as a result of the non uniqueness of the solution to the minimization process due to the conjunction of measurement noise and the biharmonic nature of the model function. In the 3 beam case two aliases 150 to 210 degrees apart are expected due to measurement noise only except in some very particular wind/antenna geometrical configurations. Contrary to the 2 beam case the residuals to the fit contain valuable information for objectively dealiasing 70 % of the winds. Yet one has to come up with the 30% left, lest error clustering preclude the impact of scatterometry onto the monitoring of rapid mesoscale developments missed by models and poorly sampled by other quantitative observing systems.

The previous discussion stresses the possible negative impact of the combination of instrumental noise, C band wind sensitivity varying across the swath and with the actual wind and poor dealiasing capability of wind retrieval algorithms onto the mapping of mesoscale wind structures of dynamical significance. As an attempt to overcome these difficulties through an integrated approach, Roquet and Ratier [ref 19] developed a variational surface wind analysis scheme, whereby the entire swath field of unique surface wind vectors $\underline{V}$ is directly retrieved from sigma naught and Kp triplets, through the stepwise minimization of a non linear cost function CF involving observation error penalty and smoothness constraint terms:

$$CF(\underline{V}) = \iint_D \left[\sum_{i=1}^{3} (\widehat{0_i 0} - 0_i 0)^2 / Kp_i^2 \widehat{0_i 0}^2 + C1 \, (DIV \, (V))^2 + C2 \, F(V) \, \sin^2(\triangle\phi) \right] dx \, dy \qquad (4)$$

$\triangle\phi$ is an angular continuity variable, C_1 and C_2 two scaling factors scaled objectively and f a function of wind speed. D is a portion of the swath.

Realistic ERS1 simulations led to promising results, as regards accuracy, implicit ambiguity removal, preservation of actual wind structures (fig. 10), and cost efficiency. No wind bias was found whatever the speed range, about 99.5% of the 10,000 wind vectors were recovered to within 20 degrees and all winds above 2.5 m/s to within 45 degrees, while the about 0.25% of 150-210 degree errors on winds below 2.5 m/s were attributed to the mentioned poor C band sensitivity. Importantly no sensitivity was found to the arbitrary initial wind field the iterative minimization process starts from, while the absence of error clustering near sharp wind gradients was confirmed by the investigation of errors induced on wind divergence by the algorithm and instrumental noise,that where found reduced by smoothness constraints (fig. 11).

Contrary to local retrieval methods this global algorithm was designed to best combine and compensate instrumental informations throughout the swath without involving external incestuous model information, in order to preserve the actual wind structures under the circumstances of maximum potential impact like extreme events ill-predicted by the numerical models, thereby meeting a key nowcasting requirement.

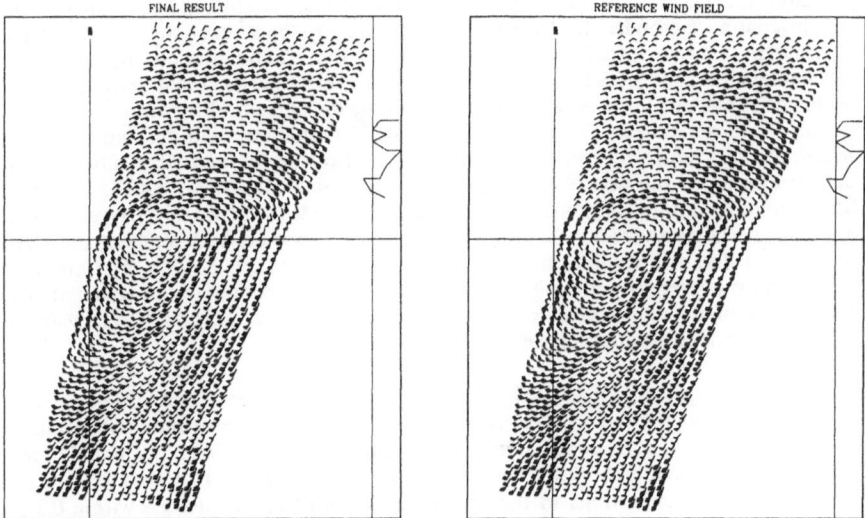

Fig.10: Variational wind retrieval: reference and retrieved wind fields for the most energetic case simulated by Roquet and Ratier (1988)

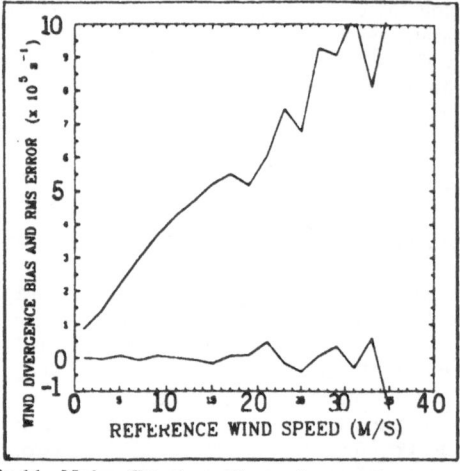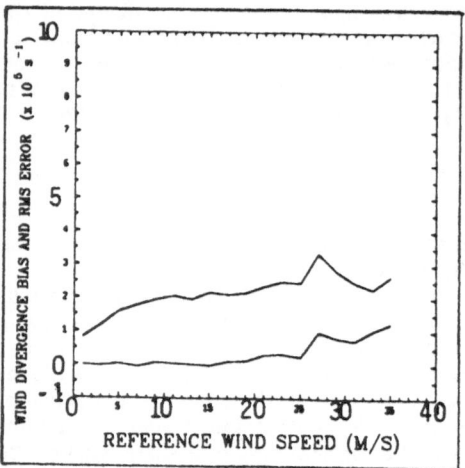

Fig.11: Noise filtering effect of smoothness constraints on divergence: Bias and RMS error on divergence are plotted as a function of wind speed for minimizations of CF with (right) and without (left) smoothness terms, both started from the true wind field.

2.2- Microwave radiometers

2.2.1- <u>AMSU sounders</u>. Infrared tropospheric sounders like HIRS2 suffer from their inability to sound beneath clouds insomuch as complete temperature and moisture soundings cannot be obtained from partly cloudy radiances, without loss of resolution and accuracy, despite indisputable progress in the physical inversion methods. Meteorologically active systems are therefore marginally documented, though thermal vorticity maps derived from the combination of the co-observing Microwave Sounding Unit and HIRS help delineate and characterize partly cloudy active zones (plate 6 a). The crude horizontal and vertical resolution achieved by the current MSU unfortunately poorly compensates the HIRS deficiencies. The 20 channel AMSU radiometer will dramatically improve satellite temperature and moisture sounding both in terms of resolution and all weather reliability, when implemented on board the next NOAA series as two separate subsystems AMSU A and B, the former dedicated to temperature sounding in the low frequency side of the 60 Ghz oxygen band at 50 km resolution and the latter sounding the atmospheric water wapor on both sides of the 183 Ghz water wapor resonant line at a resolution of 15 km. Additional channels at 23.9, 31.4 and 89 Ghz measuring integrated water vapor and flagging rain or cloud ice will complete the all-weather three dimensional mapping of the thermodynamic state of the mesoscale atmosphere by AMSU, most desirable to derive diagnostic products in the spirit of plate 6 a.

2.2.2- <u>Special Sensor Microwave Imager (SSM/I)</u>. At frequencies between 6 and 90 Ghz microwave brightness temperatures outside the 55 Ghz oxygen band are sensitive to surface temperature and emissivity, integrated water vapor, cloud liquid water content and rain. Over the weakly emitting ocean surface integrated precipitable water sensed in the weak absorption line near 22 Ghz can be discriminated from other contributions to the upwelling radiances and estimated within 0.2 g/cm2 accuracy, using simultaneously multiple wavelengths and polarizations, as demonstrated by the SMMR experiments. Surface wind speed within 2 m/s and SST within 1.5 °K and cloud liquid content can similarly be extracted as quasi-linearly related to brightness temperature below 40 Ghz. Microwave rain sensitivity is much more non linear, dominated by absorption/emission and Rayleigh scattering under 30 Ghz and by Mie scattering at higher frequencies. Lower frequencies are less attenuated along the tropospheric path and therefore directly sensitive to raindrop size distributions of heavy rains, contrary to more rapidly saturated high frequencies rather used for light rain estimations. When approaching 90 Ghz, rain sensitivity becomes indirectly determined by the scattering by anisotropic cloud ice particles aloft rain, but less affected by ground emission and hence useful over land and coastal areas.

Calibration and drift problems encountered with the SMMR should be substantially reduced by the improved hardware design of the SSM/I flown aboard the DMSP series. The dropped off SMMR low frequencies at 6 and 10 Ghz preclude SST measurements, but surface wind speed at 60 km resolution should be estimated within 2m/s according to Wentz [ref 20] using the 19, 21 and 37 Ghz channels. Sharing the same 0.65 m antenna the four channels achieve horizontal resolution varying from about 60km at 19.35 Ghz to about 15 km at 85.5 Ghz, that compare differently with scale size of weather phenomena, the high resolution at 85.5 Ghz being compatible with mesoscale widespread rains but still in excess to rain cell size. As a result dealing with partially rainy or side lobe contaminated footprints in the non linear retrieval process combining multichannel brightness temperatures is a serious problem. The deconvolution of the sensed parameters is performed using a crude model in the rain absorption/emission regime that neglects variations in the freezing level and rain layer thickness, and modelling of the scattering regime at higher frequencies is still uncertain, which makes rain rate and cloud liquid water measurements more questionable than water vapor and wind speed measurements.

Whatever these shortcomings, the extended sensing capability of the SSM/I is unique among the current operational sensors and gives simultaneous access to co-located all-weather estimates of atmospheric water and surface wind speed, close to significant weather or relevant to air/sea interactions. The high resolution and coastal applicability of the 85.5 Ghz channel consolidates the SMMR demonstrated capability of mapping widespread frontal rain and moisture [ref 21] (fig. 12) or organized convection and detecting rain at finer scales, more directly and quantitatively than via optical cloud indexing techniques and in the presence of multilayer or multitype clouds. As oulined by Liu [ref 22] integrated precipitable water and surface wind speed will moreover constrain the estimation of latent heat flux, if merged to external boundary layer data.

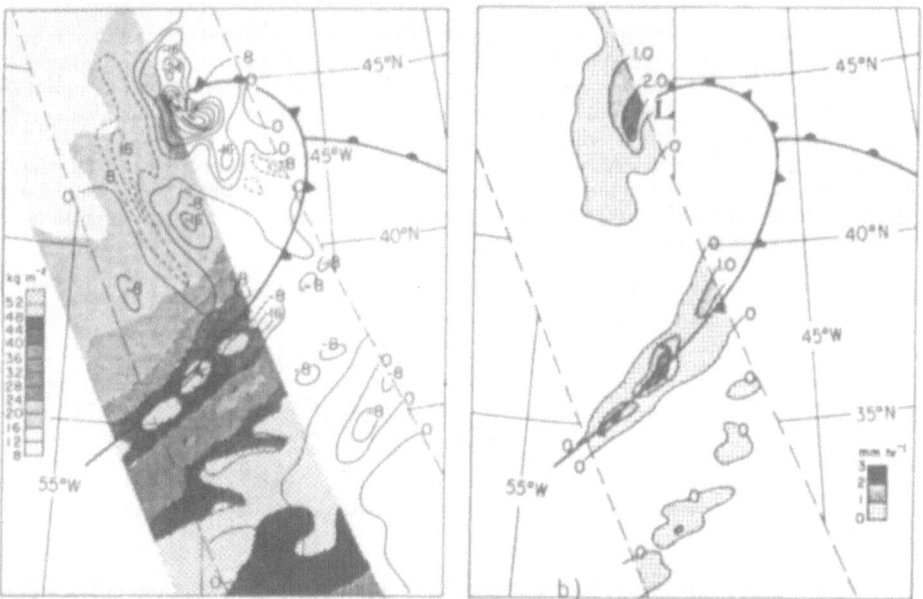

Fig.12: SMMR frontal analysis: a) integrated water vapor plotted together with scatterometer derived wind divergence b) rain rate (from Mc Murdie et al., 1987).

2.3 -Altimeters

The processing of altimeter waveforms on board ERS1 will produce quasi real time estimates of both surface wind speed and significant wave height (SWH) at an along track resolution of 7 km. SWH estimates retrieved from the slope of the leading edge of the signal have been extensively validated against buoy and model data throughout the GEOS3, SEASAT and GEOSAT missions. Based on those assessments ESA claims an accuracy of 0.5 m or 10 % whichever is greater for ERS1.

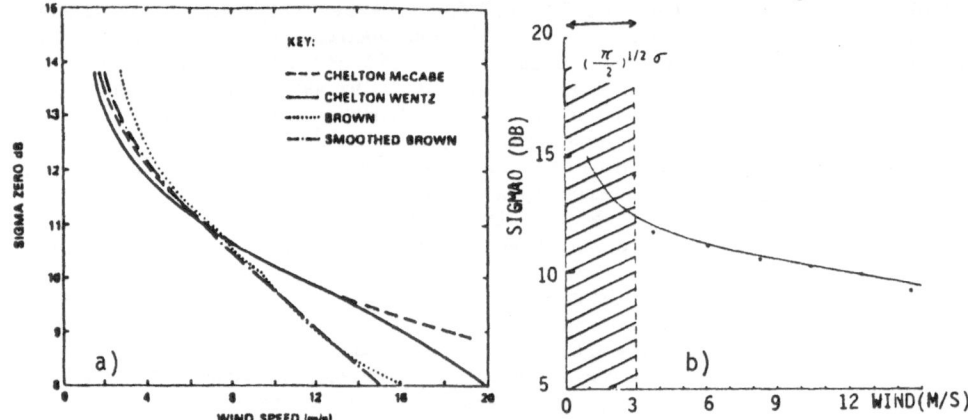

Fig. 13: a) Altimeter wind transfer functions (from Dobson et al. 1987) b) transfer function calibrated against 12 hours of SEASAT scatterometer data, with Chelton and Wentz curve superimposed. The calibration method is biased over the dashed area (from Malardé et al. 1987)

The altimeter wind sensing potential is much more disputable, as illustrated by the number of transfer functions proposed on theoretical and empirical grounds [ref 23], that differ significantly at winds above 15 m/s (fig 13a). Though continuous regional cross-calibration and quality monitoring with ERS1 scatterometer [ref 24] (fig 13b) seems feasible in real time, the key problem will remain the lack of sensitivity to strong winds, with respect to the instrumental noise of about 1 db.

Whatever the interest of surface wind speed and SWH, the coverage inherent to nadir viewing instruments is a strong limitation to the nowcasting potential of altimeters, but is somewhat balanced by the high resolution along track that enables the detection of intersected mesoscale patterns [ref 25], in spite of inaccurate high wind estimates (fig 14).

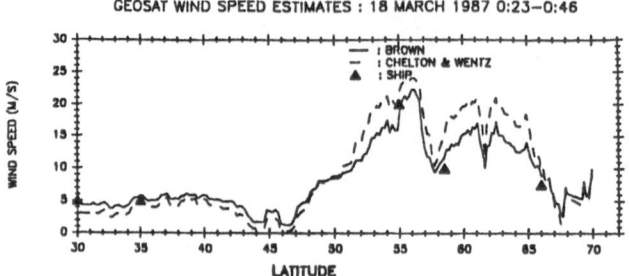

Fig. 14: (From Mognard et al., 1987) Along track wind speed variability observed from GEOSAT

Substracting a fully developed wind sea variance derived from the altimeter wind speed, from the co-located total sea state variance, yields a minimum swell estimate (5) considered as a swell index [ref 26], reliable under moderate wind and significant swell conditions.

$$H_{swell}^{1/3} > \left[(H_{alt}^{1/3})^2 - 16 \times \int_0^{\infty} S_{oo} (Valt, k) \, dk \right]^{1/2} \quad (5)$$

2.4- SAR wave mode

The interest of SAR ocean imaging has been demonstrated by SEASAT and shuttle experiments, but handling high bit rate data and extracting thematic products therefrom in real time remain difficult and seriously impede the applicability to weather nowcasting, despite substantial progress in SAR processing. A noticeable exception is the sea ice detection and tracking application based on the crude processing of raw images, not addressed in the present paper. This somewhat negative statement should not obliterate the SAR potential to depict surface signatures of ocean eddies, fronts and internal waves. The ERS1 wave mode concept circumvents the high bit rate constraint in sampling and recording on board one 'imagette' 6 x 5 km wide every 200 to 300 km alongtrack, for the purpose of extracting wave spectrum estimates over the global ocean. This attractive objective unfortunately has to face a number of obstacles. First the sensitivity is limited by random motions of scatterers to along track propagating wave lengths below a fraction of the imagette size and more critically above a value [ref 27] related to significant wave height $H^{1/3}$ and the ratio of slant range R to the spacecraft velocity V through:

$$\lambda_m \not\!/ 2 \, (R/V) (H^{1/3})^{1/2}$$

Thus the decreasing sensitivity with sea state and spacecraft altitude makes the ERS1 wave mode an azimuth swell detector at 200 to 1000 m wave lengths (fig 15a). Furthermore the non univocal transfer function from the imagette spectrum to the wave spectrum is controversial, although three imaging processes namely geometrical and hydrodynamic backscatter modulations and velocity bunching due to scatterer motion in phase with waves, are recognized to participate in it, the latter being specific to synthetic aperture and responsible for the imaging of along track travelling waves [ref 28]. The wave spectrum cannot be readily recovered under all sea state regimes, because non linearities distort the wave spectral signatures (fig 15b) [ref 29].

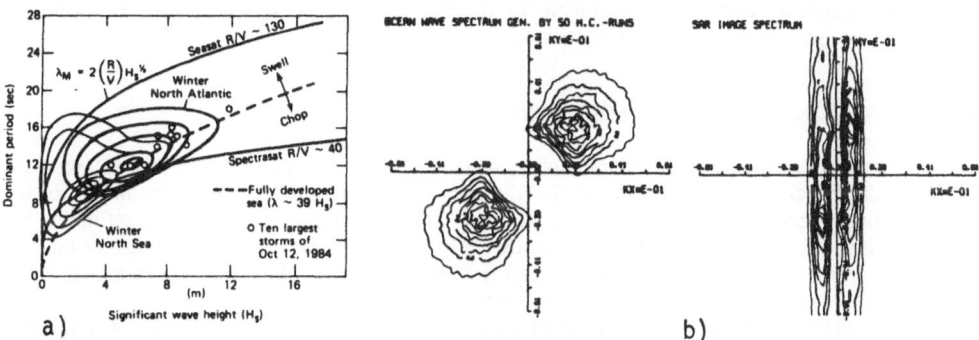

Fig.15: a) (from Beal, 1986) SAR wave detection capability from various altitudes: Joint probability densities (contours 0.1, 0.3, 0.5, 0.7, 0.9) of dominant period and significant wave height for both the North Atlantic and the North Sea, superimposed on predicted SAR azimuth wave number limits for two orbital configurations (solid lines). Also shown are the period-height relationship for a fully developed sea (dashed line) and the corresponding relashionship for ten typically large storms (dots) b): Simulated distortion of a wave spectrum by SAR imaging (From Brüning et al.)

The 180 degree ambiguity on propagation direction inherent to Fourier analysis still reduces the quantitative potential of SAR wave mode used independently of external data, which is possibly limited to estimates of near range propagating wave length, direction of propagation and spectral width of swell peaks. Consequently assimilating such dubious data into wave numerical models may be either risky or of limited interest if the only data in rough agreement with a first guess wave spectrum are accepted, in order to prevent measurement errors from propagating throughout the ocean at the high group velocity and low dissipation rate of swell.

Used as a real time swell detector the ERS1 SAR wave mode will serve some of the primary objectives of marine nowcasting. First the early detection of ill-predicted swells propagating towards a coast line will allow the timely warning and very short range prediction of their arrival at the coast, based on the group velocity diagnosed from peak frequency and local bathymetry. The crucial along track consistency and redundancy will help select the true propagation direction, by exploiting the property of a given swell component to spread rather than converge, and possibly trace back swell to its remote high wind generation zone (fig 16), when refraction is negligible, thus contributing to the early warning of weather disturbances uncovered by a single swath scatterometer, in the spirit of the interpretation by mariners of swell radiated ahead of weather disturbances. Conversely nowcasting will serve later quantitative use of SAR wave mode data, by providing a fairly reliable assessment of the sensor performances over a variety of observing configurations, as gained at monitoring simultaneously data and sea state from the very beginning of fast delivery product dissemination, and to be exploited to develop automated data selection procedures incorporated into wave data assimilation systems.

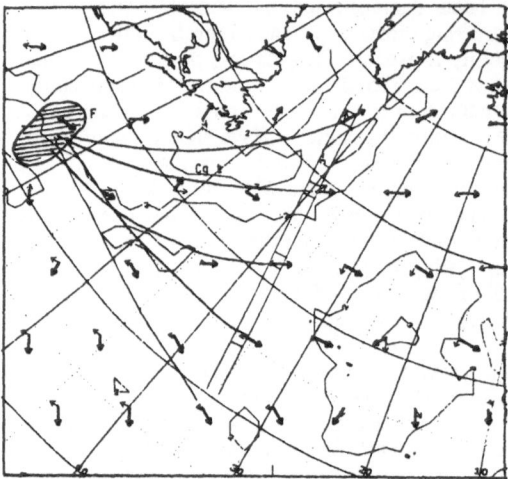

Fig.16: Reconstructing a wave generation zone F from ERS1 SAR wave mode data.

3- INTEREST OF INSTRUMENT COMBINATIONS

The mutual complementarity of microwave sensors clearly stems from the variety and wealth of their sensing capabilities summarized by fig. 17. The provision of surface kinematic and thermal variables, tropospheric moisture/temperature profiles and liquid water and sea state estimates is priceless, when attempting to understand weather systems from the pattern to details and to penetrate the complex coupling involved between the air/sea interface, boundary layer, clouds and upper atmosphere down to their implications in terms of significant weather.

Yet the nowcasting potential would be dramatically increased if such attractive measurements were co-located and simultaneously acquired by the same observing platforms, so as to effectively combine in the spirit of fig 12 a into the three dimensional multivariate atmospheric snapshot enriched by cloud and ground-based imagery needed to constrain the underdetermined nowcasting problem. This instrumental synergy also particularly helpful to constrain the estimation of air/sea fluxes possibly important in rapid developments would promisingly enhance the possibility of associating the sampling/coverage characteristics of polar and geostationary satellites and ground-based systems. In this respect the scrutiny of SEASAT, NROSS and ERS1 merits (fig 18) reveals the potential of the unfortunately cancelled NROSS and of SEASAT due to synergetic instrumental payload and swath overlap, as compared to ERS1 which rather appears a surface wind wave dedicated mission hampered in terms of coverage and reviting rate by a single scatterometer swath. This does not obliterate the uniqueness of ERS1 as regards both flight opportunity and real time capability, but should be taken into account in the design of future meteorological polar platforms and global observing systems.

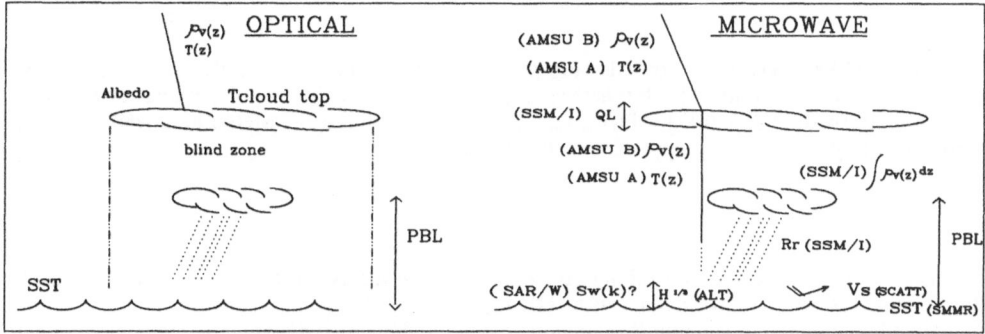

Fig. 17: Optical versus microwave measurement capabilities

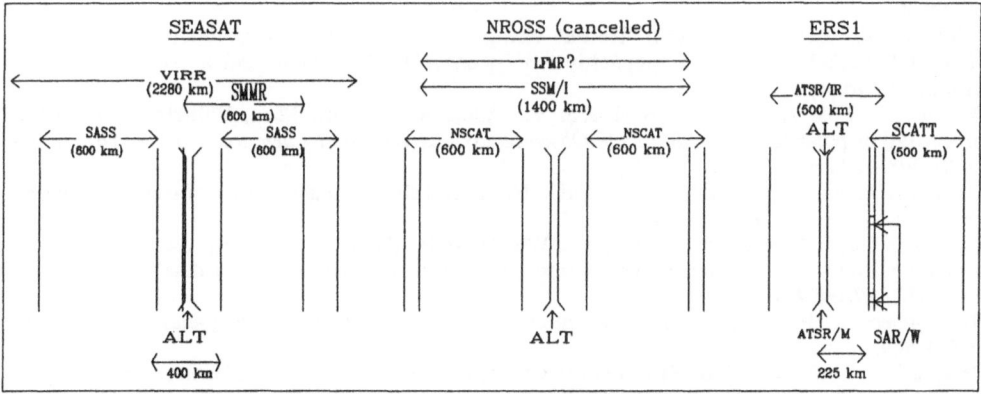

Fig. 18: Instrument swath overlap for SEASAT, NROSS and ERS1

4- CONCLUSION

The nowcasting merits of microwaves have been discussed in the prospect of operational instruments flying on board DMSP, ERS and NOAA orbiters. Contrary to optical wavelengths microwaves access boundary layer wind patterns, sea state and atmospheric liquid water in all weather conditions and at resolutions compatible with atmospheric mesoscale, and consequently document significant weather, air/sea interactions and the coupling between the boundary layer and the upper atmosphere most significant in the dynamics of explosive cyclogenesis. When merged with geostationary, coorbiting and ground-based imagery, microwave data will form comprehensive datasets increasing the number of well-documented extreme events that provide guidance for process analysis or modelling and for deriving reliable criteria for real time early warning. Conversely the possibility for nowcasting to accommodate relative accuracy, regional calibration and uncertain functions will make it a unique test bed for new operational microwave data, that should help delineate their domain of applicability for the benefit of more quantitative automated quality control and assimilation. Experience gained with ERS1 will help take the next essential step of combining microwave and optical sensors on board the same future polar platform so as to provide the wealth of synergetic measurements required to constrain the nowcasting problem and to fully exploit the potential of microwaves.

ACKNOWLEDGEMENT: The author is very grateful to H. Roquet, J.F. Geleyn, M. Jarraud, Y. Durand, F. Huynh, J. Parent , M. Hontarrède from Direction de la Météorologie and to A.Chédin from Laboratoire de Météorologie Dynamique and J.L. Tourte from Météorage, who provided material for this paper as well as appreciated encouragement.

REFERENCES

1. Parent J., Gaffard G., 1986. 'Detection of meteorological fronts over the North Sea with Valensole Skywave radar'. *IEEE Journ. of Ocean. Engin.*, OE-11, N°2, 174-179.
2. Imbard M., Craplet A., Degardin P., Durand Y., Joly A., Marie N., Geleyn J.F., 1987. 'Fine mesh limited area forecasting with the French operational Peridot system'. *Proc. of ECMWF Seminar. The nature and prediction of extra tropical weather systems*, 2, 231-269.
3. Golding B.W., 1984. 'The Meteorological Office mesoscale model: its current status'. *Met. Mag.*, 113, 288-302.
4. Geleyn J.F., Jarraud M., 1988. 'Recent changes in the French NWP operational systems EMERAUDE and PERIDOT'. To appear in *Proc. of 8th AMS Conf. on numerical weather prediction*. Baltimore, February 1988.
5. Fedor L.S., Foss A., Groenass S., Hoem V., Lystad M., Shapiro M.A., Wilhelmsen K., 1984. 'Forecasting of polar lows'. In *Mesoscale Observations and very short range weather prediction Proc. ESA SP 208*, 47-52.
6. Anthes R.A., Kuo Y.H., Gyakum J.R., 1983: 'Numerical simulations of a case of explosive marine cyclogenesis'. *Mon. Wea. Rev.*, 111, 1174-1188.
7. Anderson D., Hollingsworth A., Uppala S., Woiceshyn P., 1987. 'A study of the feasibility of using sea and wind information from the ERS1 satellite. Part 1: wind scatterometer data'. *ESA contract report 6297/86/HGE-I(SC)*.
8. Rasmussen E., 1979. 'The polar low as an extratropical CISK disturbance'. *Quart. J.R. Met. Soc*, 105, 531-549.
9. Roquet H., Ratier A., 1986. 'Storm surge statistical prediction'. In *Extended abstracts of papers presented at the WMO workshop on significant weather elements and objective interpretation methods*, Toulouse, France, 22-26 June 1987.
10. Johnson D.R., 1984. 'The inducement of planetary boundary layer mass convergence associated with varying vorticity beneath tropospheric wind maximum'. In *Mesoscale Observations and very short range weather prediction Proc. ESA SP 208*, 41-46.
11. Scott N.A., Chédin A., Breon F.M., Claud C., Flobert J.F., Husson N., Levy C., Tahani Y., Prangsma G.J., Rochard G., 1988. 'Recent advances in the retrieval of meteorological parameters through the 3I system'. To appear in *Proc of 4th Int. TOVS Study Conf.*
12. Browning, K.A. 1985. 'Conceptual models of precipitation systems'. In *ESA journal 85/2*, 157-180.
13. Chédin A., Scott N., Wahiche C., Moulinier P., 1985. 'The Improved Initialisation Inversion Method: a high resolution physical method for temperature retrievals from satellites of the TIROS N series'. *Journ. Clim. Appl. Meteor.*, 24, 124-143.
14. Wentz F.J., Peteherych S., Thomas L.A., 1984. 'A model function for ocean radar cross sections at 14.6 Ghz'. *J. Geophys. Res.*, 89, 3689-3704.

15. Long A., 1985. 'Towards a C band radar sea echo model for the ERS1 scatterometer'. *Proc. of 3rd int. coll. on Spectral Signatures. ESA/SP-247*, 29-34.

16. Woiceshyn P.M., Wurtele M.G., Boggs D.H., Mc Goldrick L.F., Peteherych S., 1986. 'The necessity for a new parameterization for an empirical model for wind/ocean scatterometry'. *J.Geophys.Res.*, **91**, 2273-2288.

17. Donelan M., Pierson W.J., 1986. 'A two scale Bragg scattering model for microwave backscatter from wind generated waves' *Proc. of IGARSS 86*, vol.1, 291-296.

18. Danard M., 1986. 'On the sensitivity of predictions ofmarine cyclogenesis to convective precipitation and sea temperature'. *Atmosphere Ocean*, **24**, N°1, 53-72.

19. Roquet H.,Ratier A., 1988. 'Towards direct variational assimilation of scatterometer backscatter measurements into numerical weather prediction models'. To be published in *IGARSS 88 proceedings*.

20. Wentz F.J., Mattox L.A., Peteherych S., 1986. 'New algorithms for microwave measurements of ocean winds: applications to SEASAT and the SSM/I'. *J. Geophys. Res.*, **91**, 2289-2307.

21. Mc Murdie L.A., Levy G., Katsaros K.B., 1987. 'On the relationship between scatterometer derived convergences and atmospheric moisture'. *Mon. Weather Rev*, **115**, 1281-1294.

22. Liu W.T., 1984. 'Estimation of latent heat flux with SEASAT SMMR, a case study in N. Atlantic'. In *Large Scale Oceanography and Satellites*, C.Gautier and M. Fieux Ed., 205-221.

23. Dobson E., Monaldo F., Goldhirsh J., Wilkerson J., 1987. 'Validation of GEOSAT altimeter derived wind speed and significant wave heights using buoy data'. *J. Geophys. Res.*, **92**, 10,719-10,731.

24. Malardé J.P., Dumont J.P., Ratier A., 1987 'Testing the feasibility of calibrating altimeter wind speed against scatterometer wind data via a 2 dimensional objective analysis procedure'. Available from the author.

25. Mognard N.M., Campbell W.J., Josberger E.G., 1987. 'GEOSAT surface wind speed estimates and comparisons in the North Atlantic for March 1987'. Paper presented at the altimeter algorithm workshop. Corvallis, Oregon, August 1987.

26. Mognard N.M., 1982. 'Apport de l'altimétrie radar sur satellite a la determination de l'état de la mer'. *Thesis*.

27. Beal R.C., 1986. 'Characteristics of a very low altitude Spacecraft for collecting global directional wave spectra with spaceborne synthetic aperture radar'. *Proc of IGARSS 86*, vol 2, 1025-1029, Zürich, 86-11 September 1986.

28. Lysenga D.R., 1987. 'The physical basis for estimating wave energy spectra from SAR imagery'. *Proc. of the symposium 'measuring ocean waves from space', John Hopkins APL technical digest*, vol 8, N°1, 65-69.

29. Brüning C., Alpers W., Hasselmann K.. 'Monte Carlo simulation studies of a 2 dimensional surface wave field by a SAR'. submitted to *Int. J. Remote Sensing*.

DESERTIFICATION STUDIES
(with emphasis on Nigeria)

D. O. ADEFOLALU
School of Science & Science Education
Federal University of Technology
Minna, Niger State
Nigeria

ABSTRACT. It has been established that five areas of
major remote sensing applications in national development
which have, hitherto, not been seriously utilised in the
planning process of most third world countries are in
activities relating to agriculture, urban development,
pollution control, weather prediction and natural resources
management. In Nigeria, which is a microcosym of most third
world or developing countries especially in West Africa, the
last 28 years as an independent nation have shown that more
problems have been created (perhaps inadvertently) through a
myraid of development projects. Studies have shown that some
aspects of poor performance may be associated with inadequate
data input which could have been better handled if remotely-
sensed data (that are now common place) had been available
for forward planning at her independence in 1960. The
present threat of desertification in the Sudan-Sahel belts of
the country has reached an alarming stage which calls for more
than tree-planting, no matter how vigorously and seriously
this approach has been handled. In this paper, the advantages
of remote sensing techniques in studying drought and problems
of geo-environmental degradation in relation to
desertification trends as confirmed by groundtruth
observations in the country are presented. The microwave
techniques in monitoring declining vegetal cover is the key to
the success of this approach.

1. INTRODUCTION

Within the last two decades at least, the Sahelian belt of
Africa defined to be between latitudes 14°N and 23°N (Glantz
1976) have experienced devastations and destruction by
repetitive droughts (Nicholson, 1981) and flood conditions
although the former has had more world-wide publicity . If
it were possible to quantify the actual costs of human,
animal and plant life losses, it would run to billions of
dollars. And yet, continous as these disaster appear to be,
no serious attempts to control or curb their effects seem
forthcoming. From obvious indications, Governments and
Organisations are contented with paliatives in the way of
relief materials and assistance by the countries that are
'well-off' with rhetorical undertones which border on
political considerations - even in the provision of such aids
whose effects, albeit genuinely motivated, can never result
in any lasting solution to the problem (Adefolalu, 1976).

R. A. Vaughan (ed.), Microwave Remote Sensing for Oceanographic and
Marine Weather-Forecase Models, 273–323.
© 1990 Kluwer Academic Publishers.

The inadequacies of relief efforts apart, there is a
definite problem of lack of data on both atmospheric
circulation patterns and geo-environmental indicators of
desertification in the entire Sahel sub-region, even if there
is the 'will' or desire to tackle the problem from basic
principles (Obasi, 1987).

Despite abundant knowledge and tested theories on
the role of atmospheric circulation and its variability in
relation to climate and its anomalous patterns of drought and
flood, Governments tend to concentrate on natural resources
monitoring to a fault (Obasanjo, 1978). The Nigerian
situation, which has been expatiated upon by Oseni (1978,
1980), clearly shows that poor execution of well defined
programmes to rehabilitate the environment may be the key
factor in poor management of the environment while reliable
data are not available at times of need. However, problems
relating to intensification of the desertification process in
recent times as postulated by various researchers have shown
that reliance on normal conventional data for future studies
will be futile (Otterman, 1974; Charney et al, 1977 and
Adefolalu 1981, 1983a). This is because the processes of
desertification are multi-faceted and inter-related to the
extent that a concerted attempt at studying them must make use
of data at different space and time scales as explained below.

2. DESERTIFICATION - A definition.

A simplified formulation representing a 'summary' definition
of desertification as postulated by Adefolalu (1981) is in
the form

$$DROUGHT + MAN = DESERTIFICATION$$

By <u>drought</u> is meant the non-availability of adequate amount
of water for man, animals and plant (growth, development and
yield or maturity) <u>as</u> and <u>when</u> needed. 'MAN' the other
'variable' on the left-hand side represents all activities of
man that bring pressure to bear on vegetal cover (leading to
'decimation'), land resources - which result in degradation
due to over-exploitation or poor management etc. These
combine in a variety of ways to alter the <u>cycle</u> of climatic
variability to the extent that some changes in total geo-
environmental resources now appear irreversible.

Thus, desertification cannot (and should not) be studied
from the context of just one of the variables on the left-
hand side of the above equation. They are inter-related by
way of some important interactive processes that arise from
the sub-components of each variable. These are illustrated
in the Organogram shown in Figure 1. Basically, drought is
composed of three types meteorological, Agricultural and
Hydrological droughts.

While meteorological drought refers specifically to
lower than normal rainfall amounts over a specific period of

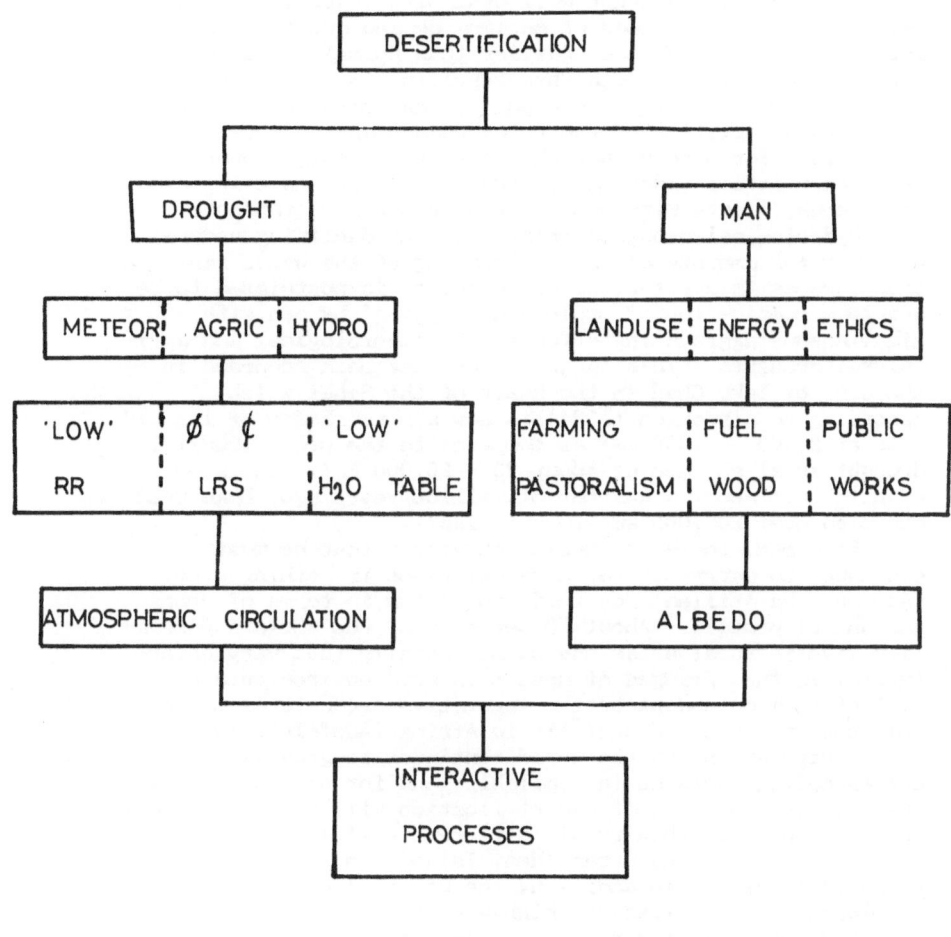

KEY

ϕ = ONSET DATE OF RAINS   METEOR = METEOROLOGICAL

$\cancel{\phi}$ = CESSATION DATE OF RAINS   AGRIC = AGRICULTURE

LRS = LENGTH OF RAINY SEASON   HYDRO = HYDROLOGICAL

Fig. 1. An organogram on component arms of desertification.

time, agricultural drought will deal with insufficient soil
moisture surplus at times of maximum demand due to either
late onset of the rains or earlier than normal cessation
dates of the rains or both thus resulting in a shorter than
usual length of the rainy season. It has been found that
these characteristics of the rainy season may be more
'stressful' for plants than shortage of absolute amount of
recorded rainfall (Adefolalu, 1986). They are more injurious
to seasonal plants with short (or near-surface) taproots.

Hydrological droughts relate more to declining underground water amounts and hence lowering of the water table. A
draw-down as this effect is often called is postulated to be
due to a combination of prolonged 'tapping' by man with
simultaneous aggravating effects of meteorological and agri-
culural droughts. This 'aggravation' has been recorded in
the case of Lake Chad in the heart of the Sahel which
according to Sircoulon (1984) is now a mere shadow of its old
size at 2,300 - 2,500 km² as compared to the pre - 1969/1973
drought areal surface of about 23 - 25,000 km². It is worthy
of note that about a quarter of million years ago, Lake Chad
occupied over 200,000 km² (Grove, 1967).

With MAN, there are basic activities that he must
continue to carry out for mere existence if nothing else.
Land must be 'tilled' for food production in terms of crops
and animal protein. About 60 per cent of the World's population live in rural areas and coincidentally (but very unfortunate) in the fringes of desert in arid environments with
most of them dependent on the vegetative products as their
sole energy source, especially in Africa (Adefolalu 1981).
As the population of the world continues to grow with no
corresponding increase in available land for development, a
third major component of desertification will be land ethics.
The dichotomy of urban/rural 'differentials' is known to be
the main cause of the urban 'Heat Island' in relation to
climatic 'State'. In addition, the use of land areas for
development (communication, highways etc) is gradually
reducing the rural environment to an alarming degree.

All of these are directly linked to increase Albedo
values either on local, regional or global scales. The
consequence of increased albedo is a direct reduction in the
effective heating power of the sun which as shown in Figure 1
is the major second arm of the Interactive Processes in the
Desertification chain of 'actions'. If any study of desertification is to have any credibility, the data input must
be such that can adequately 'capture' the role of each of the
above sub-components of the two major variables of desertification - Drought and Man. It is a complex problem as may be
visualised in Figure 2 on the inter-relationships between
climate, vegetation and man. It is believed that the interactive processes are at four different trophic levels with
man at the fourth and highest level as shown. Being at the
top implies two things - man is the largest consumer and he
'directs' all the activities at the second and third trophic
levels while he shares with mother-nature the distribution of

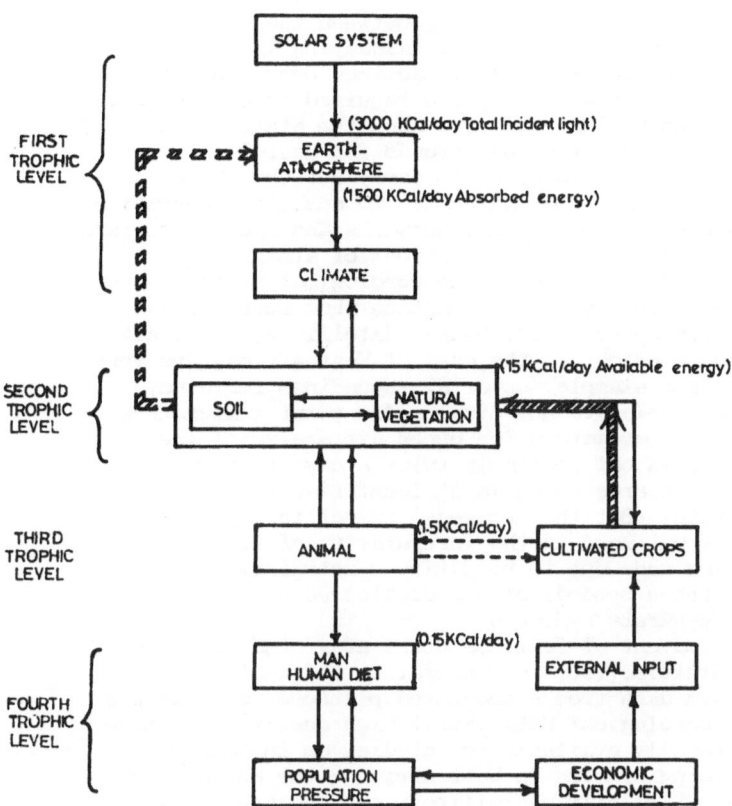

Fig. 2. Inter-relationships between climate, vegetation and man (from Adefolalu, 1983a).

the input 'energy' through coupling between the second and
first trophic levels. There is no gainsaying it therefore
that, with all these actions and reactions occurring sim
ultaneously, tremendous efforts are required to obtain useful
information on both diagnostic (to know the State-of-the-art)
and prognostic (understanding trends for useful
projections) studies. In order to carry out the above
successfully, it will be necessary to combine observations
from surface-based, conventional networks and special surface
or space-based platforms. Observations of atmospheric
variables available at Synoptic Meteorological Stations need
to be combined with similar information from such special
platforms as air-borne Radar, Buoys, Satellites, Air-crafts,
etc to fill gaps which, in the case of West Africa, are very
wide indeed. For example, upper air soundings for wind,
temperature and relative humidity or dew point temperature
(the three basic parameters for upper air analysis) are
supposedly carried out routinely twice a day at about 8 - 10
locations over an area which is at least five times the size
of Great Britain. But the 'tragedy' is not this problem of
sparse network but more in the irregularity of actual
soundings being made due to budgetary constraints and non-
availability (when needed) of expendables such as balloons or
chemicals to generate hydrogen.

With the advent of fast computers at ground receiving
stations to digitise (for interpretation and analysis)
remotely sensed data from space-based platforms such as the
METEOSAT (Meteorological Satellites) high quality data have
become more readily available for studies on the major
weather producing systems in West Africa - the Monsoon and
African wave. Similarly, satellites such as LANDSAT and
SEASAT and air-borne Radar have the capabilities for
monitoring physiographic parameters including vegetation,
soils and underground (geological and hydrogeological) structures and resources especially water. These developments are
already showing good promise in attempts to study and explain
the intricate aspects of the desertification trend in the
Sahel. In subsequent sections, recent advances on the utilisation of data generated by one of such space-based platforms
are documented while some problems that may arise from over-
dependence on them are equally highlighted.

3.   MONSOON DYNAMICS AND DROUGHT IN WEST AFRICA

3.1. Preview

In the Tropics, anomalous patterns of precipitation (annual
amounts and variable distribution in time and space, etc)
within the last two decades have aroused much interest.
Various studies have given the cause(s) of such anomalies in
West Africa as ranging from the failure of the monsoon and
tropospheric synoptic-scale features to the increasing albedo
effects due to denuded surface vegetation which will result

in lower thermal heating needed for convection (Charney et al, 1977; Adefolalu, 1984b).

The important point that becomes obvious is that the convective clouds on which tropical precipitation depends "suffered" some reverses and the role of the monsoon in this regard - be it in the supply of moisture or in enhancing sub-regional instability - needs further clarification.

In relation to the global monsoon, three aspects can be associated with precipitation dynamics in West Africa. Being a seasonal wind system, the climatological features of the monsoon suggest that there are two rainfall seasons - dry and wet. In general, it is dry from mid-October to mid-April while the wet season covers the other half of the year. Although the bulk of the rains (80 per cent of the annual total) is received during the latter season in the rainforest and Sudan-Savana zones (4° - 12°N lat.) the remaining 20 per cent is nonetheless significant, especially in the double-cropping belt, south of about 10°N.

The seasonality of West African rainfall as a function of the migratory monsoon made the climatological weather forecasting method very popular up to the sixties (Adefolalu, 1972). And, the most popular approach to weather forecasting then was the "YESTERDAY" method, as it was called (Schove, 1946), which was based on the weather zones as an indicator of the dominant weather changes to be expected (Hamilton and Archbold, 1945; Walker, 1954 and Adejokun, 1966). Since the most important phenomena then were precipitation types, their occurrence was linked with each weather zone (Aspliden and Adefolalu, 1976). However, inter-annual atmospheric circulation changes result in dramatic variations in the surface weather as quantified by total monthly rainfall in August at two locations in West Africa. Figure 3 shows that the coastal town of Lagos along the Gulf of Guinea (lat 6.5°N) has August rainfall patterns with-out-of-phase relationship with those of Niamey (13.5°N) in Niger Republic - right in the 'heart' of the Sahel. And yet, both areas are 'fed' with moisture by the same monsoon basic current. This aspect of the monsoon which results in contrasting summer rainfall patterns bring to focus its second feature - its circulation and component features in the lower troposphere (below the 700 hPa level) namely:

- the moist but rather cool southerly air with a west to southwesterly component which forms a wedge under;

- a dry and relatively hot northerly air with an east-northeasterly component.

The obvious consequence of this arrangement is the creation of a surface discontinuity which is now called the Inter-Tropical Discontinuity in West Africa (ITD). As confirmed by Lamb (1979) there has been a continuing southward decline of the monsoon influence due to an overall equatorward retreat from about 22.5°N in 1950 to about 20°N three decades later.

280

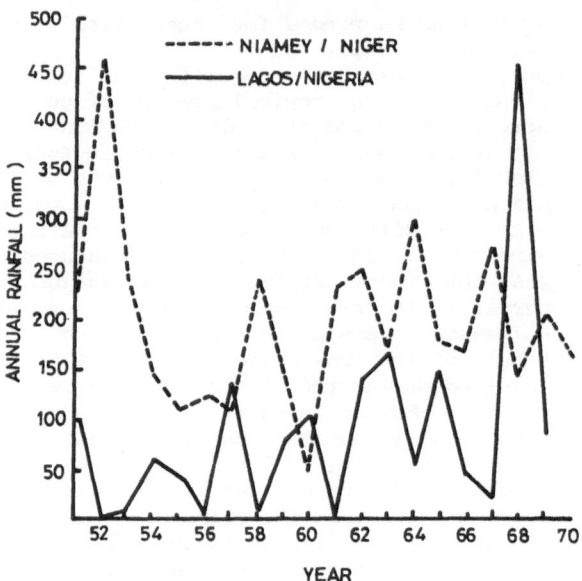

Fig. 3. Monthly rainfall in August at Lagos and Niamey (from Adefolalu, 1973).

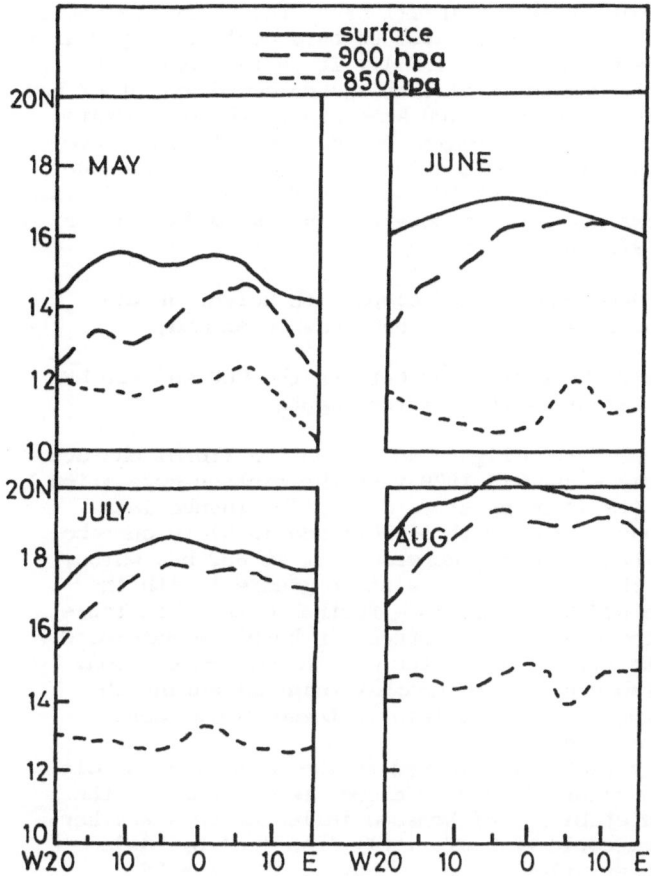

Fig. 4. Mean positions of the surface intertropical discontinuity and the monsoon trough (MT) of the 900 and 850hpa levels over West Africa during WAMEX (from Adefolalu, 1985a).

The dwindling total yearly rainfall since the early sixties in the sub-region is no doubt related to this decline, especially in the Sahel (Menekaya et al 1980). It is interesting to note that apart from the fact that the poleward extent of bad weather stops at about 200 - 400 km south of the east-west axis of the ITD, further studies by Adefolalu (1984a) have now confirmed that its wind-confluence zone at both the 900 and 850 hPa levels, known as the Monsoon Trough (MT) is also migratory. These are clearly observed in Figure 4a - d for May to August 1979 which show:

- the acknowledged equatorward slope with height of the monsoon layer from surface to 850 hPa. level and;

- the relatively sharp gradient between the 900 and 850 hPa latitudinal extent of the moisture depth.

It will also appear from Figure 4a - d on the ITD/MT characteristics in (May - August) 1979 that the earlier suggestion by Flohn et al (1974) on moisture divergence (hence less convective clouds to the north of latitude 15°N) is corroborated by the patterns of the monsoon axis at 850 hPa which, apart from May 1979, maintained a steep gradient with the position at the 900 hPa level (see further details in Table 1). This implies a definite decrease in both the pole-ward extent and depth (vertical) of the moisture-laden monsoon. Hence, the variable nature of monthly rainfall during the summer of 1979 must owe its origin to these tropospheric changes in the monsoon flow.

Were this variability limited to the acknowledged poleward decline in rainfall amounts alone, (see Figure 5) the problem of predictability of Monsoon influence as a weather-producing system could be comfortably 'handled' by basic conventional observation network. But the WAMEX - 79 experience highlighted above revealed that monthly precipitation pattern could have longitudinal aberrations as illustrated in Figure 6a - d showing departures from mean monthly rainfall in Nigeria during the months of May to August 1979. How can this 'singularity', coupled the latitudinal variations (illustrated in Figure 5), be studied on the basis of poorly (vertically) stratified data? It is obvious that in a sub-region where there are less than 10 upper air stations, monitoring of the various indicators of the monsoon influence, as reflected by the lower level tropospheric features of the monsoon trough and precipitation patterns, cannot be confined to conventional surface-based observations. There is thus the need for atmospheric monitoring by non-conventional data platforms to augment the available data by normal meteorological networks in West Africa for the following reasons:

i) Wave-types

To buttress the case for a complete 3-dimensional monitoring

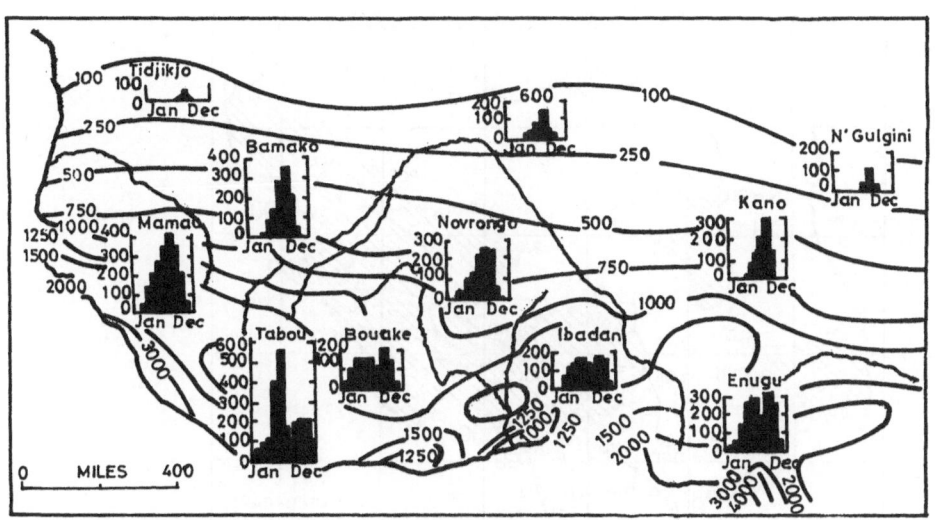

Fig. 5. Mean annual (isohyet) and monthly (histogram) rainfall in West Africa (from Glantz, 1976).

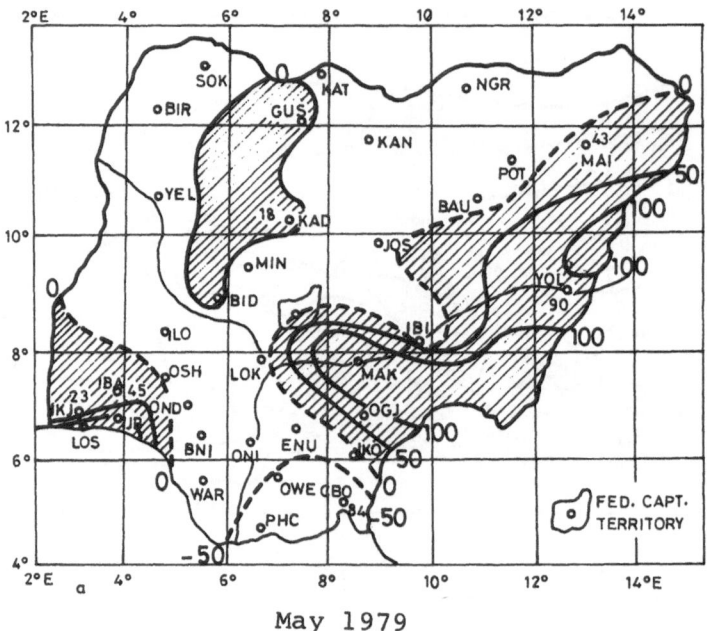

May 1979

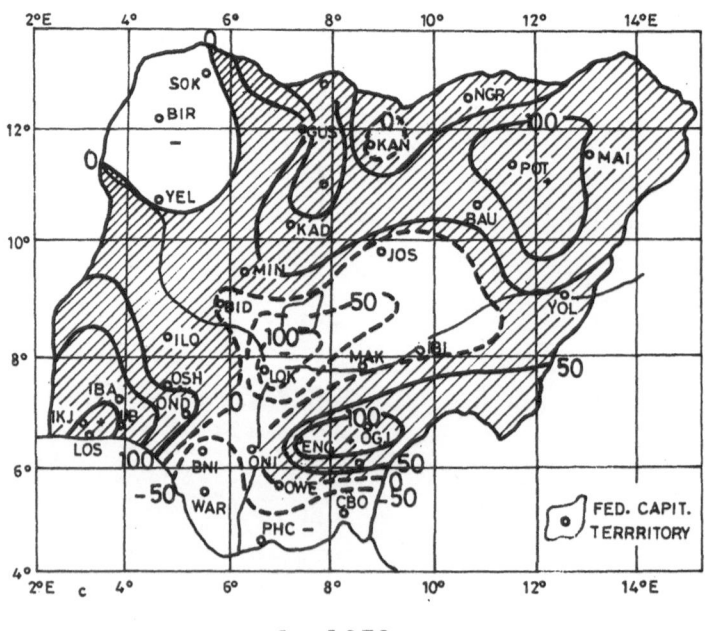

July 1979

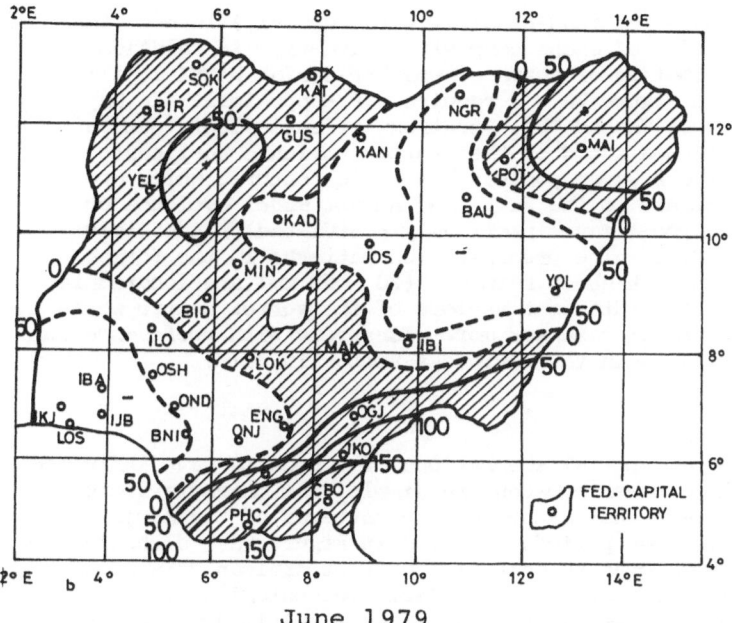

June 1979

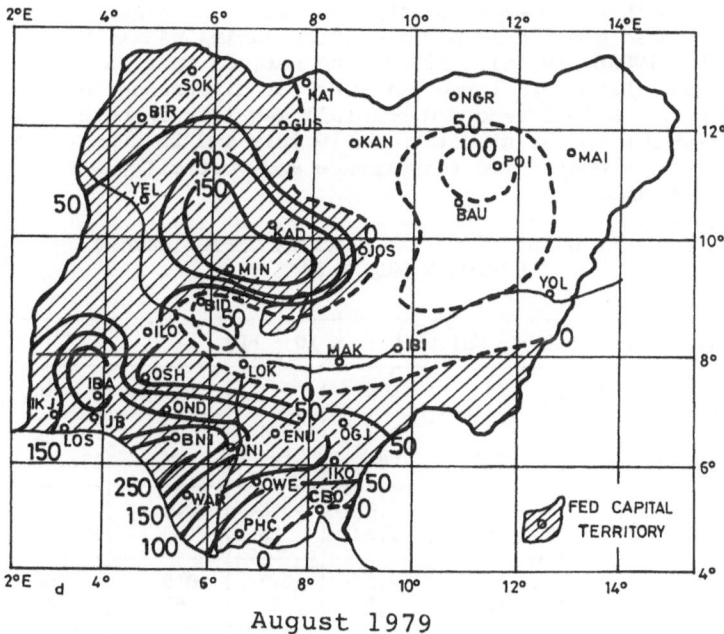

August 1979

Fig. 6. Patterns of rainfall deviation from average values based on data for 1941-80 (from Adfolalu, 1984a).

of tropospheric (and possibly stratospheric) layers, the
peculiarities of each composite wave-type in relation to
the distribution of surface weather during WAMEX as illustrated in Figure 7a - d require a re-appraisal. In May and
June (Figure 7a and 7b), centres of anticylonic outdrafts
are positioned ahead of the vortex, cylonic vortices(c).
These Type-B waves portray kinematic features which are in
sharp contrast to the July and August model cyclones or
synoptic Type-A waves previously postulated by Adefolalu
(1974, 1977). The precipitation patterns in the former
show areas of highest rainfall (20 - 50 mm) are located in
the south to south-west sectors of the vortex centre while in
July and August rain was more widespread with preference for
highest values in the southeast sector in July.

ii) Wave Character

If the above spatial aberrations of composite-wave influence
during the summer monsoon are simple, the characteristic
features of actual synoptic-hour patterns of the African wave
during the four periods selected in Figure 7a - d should be
convincing enough to confirm that West African data gathering
should take advantage of available technologies. In Figure
8a - d in four panels for (a) 15 May, (b) 12 June,(c) 19 July
and (d) 25 August, respectively, the compartments show
stream-lines, divergence, vorticity and 24-hour rainfall
associated with the passage of the four waves in Figure 7a -
d that produced widespread precipitation during WAMEX-79
(Adefolalu, 1983b, 1985a). It will be noticed that on 15 May
and 12 June, there were large patterns of divergence and
vorticity, yet actual recorded rainfall was less in both days
as contrasted with the situation on 19 July and 25 August when
these kinematic quantities (divergence and vorticity) were
less in magnitude.

TABLE 1: VERTICAL GRADIENT OF THE MONSOON DEPTH ($\Delta\Phi/\Delta P$)
DURING WAMEX - 79

(a): (1000 - 900 hPa) layer Units: Lat. x 10^{-2} hPa

Month	Gradient	20W	10W	O	10E	Mean
May	Lat.	-2.0	-2.4	-1.2	-0.7	-1.58
	km.	200	264	132	77	173.3
June	Lat.	-1.6	-1.6	-0.6	-0.0	-1.20
	km.	286	176	66	0	132.0
July	Lat.	-1.6	-0.8	-0.7	-0.5	-0.90
	km.	176	88	77	55	99.0
August	Lat.	-1.8	-0.8	-0.6	-1.05	-1.05
	km.	198	88	110	66	115.5

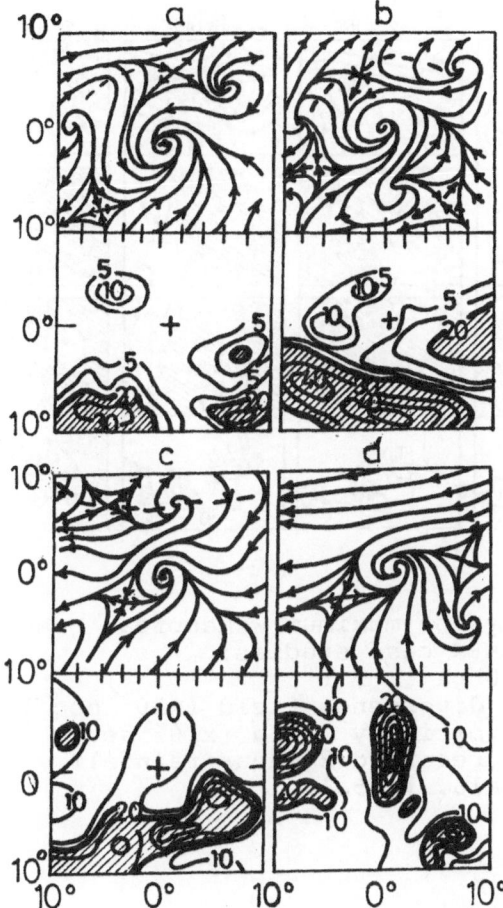

Fig. 7. Composite of waves at 850hpa and observed
precipitation for:
(a) 11-17 May 1979,   (b) 11-17 June 1979,
(c) 16-23 July 1979,   (d) 21-27 August 1979
(from Adefolalu, 1985a).

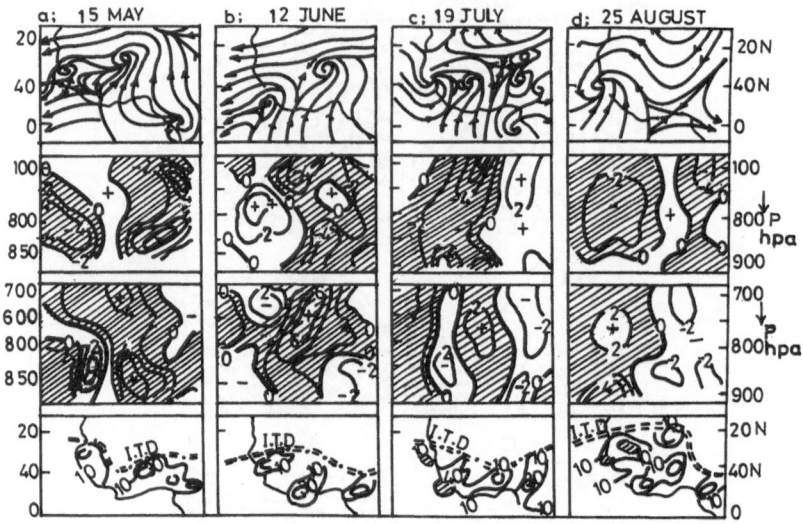

Fig. 8. Days of maximum windspread precipitation
for the 4 WAMEX case studies:
top row - 850hpa
second row - divergence field (x10^{-6} sec^{-1})
third row - vorticity field (x10^{-5} sec^{-1})
fourth row - recorded 24-hour rainfall
(from Adefolalu, 1985a).

(b): (900 – 850 hPa) layer

Month	Gradient	Units: Lat.x 2.10^{-2} hPa				Mean
May	Lat.	-0.3	-1.4	-2.3	-2	-1.55
	km	33	154	253	24	170.5
June	Lat.	-1.7	-5.0	-6.0	-5.	-4.5
	km.	187	550	660	58	495.0
July	Lat.	-2.2	-5.6	-4.3	4.6	-4.18
	km.	242	616	473	506	459.3
August	Lat.	-2.0	-4.4	-4.0	-4.2	-3.65
	km.	220	484	540	462	401.5

Note: The last column shows the average north-south retreat
of the MT. Observe in (b) that, except in May, the 850
hPa position of the MT was approximately 400 - 500 km
south of the 900 hPa location.

iii) Unstable Monsoon Basic Current

The above differences in precipitation patterns with respect
to each wave suggest that the monsoon basic current is condi-
tionally unstable. Garstang et al. (1967) defined this kind
of instability in terms of change in sign of the vertical
gradient of the mean static energy Q_t written as:

$$Q_t = C_pT + gz + Lq \qquad \qquad - \quad - \quad (1)$$

where

$$C_pT = \text{dry enthalpy}$$
$$gz = \text{geopotential}$$
$$Lq = \text{latent enthalpy.}$$

Pettersen (1946) had earlier shown that

$$C_p\theta e = C_p\theta + Lq \ (1000/p)^k \quad - \quad - \quad (2)$$

But $\quad C_p\theta = C_pT + gz \qquad \qquad - \quad - \quad (3)$
i.e. $\quad C_p\theta e = C_pT + gz + Lq \ (1000/p)^k - \quad - \quad (4)$

It has been established that the coefficient $(1000/p)^k$ is nearly unity at lower levels where q is large but is large at higher levels where q is zero (Garstang et al, 1967). Hence, for all practical purposes

$$C_p\Theta e = Q_t = C_pT + gz + Lq - \quad - (5)$$

Thus, the criterion for conditional instability is usually written in the form

$$\frac{\partial \Theta e}{\partial P} < 0 \; ; \; cp \text{ is a constant}$$

Further analysis has confirmed that the troposphere over West Africa fulfils the above criterion as illustrated in Figure 9 on latitude - height cross-section of mean Θe values in which, at lower levels (below 700 hPa approximately) and south of latitude 12.5°N

$$\frac{\partial \Theta e}{\partial P} < 0$$

Since changes in the vertical profile of mean Θe is indica tive of potential convective processes, bad weather should be associated with large vertical Θe gradients while fair weat- her may be linked with weaker gradients. Indeed, for mean seasonal rainfall, drought conditions during northern summer in West Africa appear to be associated with the above charac- teristic feature of the mean Θe structure as illustrated in Figure 10 a - c for Niamey (Adefolalu, 1972, 1973). In both 1961 and 1967, the summer was normal with average monthly rainfall but in 1963, it was a relatively drier summer season (see Figure 3). It is therefore possible to explain droughts in West Africa through the mean-state characteristics of the monsoon basic current especially since $(\partial \Theta e/\partial P)$ at lower levels (which determines conditional instability) is very much dependent on 'Lq' - the latent-heat component with 'q' representing the moisture content.

Apart from this significant thermodynamic feature of the monsoon, it has also been confirmed that it is both barotropically and baroclinically unstable due to the existence of horizontal and vertical wind shear associated with the African Easterly Jet (AET - see in Figure 11 a - e). Calculations of the meridional gradient of absolute vorticity $(\partial \bar{p}/\partial y)$ written in the form:

$$\frac{\partial \bar{p}}{\partial y} = \frac{\partial f}{\partial y} - \frac{\partial^2 u}{\partial y^2}$$

have been carried out in the latitude band most favourable for wave formation in West Africa. Results showed that the

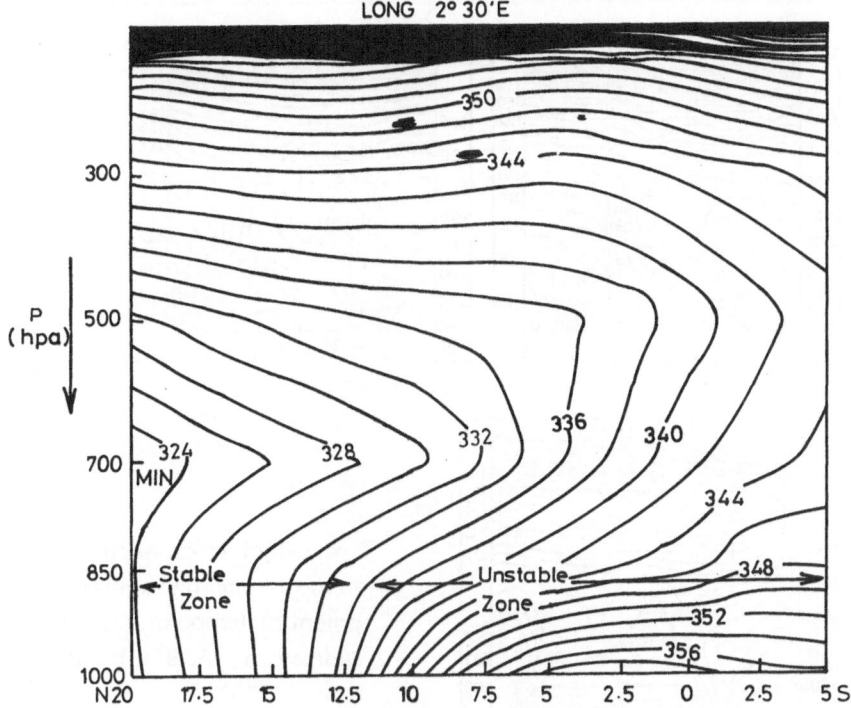

Fig. 9. Latitude–height cross section of mean equivalent potential temperature ($\bar{\theta}$ e) along longitude 2° 30 in West Africa.

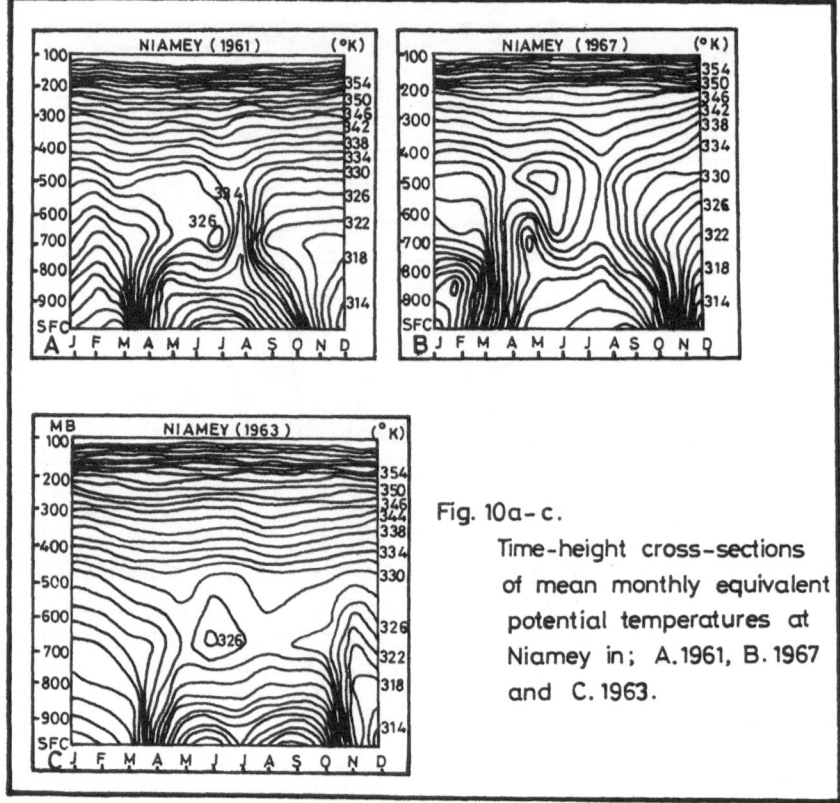

Fig. 10a-c.

Time-height cross-sections of mean monthly equivalent potential temperatures at Niamey in; A.1961, B.1967 and C.1963.

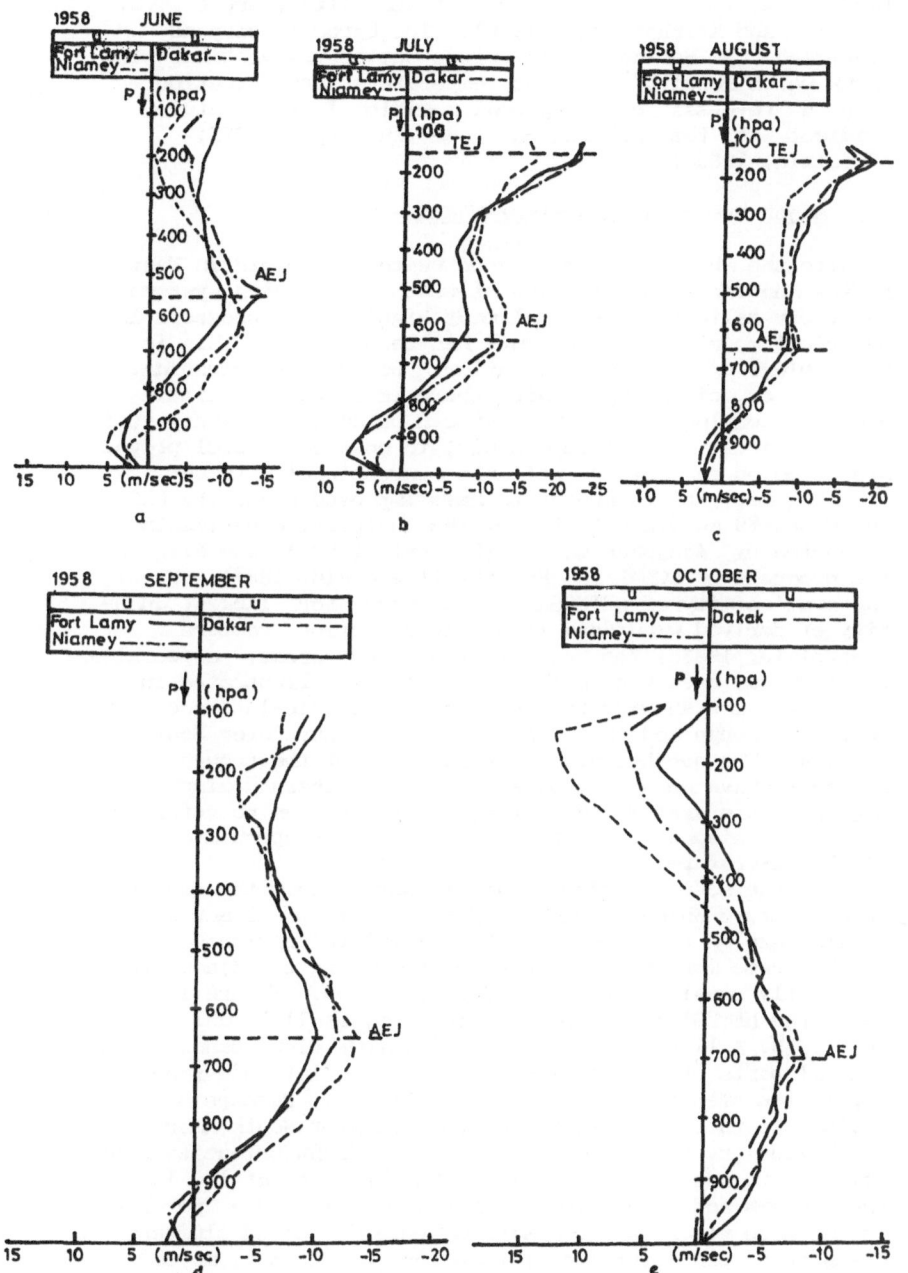

Fig. 11. Vertical profile of mean zonal component of the wind speed (u) over West Africa (Adefolalu, 1974, 1983d).

zonal wind velocity profile satisfies the necessary condition
for the existence of barotropic instability (Burpee 1972;
Pedgely and Krishnamurti, 1976). The baroclinicity as
confirmed by Hassan (1971) arises out of the vertical wind
shear which results from the temperature stratification
with a reversal of the global meridional temperature
gradient in the sub-region (see in Burpee, 1972;
Adefolalu, 1985a).

3.2 Major Break-through Since WAMEX - 79

Despite the abundant work done on weather synopsis in West
Africa and the identification of weather producing systems
and their salient features already highlighted, progress in
numerical modelling experiments has been minimal. This is due
to problems of data initialisation. Because of sparse data
network, especially upper air, there is often very little
vertical continuity in time that can be assured. Thus, it is
very difficult, if not impossible, to do a successful prognosis based on the not-so- reliable conventional data-set.

However, things appear to have improved since the FGGE
and WAMEX-79 as demonstrated by the numerical experiments
conducted by Adejokun and Krishnamurti (1983), and diagnostic products of ECMWF in Reading (Adefolalu, 1984). First,
the availability of METEOSAT data made for largest quantity of derived wind data from cloud motion vectors
(Adefolalu, 1983c) for example the ocean areas, especially
the tropical area was well represented as illustrated in
Figure 12a & b showing the intrusion of middle-latitude
westerly trough and the ridge of high pressure over South
America. The north-eastward translation of the trough appeared to have been controlled by this southern hemispheric
high pressure system. Hitherto, there has been no reference
to inter-hemispheric interaction of the type speculated here
in the literature.

On the actual surface weather associated with such upper
level feature which strengthened the sub-tropical anticyclone
at its wake on the 15th May, 1979, low-level cyclonic vor-
tices became more vigorous and the resulting cloudiness was
distinctly squally with heavy precipitation. Figure 13a
shows the METEOSAT (IR) picture (copied) at 11.30 GMT on 13th
May, 1979. A link between the low latitude cloud bands (in
central parts of West Africa) and a middle latitude frontal
depression over southern Europe is obvious. Perhaps not so
obvious is the link with the cloud band over South America
with cloudless sky in-between. But the NE/SW orientation of
all cloud bands should be noted. The response of low latitude systems and convection to this middle latitude "forcing"
became more vigorous from 14th May onward. Cloud clusters
which developed later (see Figure 13b on 15th May) moved
westwards as disturbance lines suggested by the sharp front
edges of the cloud echoes until typical monsoon cloud patterns were established on 17th May (Figure 13c). Comparison
of Figure 13c with Figure 13d for 18th July, 1979 shows

18d

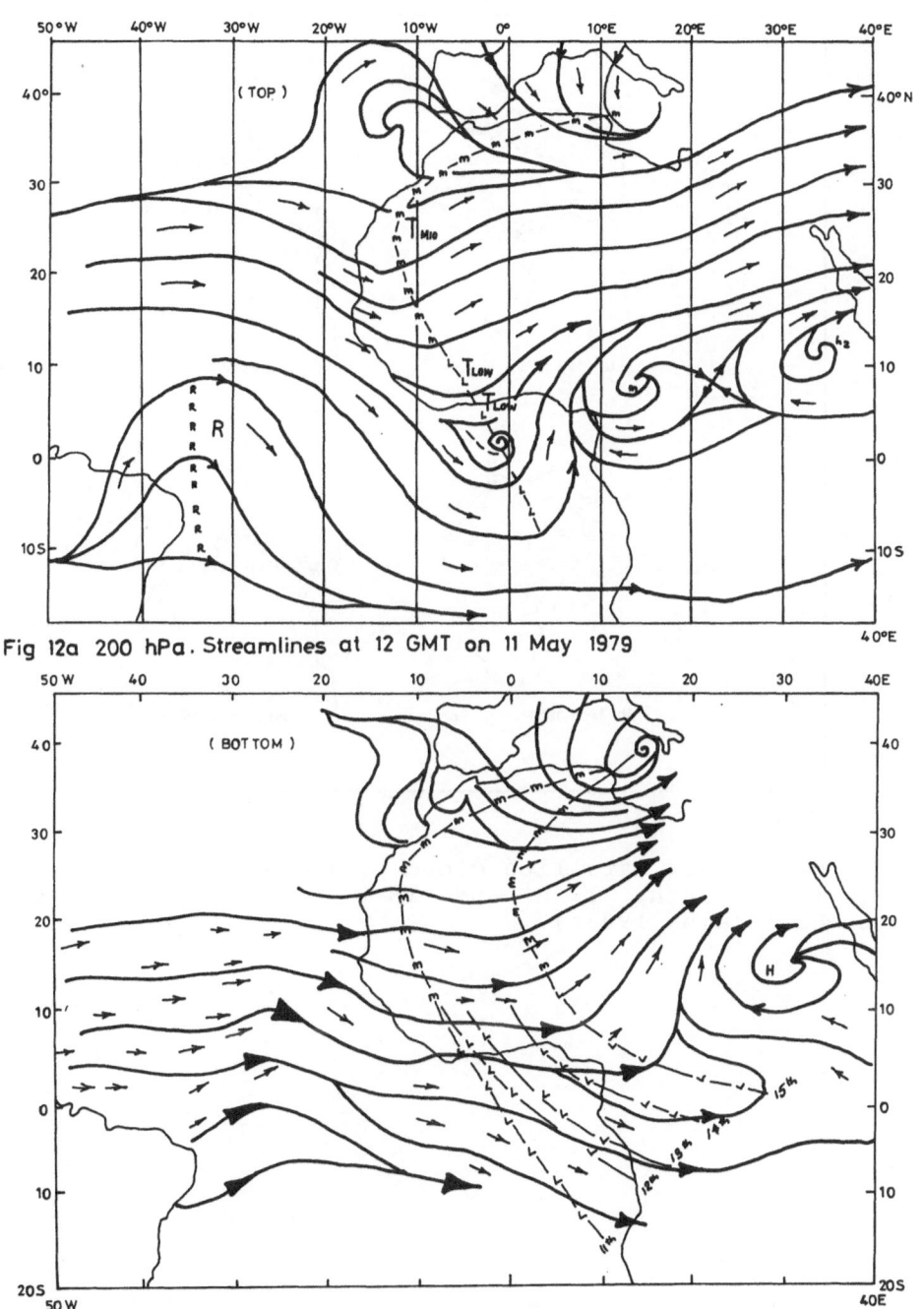

Fig 12a 200 hPa. Streamlines at 12 GMT on 11 May 1979

Fig. 12. 200hpa streamline at 12 GMT on 11 May (top) and 13
May (bottom). Positions of the Trough axis (TLOW) are marked
from 11-15 May (from Adefolalu, 1983c).

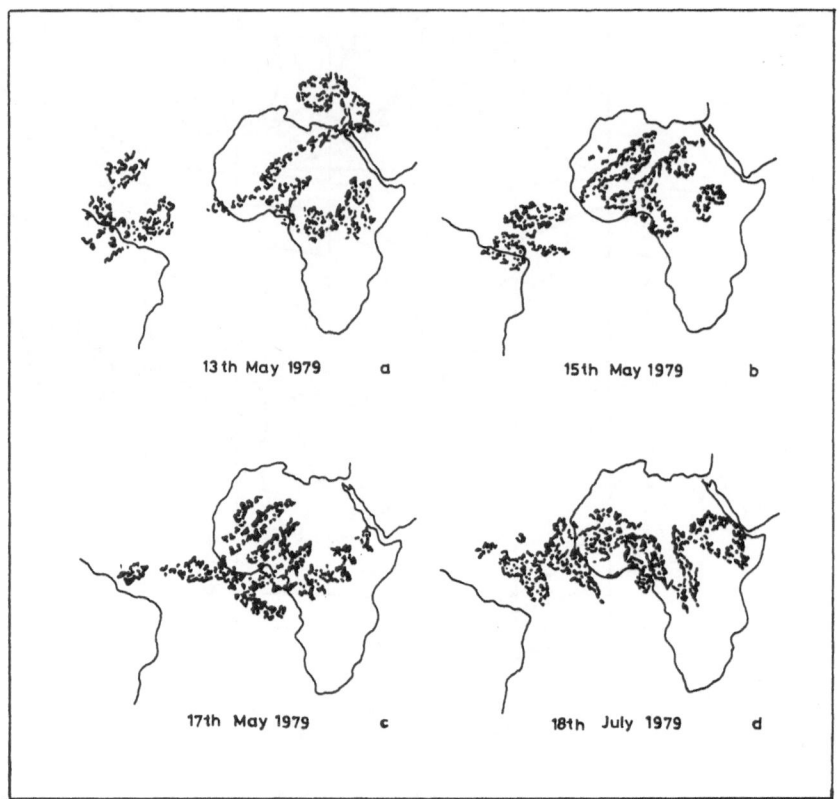

13th May 1979   a

15th May 1979   b

17th May 1979   c

18th July 1979   d

Fig. 13. Satellite cloud pictures at 1130 GMT on 13th, 15th, 17th and 18th May 1979 (from Adefolalu, 1983c).

similar patterns in areal extent and east-west orientation of cloud clusters the latter being typical of peak monsoon cloud pattern in West Africa.

The significant aspect of these cloud clusters is that they developed in the area of upper level diffluence over South America and West Africa. The low-level convergence associated with these upper-level anticyclonic circulation patterns is illustrated in Figure 14 for the 850 hPa level on 15th May 1979. The continued and widespread convective cloud build-up occurring after the upper-level diffluent features have moved out of the West African sub-region suggests that it is the initial stage of monsoon 'surge' that requires a 'trigger' and convective instability is ensured in the presence of strong diabatic heating and abundant moisture content.

Such a trigger may be related to other features apart from large-scale synoptic systems as highlighted above. Two such features are the sub-regional AEJ and the Tropical Easterly Jet (TEJ). It was observed during WAMEX - 79 that:

i) while the latitude of maximum rain increased as the latitude of both the TEJ and AEJ increased, the inverse relationship is observed between the latitude of maximum thunderstorm and the AEJ (see Figure 15a, b and c).

ii) when the vertical wind shear is related to mean rainfall and latitude band of thunderstorms, the curves are complex as may be observed in Figure 15d.

3.3 Problems and Prospects

Thus, it is obvious that there still are very many 'unknowns' in the balance of 'forces' operating in West Africa and if the problems of droughts (and also flood) are to be solved, there is need to acquire high-resolution data that will rely on appropriate technology - especially remote sensing. It is with respect to these non-conventional data acquisition platforms that the micro-wave technique leads itself to further exploitation, especially in relation to sea-surface temperatures (SST) which determine the buoyancy of the moist layer in the ocean - air boundary.

Although results are not yet conclusive, the role of moisture convergence in the lower layers during northern summer, is decisive in the occurrence of either drought or flood conditions in the Sahelian belt of West Africa. The devastations caused by the 1972 - 73 peak drought years compare with the destructions due to flood in the same belt in 1988. Hence lack of or abundant rainfall must be studied as a single phenomenon and paliatives in the way of relief aid should stop. If this is to be the case, there is no alternative to the availability of observational data platforms to monitor the environment.

As already alluded to, during the 1979 Global FGGE Observation year with the WAMEX monsoon sub-programme from

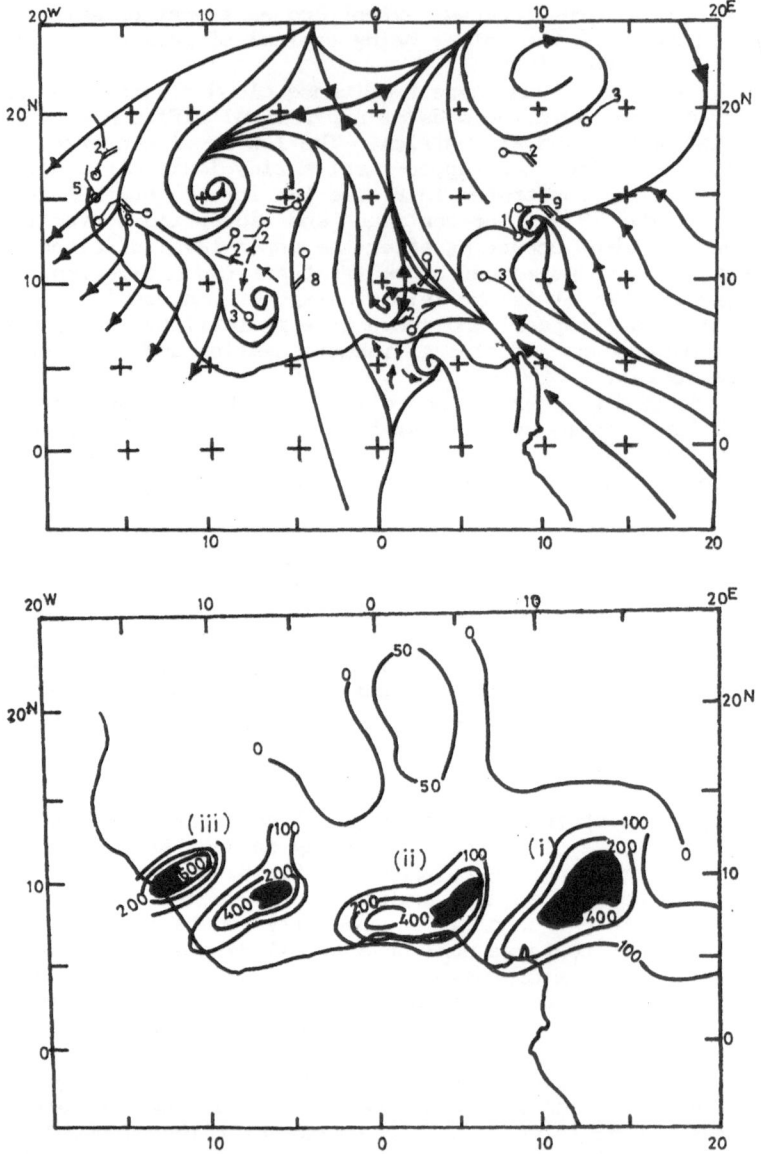

Fig. 14. Top - 850hpa flow at 06Z, 15th May 1979.
Bottom - Rainfall (x10⁻¹ mm) on 15-16 May 1979. Notice the
three "centres of action" (i) Cameroun/Adamawa Mountains,
(ii) Nigeria Highlands and (iii) Futajalon Highlands.

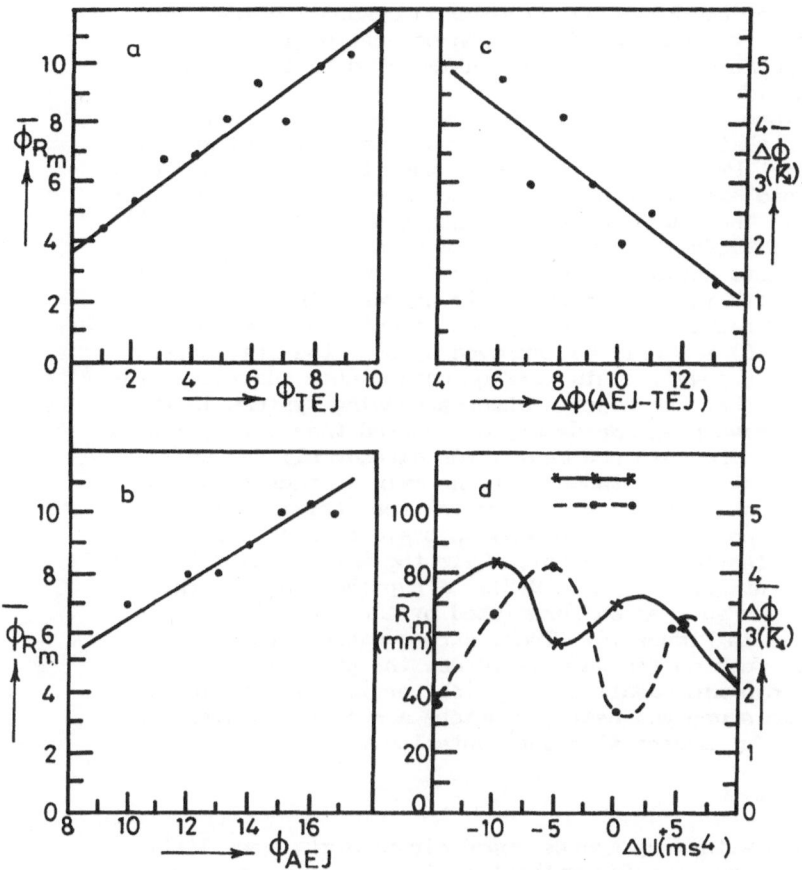

Fig. 15. Tropospheric jets/surface weather relationships in West Africa during WAMEX.
a - mean latitude of maximum rainfall against latitude of TEJ
b - mean latitude of maximum rainfall against latitude of AEJ
c - mean latitudinal width of thunderstorm band against latitude difference of the AEJ and TEJ
d - Vertical shear ($U_{max}^{TEJ} - U_{max}^{AEJ}$) against latitudinal width of thunderstorm band and maximum rainfall R
(from Adefolalu, 1984a).

May to August complementary to Global Monsoon Studies, the
special efforts made to archive non-conventional data from
special platforms especially the METEOSAT data-sets have
yielded dividends.

Analyses to-date based on these and other data-set
have shown that with adequate observational network, Numerical Weather Prediction (NWP) in the tropics is not only
feasible and expedient but could become reliable on operational basis. The numerical experimentation of Adejokun (1985)
and Assamoi (1985) hold promise in West Africa while the
ECMWF products could form the basis for inter-regional exchange. But with inadequate upper air network, the present
forecasting technique employing the 'Yesterday' method
(Schove, 1946) may only be improved upon by the statistical
method of 'analogy'. This will be unfortunate and retrogres-
sive noting that the major weather-producing systems in the
sub-region have been adequately documented (Adefolalu, 1985a)
and Data requirements are known. As already alluded to,
previous studies have shown that he main weather producing
system in West Africa during summer - the African wave -
derives its energy from the monsoon basic current which is
baroclinic (Adefolalu, 1985b), while the wave is said to have
barotropic origin (Burpee, 1972). A hypothetical energy
cycle for the sub-region formulated by Adefolalu (1985b) is
illustrated in Figure 16. Thus, a reasonable approach to
prognostic work on the wave is to use the primitive set of
the hydro-dynamic equations in which vertical stratification
of wind, moisture and heat parameters are pre-requisites for
any meaningful numerical experimentation.

3.3.1. Winds

In a region with very sparse upper air network, very little
progress can be expected without serious attempt to provide
back-up platforms to augment conventional data.

Presently, there are tested techniques on deriving
winds from cloud motion vectors (Balogun, 1985). But a major
snag is the difficulty of specifying the actual pressure
level for these cloud vectors. A simple illustration is
Figure 17 in which, at best layers (but not levels) can be
identified. For NWP models with over 10 levels at which fluid
properties are to be defined, this technique leaves much to
be desired although low- and high-level derived wind data may
be used to augment 850- and 200-hPa levels in tropical models
with considerable success (Adejokun, 1985).

It seems appropriate therefore to examine the utility
values of methods that will give precise information such as
the Range-Height Indicator (RHI) radar technology in which
micro-wave 'transmission' and 'reflection' of signals form
the basis of indentifying targets such as clouds. But perhaps the best alternative to convetional data in weather
prediction is to describe the future state of the
atmosphere by monitoring 'trends'of clouds development -
especially precipitation rates.

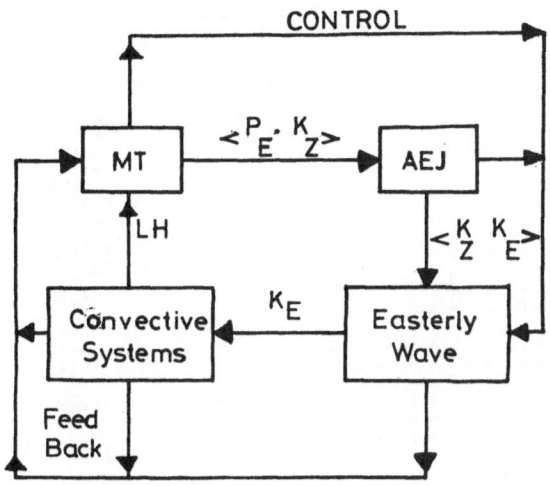

Fig. 16. The energy cycle in West Africa:
MT - meridianal oscillation of the monsoon trough
AEJ - African easterly jet
P_E - Eddy potential energy
K_z - Zonal kinetic energy
K_E - Eddy kinetic energy
LH - latent heat release at high levels
(from Adefolalu, 1985b)

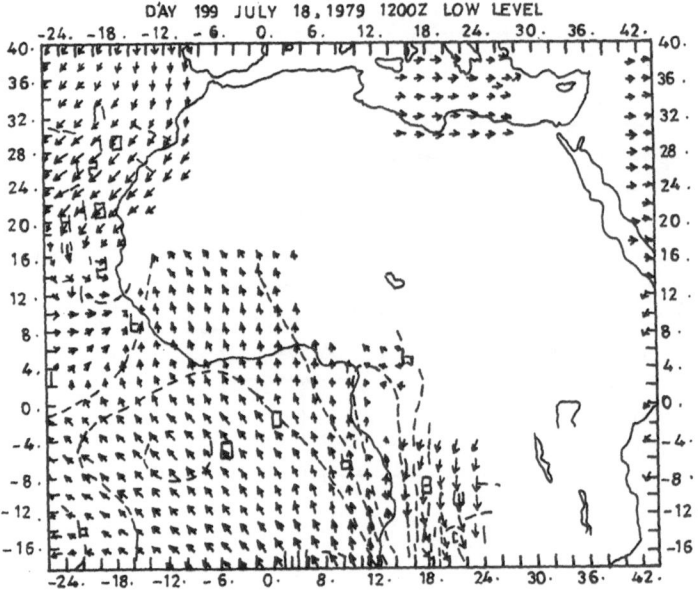

DAY 199 JULY 18, 1979 1200Z LOW LEVEL

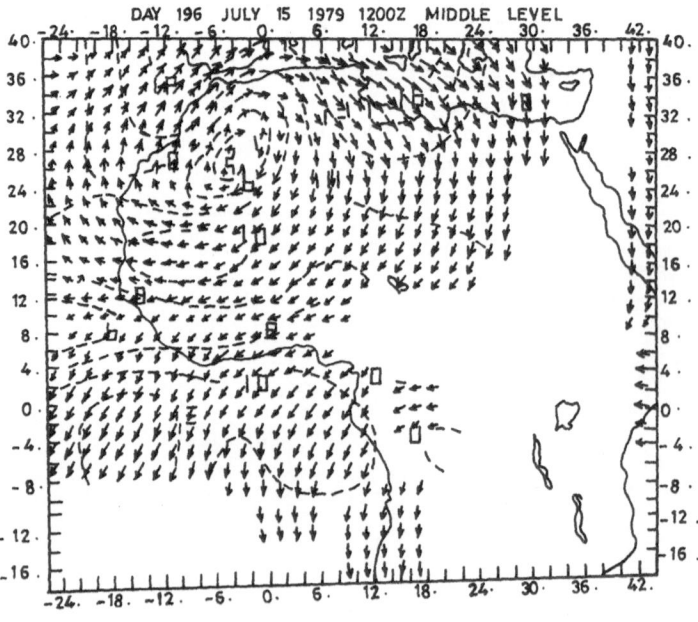

DAY 196 JULY 15 1979 1200Z MIDDLE LEVEL

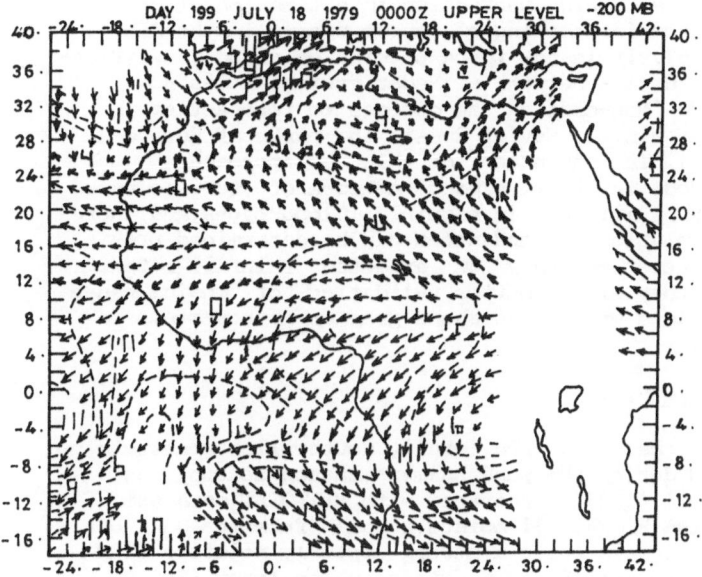

Fig. 17. Derived winds from satellite cloud and moisture motion vectors during WAMEX-79.

This approach is nearing perfection in estimation of precipitation rates and forecasting of severe storms in the Sahel
(Dugdale, 1986). It is now to be precise on cloud identification. With a decent network of Radar system, it is now
possible to do 'Now-casting' as opposed to forecasting in
such places as U.S.A., Soviet Russia and parts of Europe
(Battan, 1982). The advantage of this new approach
over NWP methods is that a good network of surface-based
Radar network is all that is required. Cloud types,and therefore their evolutionary tendencies can be determined precisely with better accuracy on expected rainfall rates.
Furthermore, they give a better validation of medium range
weather forecasts which have shown greater improvements in
recent times.

3.3.2. Moisture Flux

At low levels (below 700 hPa level), southerlies penetrate up
to about 20°N during the peak summer monsoon season. As
Cadet and Nnoli (1987) discovered with METEOSAT data set
analysed at the ECMWF, all moisture comes from the Gulf of
Guinea at low levels while a strong easterly moisture flux
(associated with the AEJ at mid-troposphere layer (700 to
500 hPa) originate from Central Africa. It is shown that
this is divergent and may be of little consequence in the
precipitation process. Another interesting finding by the
group in relation to budget studies is that evapotranspiration over the belt north of the Gulf of Guinea is very
important in the re-cycling and poleward transport of water
vapour in West Africa.
    Thus, to further understand the dynamics of rain-
producing systems in West Africa, which will lead to
successful simulation of drought or flood situations, a
network of micro-wave weather radar will be invaluable in
tracing the evolution of the monsoon in West Africa through
continuous monitoring of cloud types and their genesis. So
far, convective adjustment procedure adopted in tropical
models have not been able to successfully account for total
precipitation although cloud areas are properly defined.

3.3.3. Heat

The estimation of sensible heat in the primitive equation set
of any NWP model which includes diabatic heating is problematic
because of poor estimations of surface temperature. Except
over water, the use of the Bulk aerodynamic formula in any
form (no matter how well modified) will remain unsatisfactory
(Adefolalu, 1985a; Pearce, 1985). It seems the non-approximate
dignostic wind profile technique may hold promise as
suggested by Adefolalu (1984c).

4.   DESERTIFICATION OF THE SAHEL

Aspects of rainfall characteristics has been covered adequately under section 3. The direct link between it and declining availability of surface and underground water will be high-lighted. Although, both have a 'hand' in the deterioration of rangelands with regard to nomadic and sedentarized pastoralists as established by Berry (1984), it is only of recent that an Inter-United Nations Agency commissioned the preparation of a comprehensive handbook on 'Desert rangeland improvement techniques (see Orev, 1986).

Thus, the main thrust of the presentation in this section will be on land-use and decimation of forest resources. Remote Sensing applications in both have been demonstrated as documented by Adefolalu (1986) reports on the International Satellite Land-Surface Climatology Projects (ISLSCP) and the use of Side-Looking Air-Borne Radar (SLAR) in the Nigerian situation is fully highlighted here while results of a Ground-Truth technique to corroborate space - based results designed at the Federal University of Technology, Minna are briefly documented.

4.1. Trends in the Desertification Process

While lack of rain is generally and correctly referred to as drought, under certain conditions drought effects may be due to encroaching deserts. Writing on the theme 'doubtful dogma about deserts' Warren (1980) expatiated:

"The problem with desert is at their edges where rain, though scanty, can sometimes support pastures, wells and here and there, some agriculture".

When such highly fragile lands are over-exploited for any or all of the above purposes, the danger of expanding desert conditions is imminent and drought can only aggravate the situation but may not be its cause. It is known that the Sahara desert of Africa has taken over in marginal areas where the soil has been overworked. This problem had earlier been speculated upon (Bovill, 1921) in a study titled 'The encroachment of the Sahara in the Sudan'. Another dimension to the problem of drought and desertification is the change in some atmospheric variables like ozone which is an unstable form of oxygen. Although it is in very small quantities in the 8-50 kilometer layer above the earth's surface, it is capable of absorbing Ultra-Violet radiation (UV) from the sun. As reported by Ogunseitan (1985), such absorption of UV radiation makes it possible for plant (and perhaps animal) life which could have otherwise received extreme 'sun-tan' leading to scorching. Thus, if the arrangement of atmospheric gases suffer from variable concentrations, it is possible that vegetation will be affected and the effect on the component of solar radiation which generates convection is an obvious decrease due to higher surface reflectivity known as albedo. There are also now serious debates on Greenhouse effects of CO_2 re-distribution in the atmosphere (USEPA, 1986). All these have been tagged 'drought-related' Desertification.

If the process of desertification continues with time, the threat of desert conditions spreading further into arable lands is imminent and this is already affecting the programmes of food production by African countries. As rightly put out by Ajayi (1985), farming systems in the tropics are extremely diverse and complex in an environment where highly weathered soils of tropical Africa are characterised by low nutrient and water-holding capacities - factors which limit the growth and yield of crops. The immediate social impact of such reduction in food production needs little qualification here as it has also become a global socio-political problem.

Presently, it is being postulated that the 'rate of change of desertification' is concerned with two sets of variables. These consist of the physical variables such as rainfall, soil condition, ground-water levels, vegetative cover and character, evapotranspiration and erosion rates on the one hand and the social variables on the other hand. While the former are reasonably being monitored well, especially since the advent of Satellites - meteorological (METEOSAT) land-surface types (LANDSAT) - it is quite a different proposition with regard to the latter which is concerned with how human populations are affected by aridity and water availability, which in turn, generate a 'feed-back' (through intensive use) on the capacity of arid and semi-arid ecosystems.

When socio-economic variables work in concert with natural variables, the problem of combating or controlling desertification becomes difficult if not impossible to tackle. Adefolalu (1983a) identified four major indicators of the rate of change in desertification taking account of combined geo-environmental and socio-economic factors. These are:

i)   - deterioration of both rainfall and conditions of irrigated agricultural lands. The latter is the most critical aspect in the Sudan-Sahelian at present regions

ii)   - declining availability of surface and underground water due to uncontrolled and discriminate use

iii) - deterioration of rangelands with regard to nomadic and sedentarized pastoralists.

iv)   - deforestation with respect to use of fuelwood as main source of energy.

While the impact and effects of (i), (ii) and (iii) in the desertification process are serious, the confirmation of real crisis as induced by man directly is in relation to (iv) - fuelwood as main source of energy. It is to be realised that

70 per cent of the population in Sahelian countries are rural
dwellers whose domestic energy needs are dependent on available fuelwood.

As reported earlier by Matheson (1978), about 15 hec-
tares of forest vegetation is disappearing every minute due
to fuelwood consumption to satisfy rural energy requirements.
This became so serious that, even in Nigeria, which was then
not classified as a 'Sahelian' country, the scarcity of
fuelwood called for importation of charcoal (Morgan, 1978).
It is therefore pertinent that in conducting an impact-
assessment study or survey in the Sudan-Sahel and Guinea-
Sudan belts of West Africa, declining vegetal cover must be
appraised. Figures 18a and b clearly demonstrate the extent
to which forest vegetation has declined and must be thoroughly
appraised. In particular, quantification of the deforestation process must be attempted. The old transect method
gained ground in the late fifties and sixties when air-borne
observation platforms were non-existent. Since the 'arrival'
of air-borne Radar and Satellites, monitoring of land-surface
changes has improved tremendously. But the problem of
ground-truth observation as a way of validating data derived
from space-based platforms still require (even today) a lot
of 'leg-work'. In subsequent sections, discussion focusses
attention on the Nigerian situation (based on microwave technique employed in air-borne radar) and a new approach to
ground-truth observation.

4.2. Vegetation decline in Nigeria

As already highlighted by Adefolalu (1986), in Nigeria before
the turn of the twentieth century, most of the humid Guinea-
Sudan was a pure rain-forest belt. But these have almost been
completely replaced by Cocoa, Oil Palm, Kola and Rubber and
other economic trees while subsistence agriculture involving
small scale, peasant farming to raise tuber and some cereals
resulted in further decimation of the forest zone. As a
result, this rainforest belt which extended up to the catchment
of River Niger in the West has receded further south. In the
rest of the country the tall grass savanna dominated (see in
the Atlas of Espendale and McNally, 1960).

The transition zone between the rainforest in the south
and the tall-grass Savanna in the north was either not too
obvious at that stage or it was not of 'global' dimensions
to necessitate differentiation under the very rural equilibrium
state. However, regional studies such as those of Davies
(1973) and Menekaya and Floyd (1980) were more comprehensive
and finer details showed up. These are associated with the
consequence of plantation agriculture under intensive and
extensive farming of the 'slash and burn' type. In terms of
forest decimation, the major changes have therefore occurred in
the middle belt (north of 7° to 10°N).

As the forest lands were cleared, the tsetse fly
belt moved further south giving rise to southward extension
of grazing pastures for cattle while Benniseed was introduced

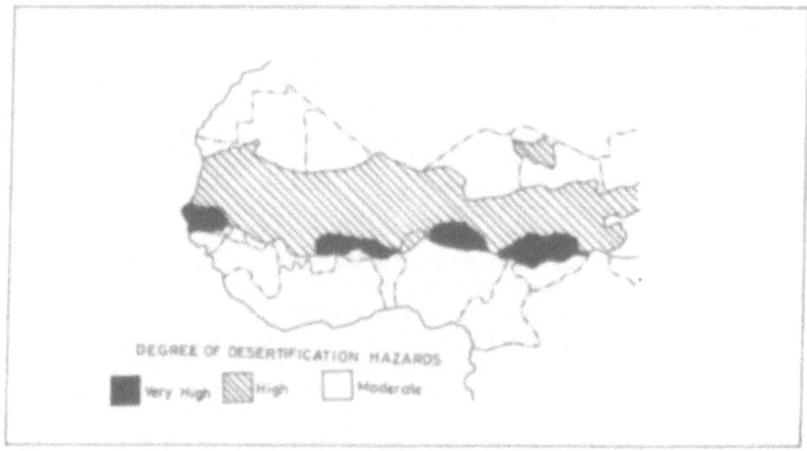

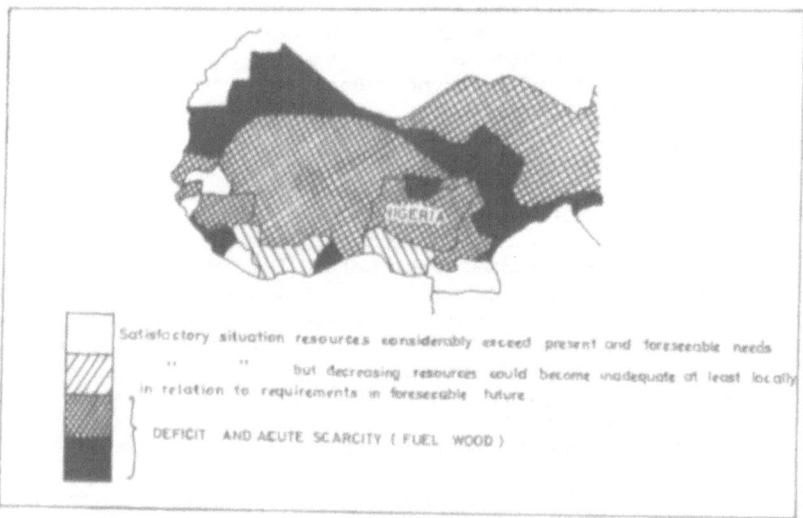

Fig. 18. Top - Trends in the desertification of
the SSR countries of West Africa
Bottom - Desertification and fuelwood crises (both
adapted from Berry, 1984).

from about 7 degree north in the Benue valley to the east and
further west, rice cultivation replaced groundnut in the
Niger Valley. The combined effects of these various activities resulted in the modified version of the ecological belt
shown in Figure 19. Up to about 7° - 8°N the forest/savanna
mosaic (humid) represent the transitions between the rainforest and savanna while the mixed woodland and savanna dominates the landscape from about 8°N to 10°N. This is the
Central Savanna. Except in the extreme northeast sector, the
rest of the country belongs to the tall-grass 'Sudan' (dry)
Savanna. In the northeast sector which falls into the Sahel,
it is mainly composed of shrubs with short grasses. It is
called wooded steppe because of the dominance of plants of
Acacia and Comiphora families while real short grasses are
highly seasonal. The dry Grassland Savanna which separates
Central woodland Savanna from the Sahel-shrub-type belt (and
presently occupying the latitude band 10° - 12°N in general)
is now being threatened by imminent, and perhaps inevitable
desert encroachment. This will result in the spread of the
shrub-type vegetation with the attendant consequence of reduced
convection due to higher albedo (Herman Otterman, 1974; Charney
et al, 1977). If the trend is not arrested drought effects
will become more 'biting' and less vegetal cover will
continue unabated.

4.3. Trends in the Vegetal Cover

The brief description above suggests that the areas greatly
endangered by encroaching desert conditions are to the north
of latitude 10°N in Nigeria.
   The Histograms of vegetal cover in the 10 northern
states threatened by imminent desertification are illustrated
in Figure 20 based on SLAR data as compiled by the Federal
Department of Forestry, Ibadan, Nigeria. The analysis recognised five major vegetal cover specifications as follows:

   1. Forest
   2. Woodlands
   3. Grasslands
   4. Shrubland and
   5. Farmland.

Under each category, further differentiation were carried out
for the 10 states. Extraction from these raw data-base was
carried out and, as shown in Tables 2a & b and 3, the
percentage areas covered by each category in the States which
fall within the aforementioned belt between latitude 7°N and
further north have been tabulated. The percentage cover of
rivers, lakes and built-up areas have been left out (they are
however very small in comparison).
   Table 3 and Figure 20 show that the two states with the
largest area; Borno and Sokoto, are presently experiencing
the harsh effects of desertification with arable lands

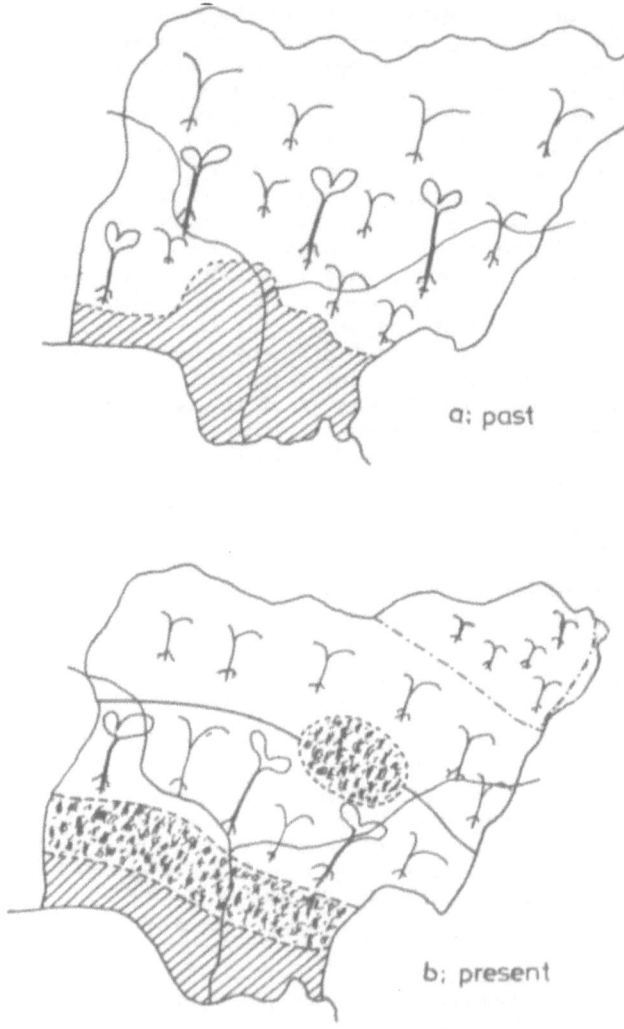

a: past

b: present

Fig. 19. Vegetation of Nigeria (from Adefolalu 1986).

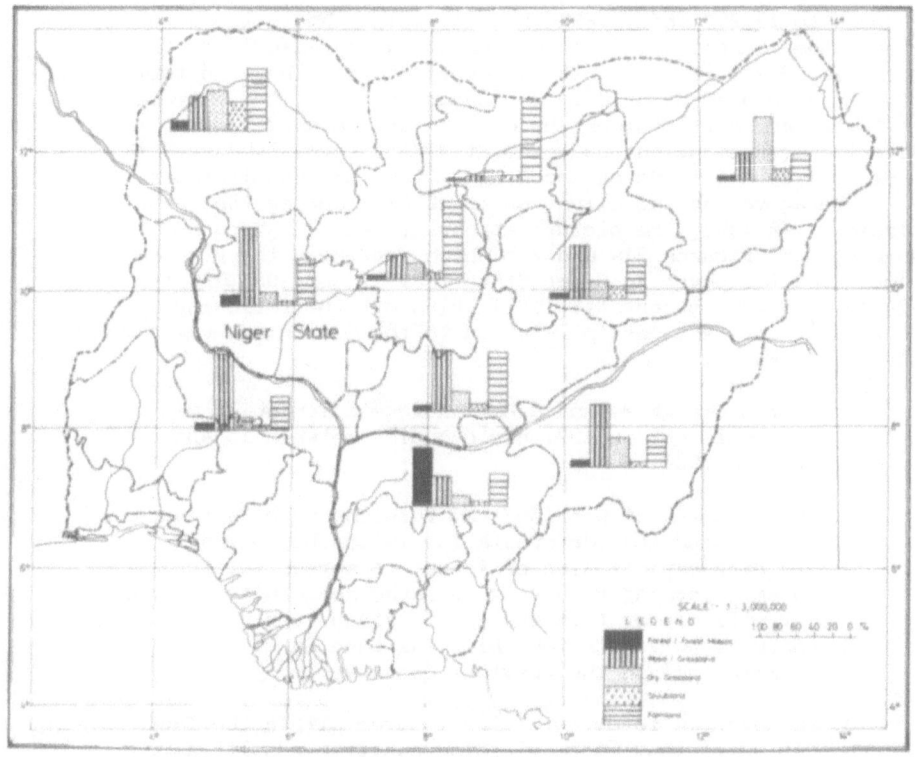

Fig. 20. Vegetal cover in the Sudan Sahel belt.

reduced to 19.29% and 41.89%, respectively. While the total
areas under Sahelian influence (shrub and dry grassland) for
Borno is 59.97% it is 38.36% in Sokoto State. Thus, in terms
of natural causes of aridity and human pressure, the extreme
northwest and northeast of the country are in danger of
desert encroachment and up to about 80% of total land area is
presently on the verge of becoming part of the Sahel-Proper.

The above was the situation in 1976/77 during the SLAR
mission. In 1987, the picture appears to have changed as
suggested by Figures 21a and 21b which represent the proportional vegetal cover in Niger State in 1977 and 1987, respectively. A close observation will bring out the following
most important features in relation to the declining vegetal
cover in Nigeria.

1. Shrub/dry grassland Sahelian vegetation which was
   non-existent in 1977 now occupies between 15 and
   20% in 1987.

2. Percentages of woodland and farmland have decreased
   suggesting that deforestation is on the increase due
   to over exploitation of forest resources for fuel
   arable forming resulting in large-scale defoliation.
   It will appear that these have given 'birth' to the
   Shrub-type vegetation which was pure Sahel
   'exclusive' in the previous decade.

3. However, a forestation initiatives within the last
   10 years seem to be yielding positive results as
   total tree-forest environment has appreciated
   somewhat.

TABLE   2a

| States | SHRUBLAND | | | GRASSLAND | | | |
	Non-Thorny	Thorny	Shrub & Thickets	Grass only Grass	Shrub	Wooded Shrub	Transition to pure shrub
Bauchi	.34	.64	–	.24	.65	8.27	2.29
Benue	–	–	–	1.34	0.50	1.13	–
Borno	.01	.07	5.21	5.0	13.30	23.12	13.26
Gongola	–	–	–	6.52	0.40	7.87	6.00
Kaduna	–	–	0.66	.59	2.54	11.25	.24
Kano	.22	.30	–	.86	3.36	5.28	–
Kwara	0.01	–	1.02	–	3.54	3.54	–
Niger	–	–	–	1.41	6.26	0.02	–
Plateau	.11	0.11	–	0.87	0.64	13.24	–
Sokoto	13.90		2.52	1.67	4.00	16.27	–

313

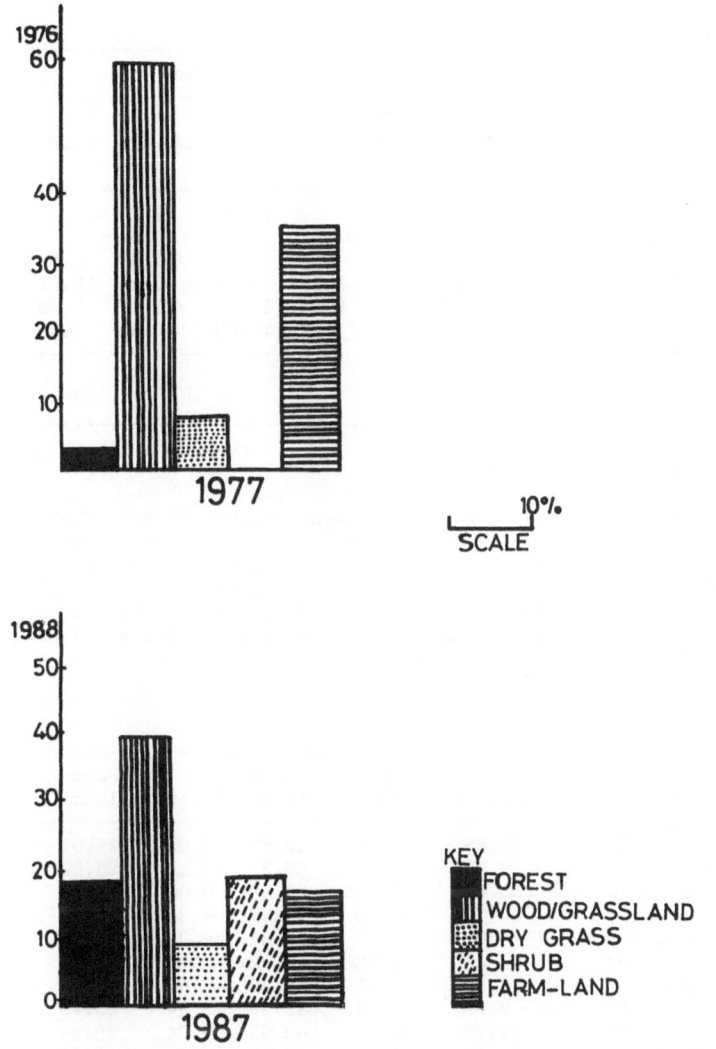

Fig. 21. Averaged vegetal cover in Niger state.
Top - based on 1977 SLAR data, bottom - ground
truth (August 1987).

TABLE 2b

States	Woodlands		Forest		Farmlands		
	Grass/Shrub Transition	Broad-Leaved	Riparian	Others	60% Intensity	30-60%	<30%
Bauchi	42.99	4.64	1.02	-	18.86	16.05	-
Benue	<------ 0.94 ------>		11.83	30.46	4.29	21.33	
Borno	17.15	3.13	-	-	7.24	12.02	0.53
Gongola	32.05	14.90	3.63	1.10	3.41	6.04	15.61
Kaduna	24.83	1.48	0.01	0.34	31.68	19.00	8.25
Kano	7.67	0.15	-	-	49.69	31.00	1.50

States	Woodlands		Forest		Farmlands		
	Grass/Shrub Transition	Broad-Leaved	Riparian	Others	60% Inten-	30-60%	<30%
Kwara	63.23	1.97	2.54	0.07	1.48	2.76	22.84
Niger	52.14	7.32	2.45	1.19	1.97	19.46	24.30
Plateau	31.55	9.08	2.55	0.06	13.68	21.14	6.12
Sokoto	18.80	0.25	0.13	-	21.59	18.64	1.65

TABLE 3

States	Total area *	Grassland	Shrub-land	Wood-land	Forest	Farm-land
Bauchi	6,548,735	11.45	0.98	47.63	1.02	34.91
Benue	4,552,817	2.97	0.00	23.31	42.29	25.62
Borno	11,913,214	54.68	5.29	20.28	0.00	19.29
Gongola	9,452,854	20.79	0.00	49.07	4.23	25.06
Kaduna	6,939,285	14.62	0.66	26.31	0.35	68.83
Kano	4,374,394	9.62	0.52	7.82	0.00	82.19
Kwara	6,011,308	4.56	0.01	65.20	2.61	27.8
Niger	6,728,658	7.69	0.00	49.46	3.64	35.73
Plateau	5,529,152	14.75	0.22	40.63	2.64	40.44
Sokoto	9,196,006	21.94	16.42	19.04	0.13	41.89

*Including water, rivers and lakes

NOTE: Values shown are in percentages based on the total area of each State.

4.4. Prospects for controlling desertification

At present, the most obvious threat to a stable ecological balance in the Sudan-Sahel belt of Nigeria is the emergence of Sahel-type shrub vegetation due to deforestation for two purposes (in the case of Niger State)

1. farming activities (illustrated)
2. rural fuelwood use (to be assessed)

It is equally obvious that this trend may not be reversed as other States in the country also pursue uncontrolled large-scale agricultural activities. A third cause of the deteriorating landscape is overgrazing by nomadic pastoralism - a problem which has not been given sufficient attention, to-date, in Nigeria.

But it is obvious that, even based on the extent of farming activities alone, a general southward extension of Sahel-type vegetation is imminent as illustrated in Figures 22a & b to the extent that areas north of latitude 10°N may become proper sahel with predominant shrub-type vegetation before the year 2000 if no immediate action is taken.

A way out is therefore the proper delineation of the various States into ecological zones based on proven techniques such as the SLAR data gathering approach corroborated with reliable ground - truth observations (The approach adapted in the case-study in Niger-State of Nigeria is briefly described in the Appendix). With LANDSAT data becoming readily available and stored data-sets of other satellites such as SEASAT and SPOT easily accessible, a good beginning could be made in data assimmilation.

There are obvious limitations in the SLAR technique on which most of the discussion in this section is based. It is known that the quality of the returned signal by the radar transmitter fixed under the fuselage of an aircraft is a function of the following:

i) - the surface materials such as water, soil, vegetation etc

ii) - moisture content of the material and related electrical properties

iii) - the slope of the terrain and associated shadow casting

iv) - the roughness of the surface with its attendant back-scattering

The third and fourth factors are very serious since shadow casting and back-scattering may lead to 'wrong' classification especially since the tone of photos are affected. However

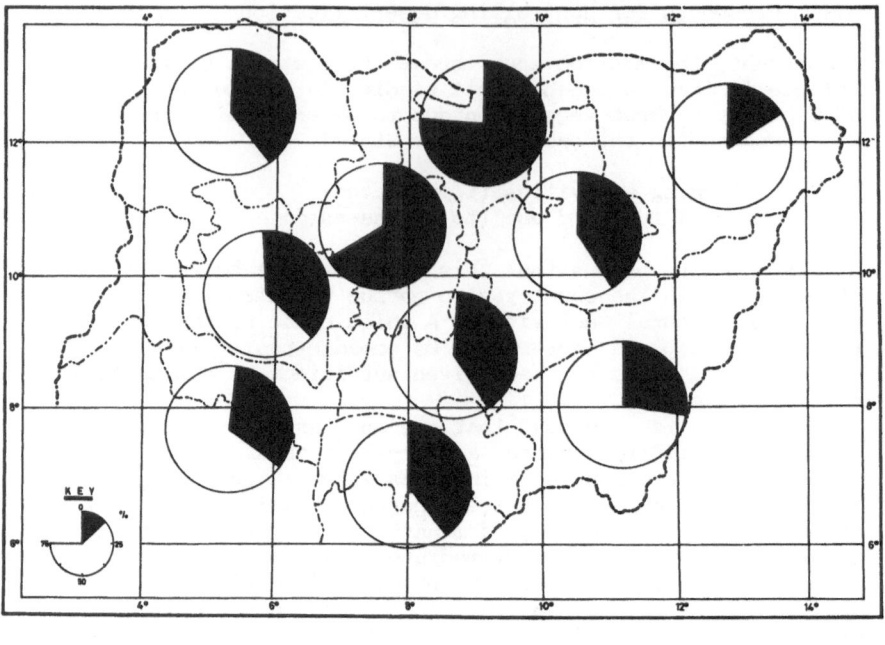

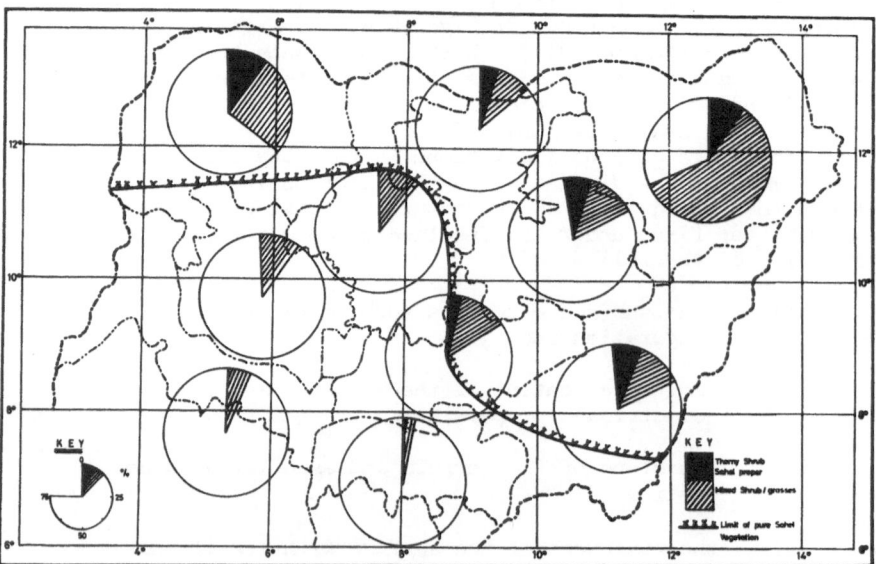

Fig. 22. Top - Percentages of farmlands in Sudan Sahel.
Bottom - Percentages of Sahel-type vegetation.

with digital techniques 'spurious' colors can be eliminated.

It is obvious from the above that side looking Air
borne Radars can be put to advantage in studying or
monitoring the trend in desertification in Sahelian
countries from

- studies on soil moisture content and underground water
  depletion and

- degradation of the environment through vegetation
  mapping.

It is hoped that, with improved data on ecological
mapping now available from special observation platforms
like Satellites with microwave sensors and reliable methods
for ground-truth observation it will be possible to
corroborate such remotely sensed data as analysed and
highlighted in this section.

5. CONCLUSION

Two problem areas in geo-environmental degradation in Sahelian
Africa, with particular reference to Nigeria have been discussed in this paper. These are on synopsis of weather and
climate in relation to drought in West Africa as it has a
bearing on desertification processes.

It is obvious that in weather prediction, numerical experimentation will benefit from micro-wave transmission techniques in data acquisition especially in relation
to the baroclinic monsoon basic current with its origin in
the Atlantic Ocean. Surface conditions which are decisive
in boundary-layer convergence of moisture are dependent
largely on good estimates of surface temperatures. Illus-
trations show that poor results of numerical experimentation
techniques cannot be reversed without a re-appraisal of the
present (even modified) bulk aerodynamic approach in use.

On desertification and monitoring its trends in the
Sahel, a combined ground-truth and space-based techniques
will enhance data gathering for decision-making process,
especially in relation to decimation of forest areas due to
increased land-use practice. It is clear that Satellite
imageries on land-surface resources can be ehanced by microwave techniques and the advantages should be further explored
and exploited.

It is hoped that improved data on ecological mapping
will become more readily available from other special platforms.
With reliable methods for ground-truth observations, it will be
possible to corroborate such remotely-sensed data as highlighted in this section.

APPENDIX - Method for Estimating Vegetation Types

Previous attempts for corroborating space-based data on natural resources by groundtruth observations relied largely on the transect method. It consists of identifying along either series of latitudinal or longitudinal lines running through the specific areas covered by remotely-sensed spaced based data (radar, satellites etc).

The problem with the above method in relation to vegetation is that since patterns and distribution are irregular and heterogeneous, some "gaps", which are naturally created in transect sampling are usually smoothed out by interpolation.

Although the method described below also employs interpolation technique in analysing data in - between locations, there is less subjectivity as all sectors of a 4 - quandrant 'pixel' are monitored.

The "4 - QUADRANT PIXEL" method for groundtruth observation

The pilot project covered Niger State of Nigeria (see figure A-1). In it, 622 sampling locations were used. Each location is treated as a point-focus. An imaginary circular path (360°) is drawn at the viewing horizon and this is bisected east-west and north-south into four quandrants. The pixel size is thus the area of the circle with the location (point-focus) as centre (illustrated in Figure A-1)

The observer monitors and estimates the percentage cover of each vegetal type: trees, wood/grassland, grasses, shrub (sahel-type of short grasses/shrub), farm land and bareground in each of the quandrants. Averaged values of the 4 - quandrants for each vegetal type is computed as follows:

Let: P_v, i, q, represent the percentage of a particular vegetal type at a location:

v = Vegetal type
i = Pixel number
q = quandrant

$$\bar{P}_{v,i} = \frac{1}{4} \sum_{q=1}^{q=4} P_v, i, q$$

For the State averages illustrated in Figure 20b, the estimate of each vegetal type over the entire state is computed thus:

$$\bar{P}_v = \frac{1}{N} \sum_{i=1}^{i=N} \frac{1}{4} \sum_{q=1}^{q=4} P_{v,i,q}$$

where N is total number of point foci   (622 in this case).

Fig. Al. Research domain with the local government
areas in Niger state.

R E F E R E N C E S

Adefolalu, D. O.   (1972):   On the   mean   structure and lower
                             tropospheric features   over   West
                             Africa   MSc.   Thesis   FSU., U.S.A.
                             321 pp.

Adefolalu, D. O.   (1973):   The mean troposphere area West
                             Africa. Nigeria Quam., Vl. 3 No.2
                             31-72.

Adefolalu, D. O.   (1974):   On Scale Interaction and the Summer
                             lower tropospheric easterly pertur-
                             bation over West Africa Ph.D.
                             Dessert.   FSU - Tall; U.S.A. 276 pp.

Adefolalu, D. O.   (1976):   "Extension of the Asia MONEX
                             Experiment to West Africa".   A
                             working paper (unpublished) delivered
                             of the   WMO   Inter-Governmental Panel
                             on   FGGE   (February 1976, General)
                             pp.

Adefolalu D. O.   (1977):   'A study of synoptic Type   2   the
                             African easterly perturbation' In
                             lectures on Forecasting of Tropical
                             Weather, including tropical cyclones
                             with particular relevance to Africa
                             (Ed. R. P. Pearce) WMO   Pub. 492,
                             255 - 270.

Adefolalu, D. O. (1982):   Acantific   Trends   in   the Tropics:
                             The   role of human interference.
                             Climatological notes,   Vl. 30
                             174 - 190.

Adefolalu, D.O. (1983a):   Desertification of Sahel.   In
                             "Natural   Resources of Tropical
                             Countries" (Ooi Jim Bee, Ed.)
                             Singapore Univ. Press, Chap. 12,
                             402 - 438.

Adefolalu, D. O. (1983b):   The West African Monsoon Dexperiment
                             ment - Field-Phase Report.   WMO-GARP
                             Pub. Series, Vol. 10, 69 pp + 10
                             Annexes.

Adefolalu, D. O. (1983c):   Monsoon   On-set   in   West   Africa
                             Application of Satellite Imagery.
                             Achieves for Met. Geophy and Biokl.
                             Ser. B, 32,   219 - 230.

Adefolalu, D. O. (1983d):   Weather Forecasting and the Model
                             of Scale Interaction in West Africa,
                             Archieves for Met. Geophy Buil
                             Ser. A 32, 103 - 116.

Adefolalu D. O. (1984a):   WAMEX: Observational Studies since
                             Field Phase (May - August, 1979).
                             Tallahsee Conf. October, 1984
                             11pp + XV figures).

Adefolalu D. O. (1984b):   The Monsoon and Precipitation in West
                             Africa (Ext. Abst.) WMO Conf. on
                             GATE-WAMEX and Trop. Met., Dakar;
                             131 - 139.

Adefolalu, D.O. (1984c):   Computation of Boundary - layer
                             Characteristics using the Gram -
                             Schmidt Orthogonalisation techniques.
                             Nig. J. Sci. No. 17.

Adefolalu D.O. (1985a):   "WAMEX - Significant contributions to
                             the understanding of monsoonal circu-
                             lation in West Africa" In "proceedings-
                             The Global Weather Experiment"
                             (Sir John Mason - Editor). WMO Pub.
                             series.

Adefolalu, D.O. (1985b):   On transport of zonal momentum and
                             energy exchange processes in West
                             Africa. Arch. Met. Geoph. Biocl.,
                             Ser. A 33, 277 - 287

Adefolalu, D. O. and   Mean State during the Onset Phase
others (1985c):   (May 1 - 15) of the West African
                             Monsoon in 1979. Archieves for Met.
                             Geophy and Bioclimat. Serv. A. 33,
                             327 - 343.

Adefolalu, D. O. (1986):   Regional Studies with Satellite data
                             in Africa - on Desertification of the
                             Sudan-Sahal both in Nigeria. In:
                             Proc. ISLSCP Conf. Rome - Italy
                             (ERoIfe, Ed.); 429 - 439.

Adefolalu, D. O. (1988):   On Characteristics of Precipitation
                             and potential evapotranspiration in
                             relation to climatic zones in
                             Nigeria. Water Management Athens
                             (under review).

Adejokun, J. A. (1985):   Numerical Weather Prediction for the
                             West African Monsoon Experiment;
                             Dakar Conf. on WAMEX. WMO. Trop. Met.
                             Pub. No. 28, 125 - 126.

Ajayi, O.   (1985):   'Improving farming Systems in Africa.
                                    Laily Times of Nigeria, August 30,
                                    1985 p.5.

Assamoi, P. (1985):   Numerical Study of Perturbations over
                                    the West African region using GATE
                                    data.   Dakar Conf. on WAMEX. WMO
                                    Trop Met. Pub. No. 28, 244.

Battan, L. J. (1982):   The Use of radar in modern
                                    Meteorology, Sem. on Radar Met., WMO
                                    Pub. No. 626, 1-24.

Balogun, E. E. (1985):   Significant features of the motion
                                    fields over West Africa during the
                                    special observation period of WAMEX.
                                    Dakar Conf.   on WAMES., WMO Trop.
                                    Met. Pub. No. 28, 106.

Berry, L.   (1984):   Assessment of Desertification in the
                                    Sudan - Sahellian Region, 1978 - 84
                                    UNEP (GC. 12) Paper, 146 pp.

Bovill, E. W. (1921):   'The Encroachment of the Sahara on
                                    the Sudan'.   J. Roy African Soc., 20,
                                    259 - 69.

Burpee, R. W. (1972):   The Origin and Structure of easterly
                                    waves in the lower troposphere or
                                    north Africa.   J. Atmos. Sci., 31,
                                    1556 - 1570.

Charney, J.C., W.U. Quirk,   A comparative Study of the Effects of
S. Chow and J. Korn Field   Albedo Change on Drought in Semi-arid
(1977):   Regions. J. Atmos.   So. 34,
                                    1366 - 1385.

Flohn, H. D. Henning and   Possibilities of a and limitations
H.C. Korff (1984):   of a large-scale water budget
                                    modification in the Sahel-Sudan belt
                                    of Africa.Met. Rdsch., 27, 97-100.

Menekaya, J.C. and   School Atlass for Nigeria 6th
B.N. (ed) 1980):   edition, Macmillan Publishers.

Ogunseitan, S. (1985):   'Climate experts raise alarm on use
                                    of sprarys', The Guardian, Nigeria,
                                    August 10 1985 p.2.

Orev, V.   (1986):   A practical Handbook on Desert range
                                    improvement techniques, Inter-Un
                                    Agency Pub., Geneva 165 pp.

Pearce, R. P.   (1985):   The Global Circulation and the West

	African Monsoon. WMO Trop. Met. No. 28, 19 - 28.
Schove, D. J. (1946):	A further contribution to the Meteorology of Nigeria, Q. J. Roy. Met. Sco. 72, 105 - 112.
Warren, (1980):	Doubtful dogma about the deserts. <u>London</u> <u>Times</u> June, 5 1980 p. 25
Usepa, (1986):	Greenhouse effect, Sea-level rise and salinity in the Delaware Estuary. USEPA Pub. No. 230 05-96-010 88pp.

REPORTS

NIRAD Report on the SLAR Project for Nigeria. Vol. I and II September, 1978.

INTEGRATION OF SATELLITE-DERIVED PRODUCTS

John A. Leese
Institute for Naval Oceanography
Stennis Space Center, MS   39529-5005 USA

ABSTRACT.   The processing and applications of remotely sensed data
from satellites have undergone many advances during the past 30 years.
The evolutionary development of the meteorological satellite from the
early TIROS experimental satellites to the current global network of
geostationary and polar-orbiting operational satellites best exempli-
fies these changes from the viewpoint of integrating the satellite
derived product with available conventional observations and products
for both research and operational use.   Quantitative information
derived from the satellite remotely sensed data are now exchanged
internationally and assimilated with other types of observations in
numerical weather prediction models.   Validation and verification of
satellite data and derived products depends heavily on suitable conventional information to assure that the satellite-derived information
maintains compatibility for integration with the other data sets
available.

1.   INTRODUCTION

    Observations of the Earth have been obtained from satellites
since the first meteorological satellite was launched in April 1960.
The capability of an orbiting satellite to produce raw data at a pro-
digious rate is unprecedented in the history of collecting geophysical
observations.   The mass of data available is of itself a major focal
point of the data management problem for remotely sensed data from
satellites.   Advances in applying satellite data and services have
benefited, in a very large measure, from the rapid developments in
computer systems.   These developments, during the past 30 years, have
continued to provide increased data processing capability at a lesser
cost.   The developments of the meteorological satellite applications
from the experimental uses of the early TIROS data to the daily operational use of satellite data and derived products from the current
global network of meteorological satellites presents some outstanding
examples of how the problems of integrating satellite-derived products
were resolved.

R. A. Vaughan (ed.), Microwave Remote Sensing for Oceanographic and
Marine Weather-Forecase Models, 325–338.
© 1990 Kluwer Academic Publishers.

The operational meteorological satellite capability evolved during the past 30 years. The decade of the 1960s witnessed the birth of the meteorological satellite as an unprecedented tool for observing broad-scale atmospheric phenomena. By the end of the sixties the meteorological satellite had grown to a highly sophisticated platform which could provide global coverage of cloud observations and was beginning to provide quantitative measurements of pertinent meteorological parameters. During the 1970s there was an evolutionary development of a cooperative international network of meteorological satellites. This effort culminated in the contribution to the Global Weather Experiment (FGGE) by a nearly complete global network of meteorological satellites. In the present decade of the 1980s we are seeing a stabilizing of the global network of meteorological satellites in terms of sensor data and services. There is now a more intensive effort in the processing and applications of satellite data to increase the information obtained.

The development of the meteorological satellites from a research tool to an invaluable operational system was greatly assisted by the Global Atmospheric Research Project (GARP) during the period 1967 through 1979. GARP consisted of a series of experimental projects in which the satellite observing system was combined with surface and aircraft observing systems to study a specific type of atmospheric phenomena or the atmospheric conditions in a particular geographic area. These experiments started with the Line Islands Experiment (1967) and included the Barbados Oceanographic and Meteorological Experiment (BOMEX) in 1970, the GARP Atlantic Tropical Experiment (GATE) in 1974 and culminated in the First, GARP Global Experiment (FGGE) in 1979. The satellite-borne sensors used during this period are described by Duggin (1987).

2. EVOLUTIONARY DEVELOPMENT OF METEOROLOGICAL SATELLITE APPLICATIONS

2.1 Image Products

The early TIROS spacecraft were spin stabilized with vidicon cameras looking out one end along the spin axis. With the spin axis in the orbital plane, this meant that the cameras saw the Earth for only a portion of each orbital pass (Fritz, 1964). The overlapping image sequences, with fore-shortening varying from horizon scenes to nadir scenes, greatly complicated the interpreters task. Familiar landmass features were often strangely deformed. The interpreter's nephanalysis first required the careful mosaicing of images into scene strips and then the tedious transcribing of pertinent cloud features from the variably distorted scene onto a standard weather map projection (Bristor and Ruzecki, 1960; Leese, 1962). The obvious first need was for latitude-longitude image locator grids, and much of the early processing effort was devoted to this requirement.

During this early period computer support was largely a reaction to problems encountered by analysts and research personnel who were spending an inordinate amount of time in the locator melding process. The real solution, however, was not to improve the hand tools but rather to automate the melding process and eliminate the waste in analyst resources. This came with the experimental activities in the TIROS Operational System (TOS).

Despite the limitations of the TIROS early exploratory effort, the potential value of cloud imagery in weather analysis and prediction was evident, and, with TOS, daily global coverage was achieved. The basic spacecraft was retained, but, with camera and horizon trigger mounted on the spinning cylinder's rim, nadir views were obtained as the satellite rolled along its orbital track. The resulting global coverage with essentially invariant distortion patterns, coupled with newly acquired large scale computing support, opened the way for a greatly expanded data processing operation. Facilities were designed and installed so as to provide for quick look analog image prints as well as for the computer manipulation of digital image data.

Although automatic grid melding simplified nephanalysis operations, it did little to broaden the information extraction opportunities. Gridded vidicon strips from one day to the next could not be conveniently juxtaposed for time trend comparisons or used in direct overlay comparisons with other coincident weather data on standard maps. Therefore, in parallel with continued support for hand interpretation efforts, software was developed by which digitized image data were transformed into global mapped mosaics in either polar stereographic or Mercator maps (Bristor et al., 1966). Once global image mapping was stabilized, attempts were made to summarize the 4096x4096 polar hemispheric arrays into a data base for other uses. After some experimentation, sequences of daily tape files were created in which brightness distributions for 8x8 sample clusters were recorded in five response ranges. This data base production continued until the end of the TOS program, and it provided the input for a four year Atlas of Relative Cloud Cover (Miller, 1971).

In order to assist those engaged in snow and ice studies, sequences of daily mapped mosaics were combined to remove transient cloud obscuration. Using five day clusters, only the minimum brightness value was recorded for each mapped image element position. The resulting composite minimum brightness image revealed the edges of snow and ice fields with little remaining cloud contamination. Such charts proved of value in several investigations (McClain and Baker, 1969).

While efforts were being made to develop applications and stabilize TOS operations, the launch of NASA's first geostationary Applications Technology Satellite (ATS) provided new opportunity for experimentation. A spin scan imaging telescope had been included as a result of the innovative efforts of Professor V.E. Suomi and his colleagues at the University of Wisconsin. Initial struggles to utilize the data somewhat paralleled the beginning TIROS effort. Earlier software was modified to produce overlay grids and the hand melding process revealed the need for more precision in navigational inputs. Again, horizon and landmark control points were used, and the several experimenting groups soon devised acceptable attitude determination procedures. With help from Professor T.J. Fujita's group at the

University of Chicago, a variety of experiments went forward in which sequences of images were examined as lapse time movie animations. All were impressed with the possibility of using moving clouds as indicators of atmospheric motion, and by 1969 the extraction of cloud motion wind estimates had become a routine operation (Young et al., 1972).

Apart from animation, ATS also provided broadscale coverage as opposed to repeated two hour time slippages obtained with strip mosaics from TOS. Since weather analysts are interested in broad synoptic comparisons, mapping codes were revised to produce both polar stereographic and Mercator ATS imagery. Such transformations were produced using ATS-1 data and later also data from ATS-3.

Many lessons were learned during these early operations with TOS and ATS. Apart from user pressures for more and improved sensors which led to ITOS (the Improved TOS series), the data processing experiences provided insight toward an improved ground processing system. It also made evident the need for more supportive processing resources. Specifically, the need was indicated for constant monitoring of incoming raw signals in the most quantitative way possible. It also indicated the importance of a proper technical staff to maintain and validate the specialized ingest, display, and other dedicated equipment. Finally, there is need for an organized effort to check the timeliness and validity of output products. Such a quality assurance group must possess a diversity of talents so as to understand both hardware and software problems and so assist under anomalous circumstances to minimize the impact on production. Further detail concerning processing and output applications has been provided by Leese et al. (1970a).

The evolutionary development of image processing to produce operationally useful products stabilized during the latter half of the 1970's with the advent of the TIROS-N and GOES series of satellites. These products are described by Dismachek, Booth and Leese (1980). These products are integrated with conventional meteorological data and the forecasts from numerical weather prediction models for many applications in providing operational meteorological services. Bader, et.al (1988) provide an assessment of the subjective interpretation of satellite imagery, in combination with radar imagery, in weather forecasting. Their paper described the results of a workshop, held in July 1987, that reviewed three areas relating to the subjective interpretation of imagery by forecasters; 1) use of conceptual models to provide a framework for understanding imagery; 2) integration of imagery, weather analysis, numerical-model products, and conceptual models by means of video workstations; and 3) training in imagery interpretation. One of the recommendations from this workshop was to organize efforts to coordinate the techniques developed by various groups and foster exchange with other groups in different countries. This could be some initial steps toward standardization which would facilitate exchange of conceptual models for image interpretation and the preparation of training materials on this topic which could be used internationally. Browning (1987) described a specific application of satellite imagery, in combination with radar imagery, for the measurement and short-term prediction of rainfall in the UK.

2.2 Initial Quantitative Operations

Data from scanning radiometers became part of the operational system in 1970. Beginning with ITOS-1, launched January 23, 1970, satellites of this series (ITOS-1, NOAA-1 through NOAA-5) provided routine measurements from a two channel scanning radiometer (0.5 to 0.7 micrometers and 10.5 to 12.5 micrometers) with a ground resolution of approximately 8 km. at the satellite subpoint.

Beginning with NOAA-2, launched on October 16, 1972, a Vertical Temperature Profile Radiometer (VTPR) provided remotely sensed measurements from which atmospheric temperature profiles could be derived. The VTPR had six discrete channels within the 15- micrometer carbon dioxide absorption region, one in the 11- micrometer atmospheric "window" region, and one in the 18- micrometer water vapor absorption region of the infrared absorption spectrum.

Data from the GOES (Geostationary Operational Environmental Satellite) also provided routine measurements from a two channel scanning radiometer beginning in May 1974. The Visible and Infrared Spin Scan Radiometer (VISSR) provided visible data in the 0.55 to 0.75 micrometer range at a ground resolution of about 1 km. and IR data in the 10.5 to 12.6 micrometer range at a ground resolution of about 8 km.

A description of the operation of the NOAA Polar Satellite System is given by Fortuna and Hambrick (1974), and ITOS products are summarized by Hoppe and Ruiz (1974). Bristor (1975) provided a detailed description of the GOES satellite together with specific descriptions of the central processing and analysis of the data received from these satellites.

Three major processing models were developed and implemented as part of the operational data processing during the period 1970-72. These models were:

- GOSSTCOMP (Global Operational Sea Surface Temperature COMPutation) provided daily global surveillance of the ocean's surface temperature structure.
- Picture Pair was an automated model for measuring cloud motion vectors using pairs of geostationary satellite images.
- VTPR Processing was a series of processing routines which provided twice daily measurements of atmospheric temperature profiles over the ocean areas of the globe.

Each of the models was initially designed to be as automated as possible. This was especially true for the early processing steps because early flaw detection with large data rates avoids waste in processor resources. Each of the models contained the basic components of an automated operational procedure. This must include a valid measurement technique, the ability to perform quality control and noise contamination measurements on the raw data, detection of valid areas for making measurements, and the exercise of quality control procedures over the numerical values obtained from the measurement process.

Hindsight indicates that the work to develop operational procedures to derive sea surface temperatures and satellite winds placed too much emphasis on the measurement techniques at the expense of other portions of the model. Work on techniques for the computer measurement of satellite winds during the period 1967-72 was highly concentrated on developing the most efficient and reliable technique. By late 1971 it became clear to this author that the specific measurement technique was not as critical as it appeared earlier. Further discussion of this evolutionary development is given in Leese et al., 1970b and Leese et al., 1971a.

The work on computer techniques for deriving sea surface temperature observations relied on a combined detection and measurement unit. The ability to eliminate cloud contamination was thought to be the most significant factor in determining the eventual accuracy of the derived sea surface temperature values. As a result of this belief nearly all of the early effort during the period 1969-73 was devoted to developing the most efficient and reliable method of measuring the surface clear radiance value. By the middle of 1973 it became very clear that the variability in atmospheric moisture attenuation was a severe problem. It was also very clear that the accuracy of the sea surface temperature value was heavily dependent on many parameters other than cloud contamination. These include atmospheric moisture attenuation, Earth location accuracy, and noise in the satellite and ground data acquisition subsystems.

Considerable effort was expended on detecting valid areas for making measurements. The subject of target selection for making valid cloud motion measurements was one of much discussion and debate between those who argued for a manual mode versus those who believed it could be completely automated. It was decided to use a fixed grid to make measurements in an operational mode, because it was less costly to make computer measurements at all locations and depend on a post editing process to delete bad measurements rather than expend the effort to analyze the target prior to making the measurement (Bradford et al., 1972). This was true for measuring low level cloud motion vectors over ocean areas. The availability of thermal IR data provided the basis for temperature slicing as a simple target analyzer, which improved the ability to make low level cloud motion measurements closer to storm systems and thus add significant information to the set of vectors derived.

The need to exercise quality control over the incoming data and the numerical values obtained from the measurement process was well recognized. A review of the technical papers describing the early operational models (Leese et al., 1971b, and McMillin et al., 1973) indicates that there was little appreciation for the scope of quality control procedures needed to sustain an operation for quantitative processing. A common weakness in all three models, at the beginning of operations, was the on-line data quality diagnostics. The quality control checks were primarily designed to reject the output products to avoid sending bad information to the users. It was possible to identify that something was wrong in the system but there was a lack of diagnostic information to identify where the problem was occurring. Searching for a problem somewhere between the spacecraft sensor and the output product from the operational processing consumed many manhours by involving operational and maintenance specialists in all portions of the data acquisition and processing system. A very sophisticated quality control program was implemented in 1972 to critically evaluate the scanning radiometer data for its suitability for quantitative processing. Thirty five parameters were examined within each frame of 25 scan lines to ascertain data quality. All 35 parameters were compared to minimum and maximum threshold values established for each parameter. Orbital summaries were produced as a quality control tool by on-line monitoring personnel. This diagnostic package also generated messages when problems were detected to assist the operational personnel in diagnosing where in the system the problem was occurring.

Quality control of the output products from the GOSSTCOMP and VTPR Processing relied very heavily on automated procedures. These automated procedures placed a heavy reliance on "first-guess" information to establish thresholds of acceptability for the product. The GOSSTCOMP made use of limits for the deviation from yesterday's analysis as thresholds for accepting today's observations. This placed severe limits on the model's ability to detect and measure areas of sudden cooling or warming. The VTPR Processing used forecasts from the Numerical Weather Prediction model to establish limits and made decisions about accepting or rejecting a specific atmospheric temperature profile measurement. Both of these suffered the drawback of providing feedback to the operational system which could cause the operational processing to go unstable. For example, if yesterday's sea surface temperature analysis inadvertently accepted bad observations into the analysis field in a certain area of the ocean, it may not be possible to accept good observations from today's processing output because the thresholds for acceptance are so heavily based on yesterday's values. Picture Pair processing always relied on manual editing as the final step in the processing, and thus the man made the final decision on accepting or rejecting a specific satellite wind observation. The manual editing made a sizable contribution to the stability and reliability of the satellite wind products.

The lack of suitable ground truth information was a severe hindrance to the development of quantitative processing techniques. GOSSTCOMP output was measured against conventional measurements of sea surface temperature made by ships. The latter represents a bulk measurement of temperature whereas the satellite derived values represent "skin" temperature measurements. VTPR Processing used rawinsonde measurements to evaluate the quality of the satellite derived atmospheric temperature profile. The rawinsonde represents an in situ point measurement as the sensor is carried vertically into the atmosphere by a rising balloon. The satellite derived profile represents an average value over a relatively large volume of the atmosphere. Picture Pair processing used the wind vectors determined from the rawinsonde processing as a measure of the quality of the satellite winds derived from cloud motion vectors. Here again, point measurements of the rawinsonde are used to evaluate the satellite wind which represents the average motion of a cloud volume over a 30- or 60-minute time span.

2.3 Global Operational Capability

The thrust of the operational meteorological satellites to date
is aimed primarily at characterizing the kinematics and dynamics of
the atmospheric circulation. The ability to achieve such objectives
was demonstrated during the Global Weather Experiment in 1979. This
capability is now part of the global operations of the World Weather
Watch. The existing network of meteorological satellites, forming
part of the Global Observing System of the World Weather Watch, regularly produces real-time weather information. This is acquired
several times a day through direct broadcast from the meteorological
satellites by more than 1000 stations located in 125 countries. There
are two major components in the current meteorological satellite network. One element is the various geostationary meteorological
satellites, which operate in an equatorial belt and provide a continuous view of the weather from roughly 70°N to 70°S. At present
there is a satellite at 0° longitude (operated by the European Space
Agency), a satellite at 74°E (operated by India), a satellite at 140°E
(operated by Japan), and satellites at 135°W and 75°W (operated by the
U.S.A.). A satellite is planned to be added by the U.S.S.R. at 76°E.
All the present geostationary satellites collect data in the visible
portion of the spectrum and in the infra-red "window" from 10.5-12.5
um. Several also collect data in other spectral intervals, especially
in the water-vapor interval from 5.7-7.1 um. Spatial resolution
varies among the different spacecraft, ranging from 1 km to 5 km in
the visible and from 5 km to 11 km in the infra-red.

The second major element comprises the polar-orbiting satellites
operated by the U.S.S.R. and the U.S.A. The "Meteor-2" series has
been operated by the U.S.S.R. since 1977. The polar satellite
operated by the U.S.A. is an evolutionary development of the NOAA
series, based on the TIROS-N system, which has been operated by the
U.S.A. since 1978. These spacecraft provide coverage of the polar
regions beyond the view of the geostationary satellites and fly at
altitudes of 850 to 900 km. Data acquired by the various sensors are
handled in two ways by the on-board processors of the polar-orbiting
satellites. Since the spacecraft are in view of their ground stations
no more than once per orbit (and sometimes not at all for several
orbits), it is necessary to record the data on tape for later playback
when the spacecraft comes into view of a ground station. At the same
time that the data are being recorded, the spacecraft is broadcasting
the data for the use of any properly equipped receiving station within
range.

Each of the satellite operators disseminate the image data in a
high resolution mode and a low resolution mode. The details of these
dissemination modes are provided by the operators through publication
of users guides or other formats. The satellite operators derive
meteorological parameters such as winds, sea surface temperatures,
cloud analysis, upper air humidity, and radiation balance, based on
the needs of the users for each satellite. The products are disseminated regularly in coded form by satellite processing centers. The
WMO adopted several codes which are used for transmitting satellite

data over the GTS (WMO, 1974).

3. MAJOR PRODUCT SUMMARY

The data from meteorological satellites are used at local user stations, at national meteorological centers, and at world meteorological centers. The data are available on analogue and/or digital formats which enable various degrees of processing based on technical sophistication and resources. The quantitative data derived from the satellite observations and processed at major central facilities are exchanged internationally via the WMO Global Telecommunication System (GTS).

It is common to evaluate the quality of quantitative atmospheric data derived from satellite observations by comparing them with in situ observations. Satellite cloud-drift winds are compared primarily with radio-wind observations, satellite temperature soundings with radiosonde observations, and sea surface temperatures with those measured by ships and buoys. It also is usual to refer to observed differences between satellite and in situ observations as "errors" in the former. However, the observed differences include the effects of the

°   different nature of the two classes of observations: satellites provide volume and time (for winds) averages (area for surface temperature) while in situ observations are close to instantaneous point readings;

°   non simultaneity in space and time of the two sets of data, often no attempt is made to even interpolate the observations to the same location and time for comparison, and

°   errors in the in situ observations, as well as the errors in the satellite data which we wish to determine.

Thus, it is important when comparing satellite performance evaluations to determine to what extent the above factors have been taken into account (Bengtsson and Morel, 1974).

Pailleux (1987) described the use of quantitative satellite data as input to numerical weather prediction and presented the results of an assessment of the impact of the satellite data on the quality of the analysis and the resultant forecast at the European Centre for Medium Range Weather Forecasts.

3.1 Cloud Wind Vector

Winds are derived from cloud movement in a sequence of geostationary satellite infra-red images, acquired at 30 minute intervals. Displacements of low clouds are determined by automated systems. Some operators are tracking upper-level clouds manually. Heights of the cloud wind vector are assigned by different methods such as cloud top temperature converted to pressure height.

The main problem in obtaining winds from cloud motions is in estimating the altitude to which each wind derivation will be ascribed. This is not as critical for the low-level vectors as for the high level, since relatively small cumulus clouds usually are used as low-level tracers, and the vertical shear in the wind through the cloud layer usually is small. Low level winds are assigned to a fixed level.

The cloud targets for the high-level winds extend over a large range of altitudes and vary as to cloud type. Wind shears can be large over the range of altitudes involved. Many of the cloud targets are thin, whose cloud top (infrared) emissivity can be highly variable and subject to large errors of estimation. The latter is important in the case of the GOES and METEOSAT high-level winds in which satellite-observed infrared temperatures are used to estimate cloud heights (Hubert, 1979).

3.2 Sea Surface Temperature (SST)

Sea Surface temperature (SST) measurements have been derived routinely from satellite remote sensing data since 1972. The basic technique for SST measurement is to measure thermal infrared radiation from the ocean in an atmospheric window where the atmospheric absorption and emission are low, and where the emissivity of the ocean is near unity.

The measurements from the polar-orbiting satellite were begun using the ITOS Scanning Radiometer (SR) data which contained only a single window channel in the 10 to 12 um range. Several different techniques were tried to correct for the atmospheric attenuation (Brower, et. al, 1976). The operational system used to derive the SST measurements during the period 1972 – 1981 was referred to as GOSSTCOMP (Global Operational Sea Surface Temperature Computation). The accuracies achievable from GOSSTCOMP were limited primarily due to an inadequate method to correct for atmospheric attenuation, and also due to residual cloud contamination.

The advanced Very High Resolution Radiometer (AVHRR) flown on the TIROS-N/NOAA series of polar-orbiting satellites provides high quality, digital measurements having a basic spatial resolution of 1.1 km at nadir in the visible (0.58 – 0.68 um) reflected-infrared (0.7 – 1.1 um) and in two or three thermal IR channels (3.5 – 3.9, 10.3 – 11.3, and 11.5 – 12.5 um). The full-resolution data are available locally by direct readout and by limited onboard tape storage. Global Area Coverage (GAC) data are available at a reduced resolution of 4 km which are recorded onboard the spcecraft and readout at the NOAA Command and Data Acquisition Stations. A Multi-Channel Sea Surface Temperature (MCSST) has been operational to provide SST measurements from the AVHRR data since 1983 (McClain, Pichel and Walton, 1985). Approximately 75,000 daytime and 25,000 nighttime observations are calculated daily, each with a resolution of 8 km. Observations are spaced every 8 km in the coastal regions of the USA and in selected research areas, every 15 km in the Eastern North Pacific and Western North Atlantic, and every 25 km. elsewhere (Strong and McClain, 1986). Comparisons of satellite-derived SST measurements with high-quality drifting buoy observations shows excellent agreement with mean differences of less than 0.3° and root mean square differences of about

0.7°C (Strong and McClain, 1984). The operationally achievable SST accuracies are estimated to be in the range of 0.5° - 1.5°C, depending on the prevailing atmospheric and surface conditions, the amount of temporal and spatial averaging involved, and the quality of the in situ data available for evaluation.

3.3 Atmospheric Temperature

The method is based on sensing in one absorption band at different wavelength, close to each other, i.e., for various absorption powers. The 15 um absorption band provides better sensitivity to the temperature of relatively cold regions of the atmosphere. The 4.3 um absorption band provides better sensitivity to the temperature of relatively warm regions of the atmosphere. The infra-red windows provide the surface temperature. When it is cloudy, microwave sensors, instead of the infra-red sensors, probe through clouds and provide the temperature. The operational method is a multiple regression statistics scheme that relies on a set of nearly coincident radiance observations and radiosondes gathered over previous days.

For a general discussion of the principles of remote sounding, see Hayden (1979) or Olesen (1987). Smith et al (1979) have summarized the general characteristics of TOVS and the data processing system utilized during the early part of FGGE. While a number of changes were introduced into the processing system during the observing year, the most significant ones were not introduced until later.

The TIROS Operational Vertical Sounder (TOVS) on the NOAA satellite includes three main components: the High-resolution Infrared Radiation Sounder (HIRS), the Microwave Sounding Unit (MSU) and the Stratospheric Sounding Unit (SSU).

McMillin et al (1983) have calculated the RMS differences between satellite retrievals and radiosondes for the same data set but using two different colocation processes. First, as in the normal operational approach, pairs of soundings within $\pm$ 3 hours and within 3 degrees of latitude are compared without space or time interpolation. The second method was that used by Phillips et al (1979) in which the radiosondes 12 hours apart are interpolated to the time of the satellite observation, and the satellite soundings for an area around the radiosonde location are interpolated to that location. The comparison indicates that with one exception the interpolation method gives a lower RMS difference. The average of the interpolated differences is 1.6 K RMS compared to 2.0 K with no interpolation.

Bengtsson and Morel (1974) also compared radiosonde reports from closely spaced stations in Europe and found variations averaging about 1.5 K, reflecting both the errors of the radiosondes in the group and the atmospheric variability over distances typical for satellite-radiosonde comparisons.

Taking all of this evidence into account, it would appear that one should not expect agreement between radiosonde observations and satellite soundings as they are normally compared to closer than about 1.5-2 K RMS and average temperatures are now in the range of 1-1.5 K except for the 1000–850 mb layer. It also reinforces the conclusion that great care must be exercised in comparing performance statistics from different sources.

In global macroscale analysis and prediction models where the higher vertical resolution of the radiosonde is not important, it also appears that the satellite soundings give much more representative observations than radiosondes in terms of both volume mean temperatures and uniformity of coverage.

There have been promising developments in physical methods of sounding retrieval which do not rely on the use of correlation coefficients with radiosondes. If this approach can be developed to a level of performance at least equal to the present operational method, it can be expected to be introduced into the operational system. This would make the satellite output nearly independent of the radiosonde observations with the latter only being used for monitoring the satellite output.

REFERENCES

Bader, M., Browning, K., Forbes, G., Oliver, V. and Schlatter, T., 1988: Towards Improved Subjective Interpretation of Satellite and Radar Imagery in Weather Forecasting: Results of a Workshop. Bulletin American Meteorological Society. 69, No. F, pp. 764–769.

Bengtsson, L., and Morel, P., 1974: The performance of space observing systems for the First GARP Global Experiment. GARP Programme on Numerical Experimentation Report No. 6, WMO/ICSU JPS, Geneva, 33, pp. and appendix.

Bradford, R., Leese, J., and Novak, C., 1972: An Experimental Model for the Automated Detection, Measurement, and Quality Control of Low-Level Cloud Motion Vectors from Geosynchronous Satellite Data, Proceedings of the Eighth International Symposium on Remote Sensing of Environment, Environmental Research Institute of Michigan, Ann Arbor, Mich. pp. 441–462.

Bristor, C.L., Callicott, W.M., and Bradford, R.E., 1966: Operational Processing of Satellite Cloud Pictures by Computer. Monthly Weather Review. 94, No. 8, pp. 515–528.

Bristor, C.L., (Editor), 1975: Central Processing and Analysis of Geostationary Satellite Data, NOAA Technical Memorandum NESS 64, National Environmental Satellite Service, National Oceanic and Atmospheric Administration, U.S. Department of Commerce, Washington, D.C. 155 pp.

Bristor, C.L., and Ruzecki, M.A., 1960: TIROS I Photographs of the Midwest Storm of April 1, 1960. Monthly Weather Review 88, No. 9, pp. 315–326.

Brower, R. Gohrband, H., Pichel, W., Signore, T. and Walton, C., 1976, Satellite derived sea surface temperature from NOAA spacecraft, National Environmental Satellite Service Technical Memorandum, NESS 78, National Oceanic and Atmospheric Administration, U.S. Department of Commerce, Washington, D.C.

Browning, K., 1987: Use of Radar and Satellite Imagery for the Measurement and Short Term Prediction of Rainfall in the United Kingdom, Remote Sensing Applications in Meteorology and Climatology (edited by

R.A. Vaughan), D. Reidel Publishing Co., NATO ASI Series, Series C, Vol. 201, pp.189-208.

Dismachek, D., Booth, A. and Leese, J., 1980, National Environmental Satellite Service Catalog of Products, Third Edition, NOAA Technical Memorandum, NESS 109, National Oceanic and Atmospheric Administration, U.S. Department of Commerce, Washington, D.C.

Duggin, M., 1987; Sensors to Record Atmospheric and Terrestrial Information: Principles of Collection and Analysis, Remote Sensing Applications in Meteorology and Climatology (edited by R.A. Vaughan), D. Reidel Publishing Co. NATO ASI Series, Series C, Vol. 201 pp. 155-172.

Fortuna, J. and Hambrick, L., 1974: The Operation of the NOAA Polar Satellite System, NOAA Technical Memorandum NESS 60, National Environmental Satellite Service, National Oceanic and Atmospheric Administration, U.S. Department of Commerce, Washington, D.C. 127 pp.

Fritz, S., 1964: Pictures from Meteorological Satellites and their Interpretation. Space Science Reviews, 3, pp. 541-580, D. Reidel Publ. Co., Dordrecht-Hollarnd.

Hayden, C.M., 1979: Remote Soundings of Temperature and Moisture. Quantitative Meteorological Data from Satellites. (Edited by J.S. Winston, WMO Technical Note No. 166, WMO-No. 531, pp. 1-32.

Hoppe, E. and Ruiz, A., 1974 Catalog of Operational Satellite Products, NOAA Technical Memorandum NESS 53, National Environmental Satellite Service, National Oceanic and Atmospheric Administration, U.S. Department of Commerce, Washington, D.C. 91 pp.

Hubert, L., 1979: Wind Derivation from Geostationary Satellites, Quantitative Meteorological Data from Satellites. (Edited by J.S. Winston, WMO Technical Note No. 166, WMO-No. 531, pp. 33-59.

Leese, J., 1962: The Role of Advection in the formation of Vortex Cloud Patterns. Tellus. 14 No. 4, pp.409-421.

Leese, J., Booth, A. and Godshall, R., 1970a: Archiving and Climatological Applications of Meteorological Satellite Data, ESSA Technical Report NESC 53, National Environmental Satellite Center, Environmental Satellite Services Administration, U.S. Department of Commerce, Washington, D.C.

Leese, J., Novak, C. and Taylor, R., 1970b: The Determination of Cloud Pattern Motions from Geostationary Satellite Image Data, Pattern Recognition, Pergamon Press, Vol. 2, pp 279-292.

Leese, J., Novak, C. and Clark, B., 1971a: An Automated Technique for Obtaining Cloud Motion from Geostationary Satellite Data Using Cross Correlation, Journal of Applied Meteorology, American Meteorological Society, Vol. 10, No. 1, pp. 118-132.

Leese, J., Pichel, W., Goddard, B. and Brower, R., 1971b: An Experimental Model of Automated Detection, Measurement, and Quality Control of Sea-Surface Temperatures from ITOS IR Data, Proceedings, Seventh International Symposium on Remote Sensing of the Environment, Center for Remote Sensing Information, Ann Arbor, Mich. pp. 625-647.

McClain, E.P. and Baker, D.R., 1969: Experimental Large-Scale Snow and Ice Mapping with Composite Minimum Brightness Charts. ESSA Technical Memorandum NESCTM-12.

McClain, E., Pichel, W. and Walton, C. 1985: Comparative Performance of AVHRR-Based Multichannel Sea Surface Temperatures, Journal of

338

Geophysical Research, 90, No. C6, pp. 11587-11601.

McMillin, L., et al, 1973: Satellite Infrared Soundings from NOAA Spacecraft, NOAA Technical Report NESS 65, National Environmental Satellite Service, National Oceanic and Atmospheric Administration, U.S. Department of Commerce, Washington, D.C. 112 pp.

McMillin, L., Gray, D., Drahos, H., Chalfant, M., and Novak,C., 1983: Improvements in the Accuracy of Operational Satellite Soundings, Journal of Climatology and Applied Meteorology, 22, pp. 1948-1955.

Miller, D.B., 1971: Global Atlas of Relative Cloud Cover 1967-70. Joint NOAA/NESS - Air Weather Service (ETAC) Publication, Washington, D.C., September 1971.

Olesen, F., 1987: Vertical Sounding from Satellite, Remote Sensing Applications in Meteorology and Climatology (edited by R.A. Vaughan), D. Reidel Publishing Co. NATO ASI Series, Series C, Vol. 201, pp. 155-172.

Pailleux, J., 1987: The Impact of Satellite Data on Global Numerical Weather Prediction, Remote Sensing Aplication in Meteorology and Climatology (edited by R.A. Vaughan), D. Reidel Publishing Co. NATO ASI Series, Series C, Vol. 201, pp. 173-187.

Phillips, N., McMillin, L., Gruber, A., and Wark, D., 1979: An Evaluation of Early Operational Temperature Soundings from TIROS-N. Bulletin, American Meteorological Society, 60, pp. 1188-1197.

Smith, W., Woolf, H., Hayden, C., Wark, D., and McMillin, L., 1979: The TIROS-N Operational Vertical Sounder. Bulletin, American Meteorological Society, 60, pp. 1177-1187.

Strong, A., and McClain, E. (Editors) 1986; Field Workshop on Intercalibration of Conventional and Remote-Sensed Sea Surface Temperature Data, WMO Marine Meteorology and Related Activities Report No. 16; WMO/TD-No.95.

Strong, A. and McClain, E. 1984: Improved Ocean Surface Temperatures from Space-Comparisons with Drifting Buoys, Bulletin, American Meteorological Society 65, No. 2, pp. 138-142.

WMO, 1974: Manual on Codes, Volume 1--International Codes, WMO-No. 306.

Young, M., Dolittle, R. and Mace, L., 1972: Operational Procedures for Estimating Wind Vectors from Geostationary Satellite Data, NOAA Technical Memorandum NESS 39, National Environmental Satellite Service, National Oceanic and Atmospheric Administration, U.S. Department of Commerce, Washington, D.C. 19 pp.

THE ERS-1 SATELLITE: OCEANOGRAPHY FROM SPACE

E. ORIOL-PIBERNAT
European Space Agency
Galileo Galilei
Frascati 00044
Italy

ABSTRACT. The European Space Agency is preparing for the launch of the first European Remote Sensing satellite, which will take place in 1990. The ERS-1 will carry on-board a set of active microwave sensors which will allow the collection of global geophysical information over the oceans as well as the imaging of selected coastal and land areas with high resolution. The main feature of this satellite is that the ground segment has been designed in order to offer a duality of services: there will be dissemination of the data within three hours of acquisition for near real time use; also, the data will be archived at specific centres for subsequent off-line precision processing.

1. INTRODUCTION

The European Space Agency is preparing for the launch of the first European Remote Sensing satellite, which will take place in 1990 using the Ariane IV launcher. The ERS-1 will carry on-board a set of active microwave sensors which will allow the collection of global geophysical information over the oceans as well as the imaging of selected coastal and land/ice areas with high resolution. The nominal life of the satellite being 2 years, it is expected to last 3 years.

The predecessor of the ERS-1 satellite, the Seasat, was operating during only 3 months in 1978. Yet, the amount of information supplied to oceanographers, meteorologists and other scientists was such, that not all of it has been completely exploited.

The Seasat demonstrated the usefulness of active microwave information for remote sensing applications, but also showed the need for a preparation of a ground segment designed to cope with the huge amount of data that will need to be handled.

R. A. Vaughan (ed.), Microwave Remote Sensing for Oceanographic and Marine Weather-Forecase Models, 339–353.
© 1990 Kluwer Academic Publishers.

During the last ten years, the scientific community has been preparing for the optimal utilisation of active microwave data. A number of centres are now operationally producing weather and sea state forecasts; the availability of almost real-time oceanographic information at places not covered by conventional measurements is of an obvious interest. The main feature of the ERS-1 satellite is that the ground segment has been designed in order to cover such demand: there will be dissemination of the data within three hours of acquisition. Furthermore, the raw data and all auxiliary information will be archived at specific centres for subsequent off-line precision processing.

2. THE PAYLOAD

The ERS-1 will use the same platform developed for the french SPOT satellite. The payload is basically formed by an Active Microwave Instrument operating at C-band (5.3 GHz) VV polarisation, a Radar Altimeter operating at Ku-band (13.8 GHz) plus the sensors embarked as a result of the Announcement of Opportunities: Along Track Scanner Radiometer/ Microwave Sounder, Precise Range and Range Rate Equipment and the Laser Retroreflector. Figure 1 illustrates all the instruments.

The satellite will fly at a nominal altitude of 785 Km in a circular almost polar sun-synchronous orbit; the local solar time at descending node will be 10h 30 '; mean nodal period 100 min 27.9 s. The reference orbit will have a repeatability of 3 days , but there are up to six foreseen demonstration orbits which can go to 76 or more days repeat cycle.

The sensors can also be grouped according to their mission: The ones in charge of ocean and ice monitoring will be used at a global scale. All these sensors are named Low Bit Rate : AMI wind (scatterometer), AMI wave (SAR in wave mode), radar altimeter and Along track Scanning Radiometer / Microwave Sounder. The relevant data will be recorded on-board for a complete orbit and damped to dedicated stations in addition of being transmitted in real time.

The AMI image (Synthetic Aperture Radar) is oriented towards an all-weather high resolution imaging over land and coastal areas. Since it is a High Bit Rate instrument, the data cannot be recorded on-board, and therefore the mission coverage is defined by the coverage of the acquisition stations.

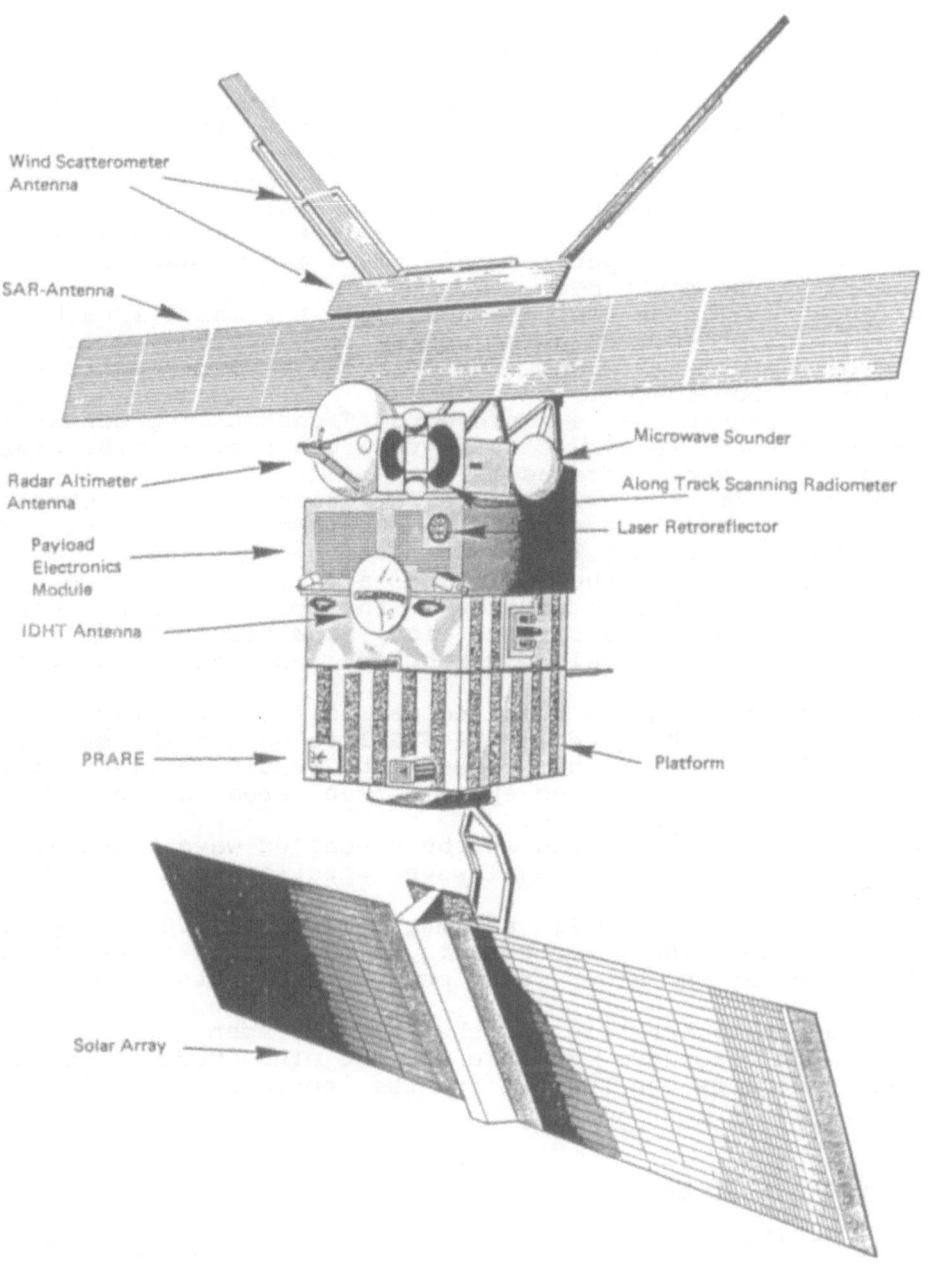

Wind Scatterometer
Antenna

SAR-Antenna

Radar Altimeter
Antenna

Payload
Electronics
Module

IDHT Antenna

PRARE

Solar Array

Microwave Sounder

Along Track Scanning Radiometer

Laser Retroreflector

Platform

Fig. 1   ERS-1 instruments

2.1 The Active Microwave Instrument

The on-board Synthetic Aperture Radar (SAR) operating in image mode will obtain high resolution (i.e. 30 m) images along a swath of 100 Km width centred with an incidence angle of 23 ° to the right of the flight direction. This angle can vary up to 35 °, for experimental applications, using the roll tilt capability of the satellite. The image of the ground depends on the scattering properties of the individual targets illuminated by the radar beam.

High range resolution is obtained from the short length of the transmitted pulse. By using pulse compression techniques on board the satellite (OBRC), the peak power requirements for the system can be reduced.

High azimuth resolution is achieved by "synthetic" processing of the overlapped return signals received from a wide beam generated from a short antenna.

The purpose of the wind mode instrument is to obtain information on the wind speed and direction at the sea surface. The scatterometer will measure the echo power of a signal transmitted by the satellite and returned from the ocean surface; this signal depends on the ocean roghness which is closely related to the surface wind conditions as well as the angle between the wind direction and the radar beam direction.

Three antennae, each one pointing at a different direction (aft, mid, aft, will be used on the ERS-1 to obtain the desired information. The on-ground processing result will be the averaged wind direction and speed for an area of 50X50 Km along a swath of 500 Km width.

The SAR can also be operated in the so-called wave mode. This in order to obtain directional wave spectra from the ocean sampled areas.

At this end, the SAR will be imaging 200 or 300 Km spaced areas of 5X5 Km. The power spectra of the small images will provide information about the ocean wavelength and wave direction.

Due to the characteristics of the AMI instrument, the use of the image mode excludes the use of the wind mode. However, wind and wave mode can operate in the so-called interleaved mode.

2.2 The Radar Altimeter

The Radar Altimeter is a Nadir pointing pulse radar which is designed to measure echoes from ocean, ice and land surfaces.

The ERS-1 Radar Altimeter will provide high precision measurements of the distance from the satellite to the surface; the accuracy over oceans and smooth ice sheets will be better than 10 cm. The instrument will also be capable of operation over land surface when the terrain is not too mountainous.

Another major objective of this instrument is to measure the significant wave height of ocean waves derived from the echo shape; additionally the wind speed can be determined from the measurement of the echo power.

2.3 The Along Track Scanning Radiometer and Microwave Sounder

The ATSR is a passive instrument consisting of an advanced four-channel infrared radiometer working at the spectral channels of 1.6, 3.7, 11 and 12 μm. It will provide measurements of the sea surface temperature and cloud-top temperatures on a swath 500 Km wide on a spatial resolution of 1X1 Km. There will be as well a two-channel microwave sounder (23.8 and 36.5 GHz) providing information on the total water-vapour content of the atmosphere for an area of 20 Km. This information will be used for correction of the atmospheric absorption.

The infrared radiometer is being designed and build by a UK and Australian consortium, while a consortium of french and danish laboratories is responsible for the microwave sounder.

2.4 The Precise Range and range Gate Equipment

The PRARE is a highly accurate microwave ranging system which will be used for orbit determination at decimetre level as well as for various geodetic applications.

The development is financed by the German Ministry of Research and Technology and built under the responsibility of the Institute of Navigation associated with the Geodetic Research Institute in Munich.

The PRARE system is supported by a network of small, mobile and low cost ground stations.

2.5 The Laser Retroreflector

A cluster of 9 corner cube reflectors is carried to permit ranging from laser ranging stations. The measurements are to be used for calibration of the Radar Altimeter to within 10 cm and orbit determination of the radial component to within 1-2m.

3. THE GROUND SEGMENT

A detailed scheme of the ERS-1 ground segment showing the
interfaces between the various elements is presented in Fig.
2.

3.1 The Ground Stations

Two data streams are provided in the X-band (at 8140 and 8040
MHz). The High Rate stream of 105 Mbit/s and a Low Rate stream
which carries the real-time data at 1093 Kbit/s or a rate of
15 Mbit/s when the recorder is playing back an orbit's worth
of data. A separate S-band link is used for other telemetry
and up-link command.
   The ESA network of ground stations will comprise the ones
funded by the Agency: Kiruna (Sweden), Fucino (Italy),
Maspalomas (Spain) and Gatineau (Canada). They will acquire
and will process a selected part in near real time in order to
disseminate Fast Delivery products to the users within 3 hours
of acquisition. The Kiruna and Fucino stations will acquire
both SAR and LBR data, the latter for regional use
(Mediterranean). The other two ESA stations will acquire LBR
data and provide also a FD processing service.
   The stations set up by countries belonging to the states
participating to the ERS-1 program are called National
Stations. So far, they are : West Freugh (UK), AUssagel
(France), Tromsφ (Norway) and Gatineau (Canada). Another
station will be placed by Germany in the Antarctica.
   The rest of ground stations upgraded to acquire ERS-1 data,
are called foreign stations. Negotiations are going on between
the Agency and a number of interested countries : Alaska
(U.S.A.), Australia, Japan (also for a station in
Antarctica), India, Ecuador, Brazil, Argentina, Kenya. Fig 3
shows the potential coverage of all the mentioned stations.

3.2 MMCC

The Mission Management and Control Centre located at ESOC
(Darmstadt, FRG) will operate and control the satellite via
the Kiruna ground station. it will aslo receive from the
Earthnet ERS-1 Central Facility (see 3.4 below) the Preferred
Payload Exploitation Plans from which to generate the Detailed
Mission Operation Plan as well as the Ground Stations
Acquisition schedule.

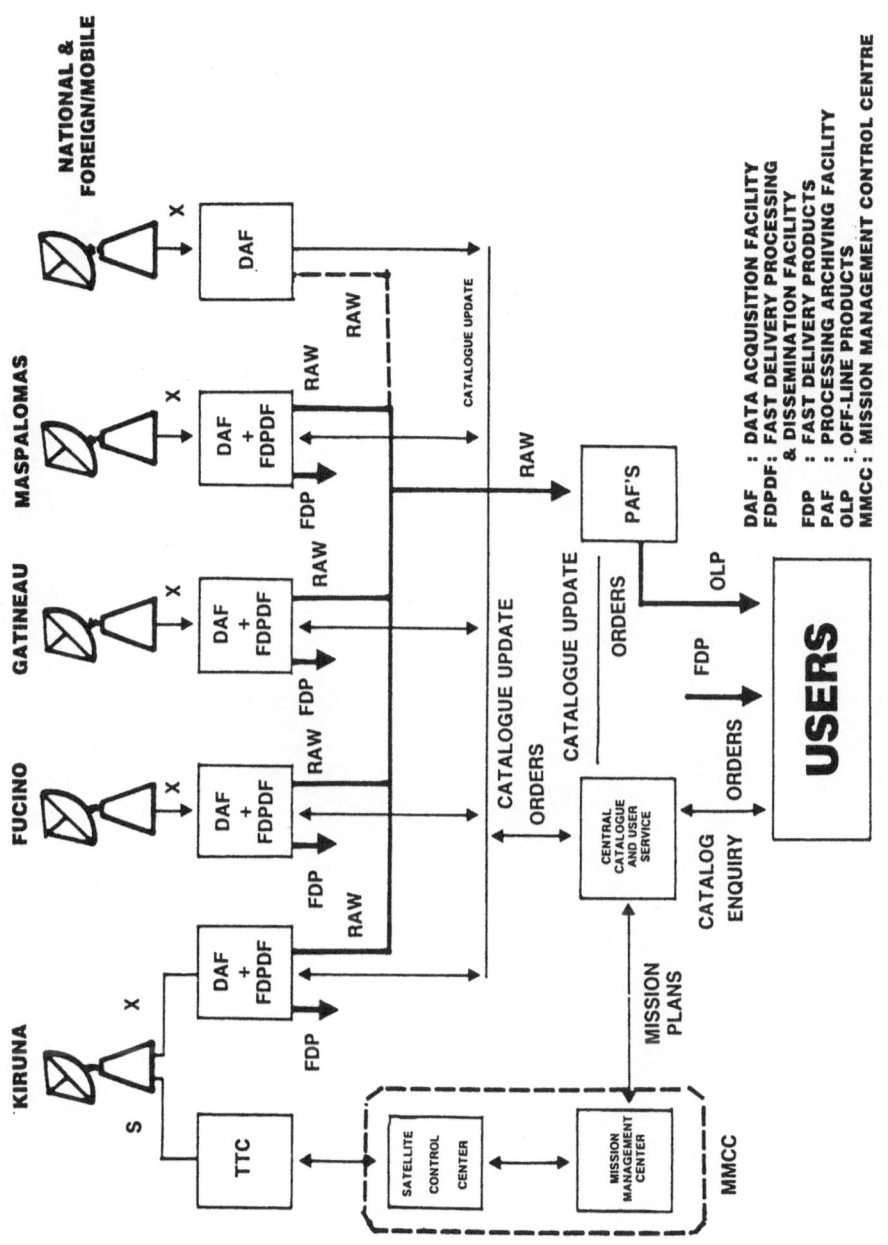

Fig. 2 The ERS-1 Ground Segment Data Flow

DAF : DATA ACQUISITION FACILITY
FDPDF: FAST DELIVERY PROCESSING
       & DISSEMINATION FACILITY
FDP : FAST DELIVERY PRODUCTS
PAF : PROCESSING ARCHIVING FACILITY
OLP : OFF-LINE PRODUCTS
MMCC : MISSION MANAGEMENT CONTROL CENTRE

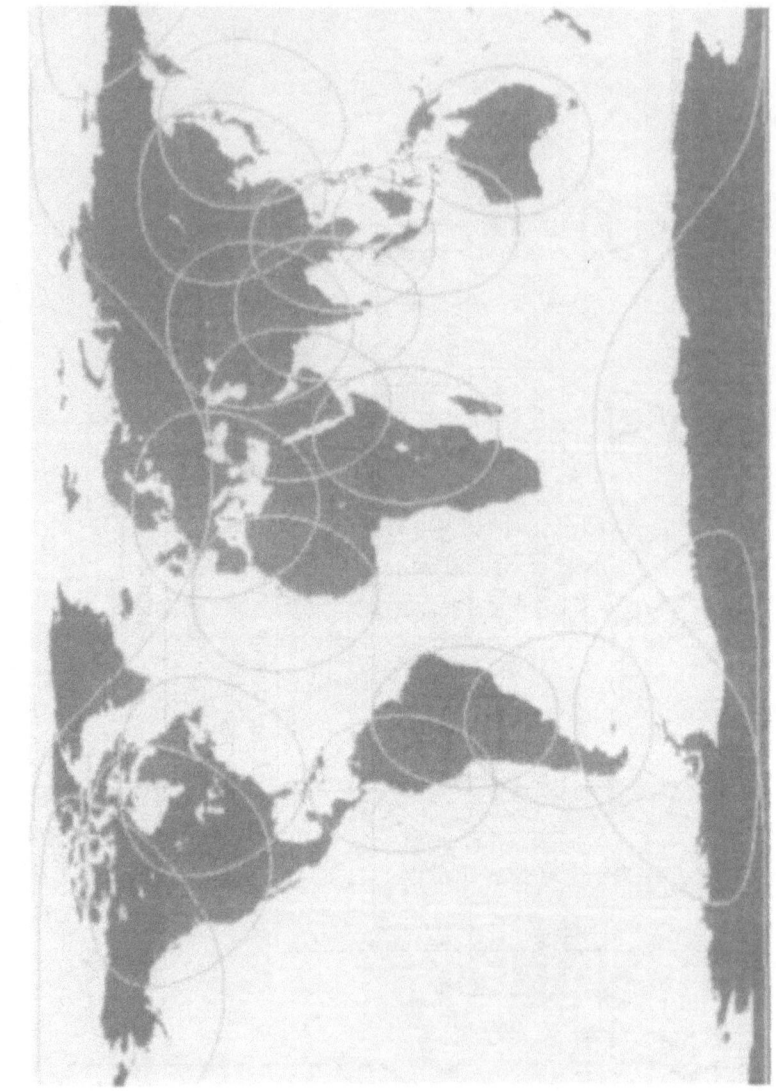

ERS-1
POTENTIAL STATIONS COVERAGE

Fig. 3

3.3 The Processing and Archiving Facilities

All ESA acquired ERS-1 data and associated auxiliary data will be archived at the four Centres which will be set up in France, U.K., Germany and Italy. The PAF operations will continue up to 12 years after the ERS-1 launch.
The scope of the PAFs is not only to archive such data, but also to elaborate precision processed products using different algorithms or more accurate auxiliary information than in the fast delivery case. The tasks have been distributed among the four PAFs according to their vocation. The PAFs are the result of a joint venture between the national governments and ESA.

The French PAF will is being established at IFREMER in Brest, an Institute specialised in Oceanographic research. Three french institutions are involved in the project: IFREMER, CNES and Meteorologie Nationale. It will be in charge of archiving LBR raw data and obtaining high precision LBR products over the oceans.

The UK PAF will be located at RAE in Farnborough. The Royal Aerospace Establishment is very experienced in SAR processing on one hand as well as in Remote sensing data management. It will therefore be in charge of archiving HR raw data and of high precision processing of SAR images as well as SAR waves and Altimeter data over ice or land areas.

The German PAF is setting up a high precision processing centre for SAR images at DFVLR in Oberffafenhofen. It will also restitute very precise orbit for the ERS-1 in co-operation with DGFI (Munich).

The Italian PAF will be a regional archiving and processing centre dedicated to all ERS-1 data acquired over the Mediterranean Sea. Its location will be Matera, in the South of Italy.

3.4 The Earthnet ERS-1 Central Facility

The EECF is being implemented at ESRIN in Frascati (Italy). It is in charge of all user interface functions: cataloguing, handling of user request, interfacing with the MMCC for payload operations, interface to users for future planning, quality control of data products and sensor performance monitoring. The EECF operations will continue up to 12 years after the ERS-1 launch i.e. well into year 2000s.

These functionalities will be distributed amongst the three EECF components: Central User Service, Browse Service and Product Control Service.

The CUS will provide a central catalogue of all ESA ERS-1 products, a central interface for user requests, the schedules for PAFs operations and the control of ESA products dissemination.

The PCS will include a Quality Assurance system whose function is to monitor and control the quality of products from all ESA processing facilities (Ground stations and PAFs) as well as initiate corrective actions when required. It will be responsible for the maintenance of Fast delivery chain application software. In addition, PCS will perform long term validation of the products and mission; will control and support the development of algorithms for product generation within the ERS-1 ground segment and will contribute to the calibration of ERS-1 instruments.

4. THE PRODUCTS

In order to be of maximum use to the Oceanographic and Meteorological communities, ESA has committed to generate and disseminate Fast Delivery products to nominated Centres within 3 hours from satellite observation. This time constraint puts a limit to both the accuracy and the coverage of the FD products; this is the one of the reasons why the PAFs will be in charge of the production of the so-called off-line precision processing products.

4.1 Fast Delivery

Most of LBR data acquired by ESA stations will be processed and disseminated trough land lines to Nominated Centres, for real-time use. The baseline is also to process 3 SAR images per orbit and disseminate them via a satellite link.

The FD products consist of :
- wind direction and speed on a grid spaced 25 Km along the scatterometer swath. Wind speed range 4-24 m/s; accuracy 2 m/s; wind direction accuracy +/- 20°.
- image power spectra on a polar-log scale; averages over sectors of 30 ° overlapped 15 °; 12 wavelengths ranging from 100 to 1000 m.
- significant wave height, satellite altitude and wind speed at nadir averaged every second.
- SAR FD 100x100 Km ground range images at resolution better than 33 m.

4.2 Off-line Products

The various PAFs have put together expertise teams which are defining algorithms which will exploit at their best the capabilities of the ERS-1 instruments together with the use of certain auxiliary information e.g. better orbits, external calibration etc not available for real time processing.

The PAFs have proposed to ESA a certain number of products for their off-line processing. ESA has accepted a subset of them as baseline products i.e. those with agreed specifications, verifiable in quality and properly validated. The rest has been put under the category of experimental products i.e. those with good potential, but whose quality cannot be adequately verified today. The latter have the possibility of becoming baseline, if and when the quality and validation requirements are met.

BASELINE PRODUCTS:
AMI image: raw data, SAR precision image one look, SAR precision image 3 looks.
AMI wave: SAR wave mode image three complex looks.
AMI wind: Sigma nought triplets, wind fields.
Radar Altimeter: sensor record, ocean product.
Orbit: Preliminary, Precise.
EXPERIMENTAL PRODUCTS:
AMI image: SAR rectified image, SAR geocoded image.
AMI wave: Wave spectrum.
Radar Altimeter: ice/land, lakes, gridded/averaged.
Orbit: ERS-1 gravity model.
ATSR: IR and MW radiances, coarse and precision SST, water vapour and liquid water content.

5. THE PREPARATORY ACTIVITIES

ESA has long ago started a number of activities in order to inform and prepare the user community for the wealth of data that the ERS-1 will provide.

At this end, ESA has organised or contributed to the organisation of a number of seminars, training courses and workshops dealing with the scientific and operational problems to be solved for a proper use of the ERS-1 products.

5.1 The Announcement of Opportunities

The ERS-1 AO was issued in 1986. The purpose was to select a number of experiments which would make use of the ERS-1 data, and which Principal Investigators would be guaranteed the access to such data.

This will allow scientific investigations in the fields of Oceanography, Climatology, Glaciology, Geophysics, Coastal and Land Processes and resources. It will support application demonstration projects to evaluate the practical utility of ERS-1 data for operational and industrial activities. It will contribute to the geophysical validation of ERS-1 data and performance verification of the ERS-1 system.

The response was very large: it came from all ESA member states plus Finland, Brazil, Senegal, South Africa, India, Japan, Australia and New Zealand, including co-investigators from Egypt, Mexico, Columbia and Greece. Plus International Organisations: NATO, EEC, ECMWF. A total of 140 proposals involving 1000 investigators covering 300 experiments.

The selected investigators (PIs) met in Frascati (Italy) for the first time in 2-5 may 1988. The next plenary meetings will take place 3 months prior to the launch, 1 year after the launch and 2 years after the launch; the last two will coincide with ESA-organised Workshops. The Co-ordinating Investigators, which amongst other tasks will act as contact point between ESA and the PIs, will have the first meeting in Frascati on 17-21 April 1989. Another three meetings are scheduled before and after the launch.

5.2 Calibration and Validation

The ERS-1 calibration and validation activities have four well-defined objectives:

Engineering calibration of the on-board instruments aimed at characterising their performances in terms of their engineering parameters. This activity is to be performed primarily between ESA and the industrial contractor.

Geophysical validation of the ERS-1 Fast Delivery products, as specified in the ERS-1 declaration. ESA will be supported on this activities by appropriate Institutes/ Laboratories/ PAFs, also via some selected A.O. proposals.

Validation and characterisation of other ERS-1 products, namely those generated off-line in the PAFs. This is a PAFs typical responsibility.

Long-term monitoring of the performances of the ERS-1 instruments and of the eventual impact of the variations on the characteristics of the derived products. This is an ESA responsibility, based on contributions from the PAFs.

ERS-1 calibration and validation will include both pre-launch and post-launch activities.

The pre-launch activities are concentrating on:

Campaigns and studies performed jointly by ESA and laboratories/Institutes to better understand geophysical phenomena and models (e.g. Airborne C-band scatterometer campaigns, Radar Altimeter campaigns).

Development of algorithms for conversion of instrument measurements into calibrated engineering quantities, plus calibrated engineering quantities into geophysical parameters, where appropriate.

Preparation of measurement campaigns to be performed during ERS-1 Commissioning Phase, as well as subsequent data processing and model/algorithm tuning.

The post launch activities will include two phases:

The Commissioning Phase of about 3 months, for the engineering calibration of sensors as well as the intensive geophysical validation of the basic FD products. At the end of this phase, production and dissemination of FD products will start on a routine basis.

The Exploitation Phase during which validation of other ERS-1 products, long term monitoring of ERS-1 sensor performances, fine tuning of algorithms etc, will be performed. Engineering calibration as well as geophysical validation will still continue but in a less extensive fashion.

5.3 Mission Planning

The objectives of the ERS-1 Mission Planning during the pre-launch period are to establish priorities and guidelines for the operations of the ERS-1 system as well as to provide a formal operation plan describing the use of the ERS-1 instruments. After launch, it will monitor the use of the ERS-1 system resources and accept, process and implement users requests.

At this end, an ERS-1 Operation Plan Advisory Group to ESA's Earth Observation Programme Board has been created. The main tasks of the EOPAG are: to consider and recommend guidelines for the definition of the Mission Operation Plan, to consider and recommend a High Level Operation Plan, to review regularly the updates proposed to the Mission Plan and to review on a regular basis the mission implementation reports issued by the ESA Executive.

The ERS-1 mission planning has to take into consideration that two missions need to be carried out simultaneously: a global ocean/ice mission for the LBR sensors and a set of regional missions aimed at imaging specific ice, coastal and land areas by SAR. Particular consideration to be given to: sensor calibration and data validation; demonstration of pre-operational applications and support to international environmental and climate research programmes.

6. METEOROLOGICAL AND OCEANOGRAPHIC OPERATIONAL APPLICATIONS

ERS-1 is devoted primarily to studies of the Earth's ocean and ice. It measures, as previously seen, surface quantities; it will therefore address only a part of the problems of the scientific and operational communities. However, the ERS-1 system will add valuable new information and in particular, will provide repeated global or regional datasets.

It is intended to take Europe as far as possible towards a fully operational system, which will be justified mostly on the grounds of the direct economic benefit resulting from the services provided. The industries which are expected to benefit from the ERS-1 are principally the off-shore industries (e.g. oil and gas exploitation), shipping and fishing. Their primary requirements are to obtain the best possible forecasts of oceanic and atmospheric conditions as far ahead as possible. Forecasting is based on the use of numerical models of the relevant natural processes, and the best predictions, therefore, require the best models as well as the best input data.

The operational oceanographic and meteorological community are preparing for the use of this wealth of data. In particular, they need to upgrade their systems for the data assimilation of single-level non conventional data.

The first attempt of assimilating scatterometer data in a model for medium range weather forecasts was done by Anderson et altri at ECMWF (Reading) by using Seasat Scatterometer data. It was proved that both analysis and forecasts were improved, in particular in areas such as the Southern Hemisphere, where conventional data are scarce.

The improvement provided by the use of the scatterometer data on the wind field analysis has an immediate impact on the sea-state forecasting models. Such models rely on the wind field provided by atmospheric models, and have proved to be very sensitive to their quality (Janssen et altri, 1987); This author also attempted with success to assimilate Significant wave height data from a Radar Altimeter in order to improve wave model outputs.

Plate 7 shows the swell wave height for the 13/9/78 (JASIN period); this is the result of a hindcast run after a wind field obtained at ECMWF after assimilation of the relevant Seasat Scatterometer data. Studies are being conducted by M.P.I. (Hamburg) in order to assimilate the power spectra to be obtained from the AMI wave instrument of the ERS-1; the general theory and first attempts of such assimilations are reported by Hasselmann et altri (1988).

7. CONCLUSIONS

The first European Space Agency Remote Sensing satellite is approaching its launching date. ESA is actively preparing the ground segment to be ready for what is a challenging task: to support the real time operations as well as a large variety of users ranging from operationel centres and industry to the isolated scientific researcher. To all of them, ESA has committed to demonstrate the operational capabilities of the ERS-1 system in order to prepare the community for future advanced systems, e.g. the Polar Platform.

8. REFERENCES

8.1 Anderson D. et altri 1987 'A study of the feasibility of using sea and wind information from the ERS-1 satellite. Part I: Wind scatterometer data'. ECMWF contract report to ESA.

8.2 Hasselman K. et altri 1988 ' Development of a satellite SAR image spectra and altimeter wave height data assimilation system for ERS-1' . MPI contract report to ESA.

8.3 Janssen P. et altri 1987 ' A study of the feasibility of using sea and wind information from the ERS-1 satellite. Part II: Use of scatterometer and altimeter data in wave modelling and assimilation'. ECMWF contract report to ESA.

8.4 ESA ERS-1 Programme proposal. Volume I: System description and mission capabilities.

WIND FIELDS AND SURFACE FLUXES

D. OFFILER
Meteorological Office
London Road
Bracknell, UK

ABSTRACT. The European Space Agency's ERS-1 satellite will carry several active radar instruments, one of which — the AMI scatterometer — will be used to derive surface wind fields. This paper describes how this instrument works, and how the winds are to be extracted from the radar data during ground processing. Plans to calibrate the satellite product, and to validate it during its lifetime will be outlined. Finally, the use to be made of these data, and derived parameters, such as surface fluxes, will be given.

1 Introduction

The European Space Agency (ESA) is planning to launch its first remote sensing satellite, ERS-1, in 1990. Its main complement of instruments are various types of radar, operating at microwave frequencies (5-15 GHz), plus a passive infra-red and microwave instrument, the Along-Track Scanning Radiometer (ATSR). Primarily, the satellite mission is to sense several oceanographic parameters, both for experimental purposes and as a demonstration that such instruments could have a role to play in operational meteorology and oceanography on the global scale.

Of these instruments, the one most applicable to meteorology is the Advanced Microwave Instrument (AMI) in its wind mode of operation. This is designed to sense the near-surface wind vector over the oceans. This paper describes the principles of operation of this type of sensor, and the processing necessary to interpret its raw measurements into familiar geophysical units. The results from some simulation experiments will be given. Any form of instrument needs calibrating and validating, and the AMI is no exception; the plans to achieve this will be outlined, and some examples using SEASAT data will be presented. Finally, we will deal with surface flux parameters, and the röle that satellite data can play in measuring them.

2 Theoretical and experimental background

As soon as microwave radar became widely used in the 1940's, it was found that at low elevation angles, surrounding terrain (or at sea, waves) caused large, unwanted echoes. Ever since, designers and users of radar equipment have sought to reduce this noise (e.g. [1]).

R. A. Vaughan (ed.), Microwave Remote Sensing for Oceanographic and Marine Weather-Forecast Models, 355–374.
© 1990 Kluwer Academic Publishers.

Researchers investigating the effect found that the backscattered echo from the sea became larger with wind speed, so opening the possibility of remotely measuring the wind (e.g. [2], [3]). Radars designed to measure this type of echo are known as "scatterometers".

The backscattering is due principally to in-phase reflections from a rough surface; for incidence angles of more than about 20° from the vertical, this occurs when the Bragg condition is met:

$$\Lambda \sin \theta_i = n\lambda/2 \tag{1}$$

where Λ is the surface roughness wavelength, λ is the radar wavelength, θ_i the incidence angle and $n = 1, 2, 3 \ldots$ First order Bragg scattering ($n = 1$), at microwave frequencies, arises from the small ripples (cat's paws) generated by the instantaneous surface wind stress. The level of backscatter from an extended target, such as the sea surface, is generally termed the Normalised Radar Cross-Section (NRCS), or σ^o. For a given geometry and transmitted power, σ^o is proportional to the power received back at the radar. In terms of other known or measurable radar parameters,

$$\sigma^o = \frac{P_r}{P_t} \cdot \frac{64\pi^3 R_s^4}{\lambda^2 L_s G_o^2 \left(\frac{G}{G_o}\right)^2 A} \tag{2}$$

where P_t is the transmitted power and P_r the power received back at the radar. R_s is the slant range to the target of area A. λ is the radar wavelength, L_s includes atmospheric attenuation and other system losses, G_o is the peak antenna gain and G/G_o the relative antenna gain in the target direction. Equation 2 is often referred to as the Radar Equation. σ^o may be set in a linear form (as above) or in deciBels — i.e. $\sigma^o{}_{dB} = 10 \log_{10} \sigma^o{}_{lin}$.

Experimental evidence from scatterometers operating over the ocean show that σ^o increases with surface wind speed (as measured by ships or buoys), decreases with incidence angle, and is also dependent on the radar beam angle relative to wind direction. Figure 1 is a plot of σ^o against wind direction for various wind speeds at C-band (5 GHz) and an incidence angle of 35°. Direction 0° corresponds to looking upwind, 90° to crosswind and 180° to downwind.

Over the past few winters, ESA have coordinated a number of experiments to confirm these types of curves at 5.3 GHz, the AMI operating frequency. Several aircraft scatterometers have been flown close to instrumented ships and buoys, in the North Sea, Atlantic and the Mediterranean. The σ^o data is then correlated with the surface wind, which has been adjusted to a common anemometer height of 10m (assuming neutral stability — see Section 8.5). An empirical model function has been fitted to this data [4] of the form:

$$\sigma^o = a_0.U_{10}{}^\gamma.\frac{(1 + a_1 \cos\phi + a_2 \cos 2\phi)}{(1 + a_1 + a_2)} \tag{3}$$

where the coefficients a_0, a_1, a_2 and γ are dependent on incidence angle and are weakly dependent on wind speed. This model relates the neutral stability wind speed at 10m, U_{10}, and the wind direction relative to the radar, ϕ, to the NRCS.

It may also be the case that σ^o is a function of sea surface temperature [5], sea state [6] and surface slicks (natural or man-made), but these parameters have yet to be demonstrated as having any significant effect on the accuracy of wind vector retrieval.

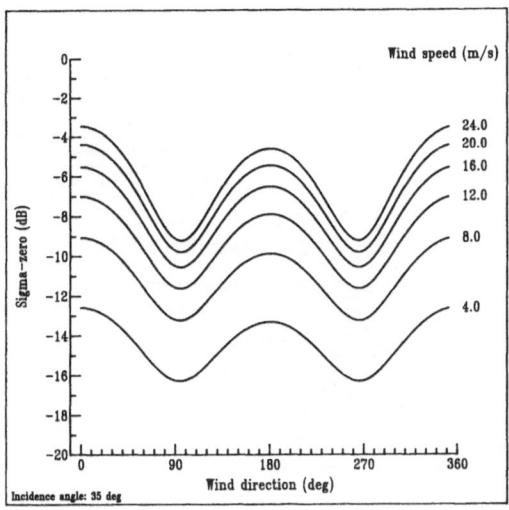

Figure 1: Modelled backscatter, σ^o (in deciBels) against relative wind direction for different wind speeds. Data for C-band (5.3 GHz), vertical polarisation, 35° incidence angle.

3 The AMI scatterometer

Since σ^o shows a clear relationship with wind speed and direction, in principle, measuring σ^o at two or more different azimuth angles allows both wind speed and direction to be retrieved. However, the direction retrieved may not be unique; there may be "ambiguous" directions.

The first multi-beam wind scatterometer to be flown on a satellite — the SEASAT-A Satellite Scatterometer (SASS) — was in 1978 and ably demonstrated the accuracy of this new form of measurement. The specification, as for the AMI, was for RMS accuracies of $2\,\mathrm{ms^{-1}}$ for wind speed and 20° for direction; comparisons with JASIN conventional wind measurements showed that these figures were met if the approximate wind direction were known, so as to select the best from the ambiguous set of SASS directions [7].

The SASS instrument used two beams either side of the spacecraft; the ERS-1 AMI scatterometer uses a third, central beam to improve wind direction discrimination, but is only a single-sided instrument, so its coverage will be less. Figure 2 shows the geometry of the ERS-1 scatterometer. The three antennae each produce a narrow beam of radar energy in the horizontal, but wide in the vertical, resulting in a narrow band of illumination of the sea surface across the 500km width of the swath. As the satellite travels forward, the centre, then rear beam will measure from the same part of the ocean as the fore beam. Hence each part of the swath, divided into 25km squares, will have three σ^o measurements taken at different relative directions to the local surface wind vector.

Figure 3 shows the coverage to be expected of the scatterometer for the North Atlantic

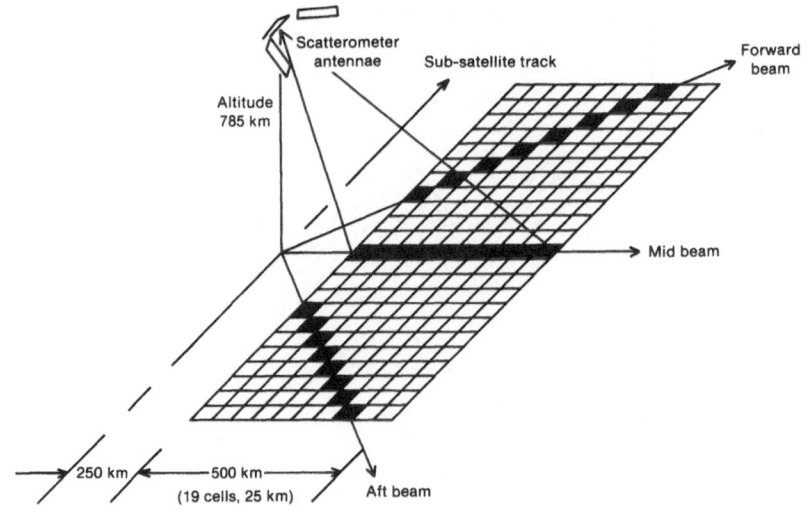

Figure 2: ERS-1 wind scatterometer geometry. For clarity, the 500km-wide
swath is shown as 10 non-overlapping cells. In reality, there are 19
cells across the swath, on a 25km grid, each one being 50x50km
in size.

over 24 hours. These swaths are not static, but "move" eastwards to fill in the large gaps
on subsequent days. Even so, the coverage will not be complete, due to the relatively small
swath width in relation to say, the AVHRR imager on the NOAA satellites. However, there
will potentially be a wind available every 25km within the coverage area, globally, and ESA
are committed to delivering this to operational users within three hours of measurement
time. The raw instrument data are to be recorded on board and replayed to ESA ground
stations each orbit, the principle station being at Kiruna in northern Sweden, where the
wind vectors will be derived.

4   Wind vector retrieval

4.1   GENERATION OF AMBIGUOUS VECTORS

As already mentioned, the AMI scatterometer principally measures the power level of the
backscatter at a given location at different azimuth angles. Since we know the geometry,
such as range and incidence angles, Equation 2 can be used to calculate a triplet of σ^os for
each cell.

It ought to be possible to use the model function (Equation 3) to extract the two pieces
of information required — wind speed and direction — using appropriate simultaneous
equations. However, in practice this is not feasible; the three σ^os will have a finite measurement error, and the function itself is highly non-linear. Indeed, the model, initially based

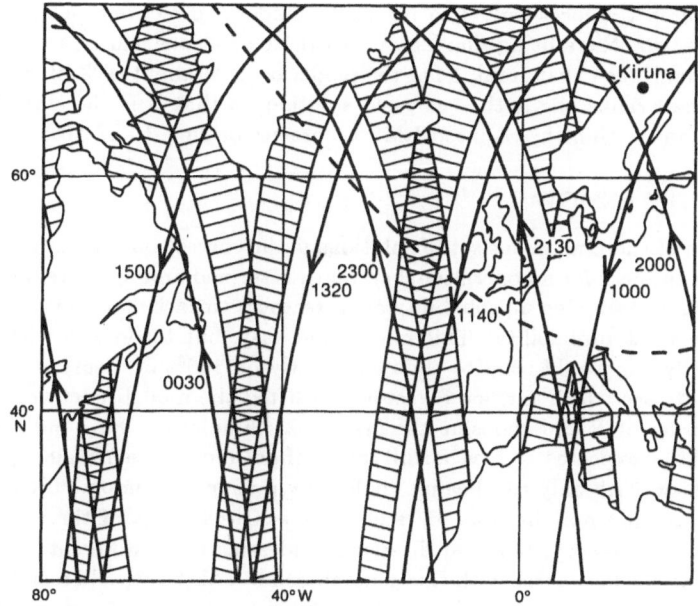

Figure 3: ERS-1 sub-satellite tracks and wind scatterometer coverage of the North Atlantic region over 1 day. The large gaps are partially filled on subsequent days; nominally this occurs on a three-day cycle. The dashed line shows the limits of reception for the Kiruna ground station in Sweden.

on aircraft data, may not be applicable to all circumstances. The wind speed and direction must be extracted numerically, usually by minimising a function of the form:

$$R = \sum_{i=1}^{3} \left(\frac{\sigma^o{}_i - \sigma^o(v, \phi_i, \theta_i)}{\sigma^o{}_i . K p_i} \right)^2 \tag{4}$$

where:

$$Kp^2 = \frac{1}{B\tau_{sn}} . \left[1 + \frac{2}{\text{SNR}} + \frac{1}{\text{SNR}^2} . \left(1 + \frac{\tau_{sn}}{\tau_n} \right) \right] \tag{5}$$

R is effectively the sum of squares of the residuals, comparing the measured σ^os to those from the model function (using an estimate of wind speed and direction), weighted by the noise in each beam, Kp_i. (Kp — Equation 5 — is related to the signal-to-noise ratio, SNR, signal plus noise integration time, τ_{sn}, noise-only integration time, τ_n and system bandwidth, B.)

The wind vector estimate is refined so as to minimise R. Starting at different first guess wind directions, the numerical "solution" can converge on up to four "aliases", or ambiguous wind vectors having similar speeds but different directions. The three-beam configuration often produces only two markedly different directions, usually about 180° apart, with the two others being duplicates. All these solutions can be ranked in order of increasing residual, such that Rank 1 fits the model best; in about 95% of simulated cases, of the two highest ranks one is the "correct" solution, in that it is the closest to the true wind direction and within the required RMS accuracies of 2ms^{-1} and 20°.

4.2 AMBIGUITY REMOVAL

Selecting which of the ambiguous set of solutions is the correct one, when the true wind is unknown (as is the case for an operational system), is termed ambiguity removal or dealiasing. This is merely assigning a probability of correctness to each solution and choosing the one with the highest probability. This may sound easy, but to do this with a consistent high skill (ideally one would want to choose correctly on 100% of cases) is a difficult task. A first stage is to see which extracted vector best fitted the model function, ranking them in increasing order of R and choosing the first rank. Simulations have indicated that this technique might be expected to have a skill rating of not more than 70% [8]. (Compare this with SASS, which, with only two beams, could only manage a random 25% skill.)

Further processing may be done by seeking objectively areas of the swath where the first rank shows some consistency in direction, and selecting a different rank for nearby cells which do not at first agree with this general trend. This may yield a skill score of 90% or so in most cases, but areas which have been consistently ranked wrongly to start with will only be reinforced, so drastically reducing our skill in picking the right solution [8] [9].

Ideally, we would like some indication of the "true" direction, such as from ships or buoys, or even better, because of greater spatial coverage, the wind analysis from a numerical weather prediction model. Unfortunately, while these may be available after a few hours, at the time the satellite winds are required, these independent data will not be ready. The best that we could have is a forecast field of say 6 or 12 hours ahead for the verification time of the satellite data. Winds from a forecast grid can be interpolated to each scatterometer data point and the solution with the nearest direction chosen [8] or used to weight the residuals R (Equation 4) during the retrieval stage [10].

Although this can result in high skill scores (over 95%), there is a major drawback; if a forecast is wrong in the positioning of a circulatory feature by 100-200km (a common occurrence), then there is an area of winds between the forecast centre and true centre which are 180° in error. Here, we will select exactly the wrong scatterometer solution. Worse, if fed back into the next analysis, these winds could reinforce the incorrect forecast which is also being used as the assimilation background. Whilst there are probably enough conventional surface data to prevent this happening in the North Atlantic region, there could be a problem in the Southern oceans and Pacific where other data are sparse.

4.3 EXAMPLE ERS-1 WIND PRODUCT

An example of a simulated ERS-1 swath of wind data is shown in Figure 4. The left hand side illustrates the data that might be available operationally; the AMI Rank 1 wind vectors cover a swath in the North Atlantic south of Greenland with the model 12-hour forecast field as a background. The right hand side shows the validation; the background is the

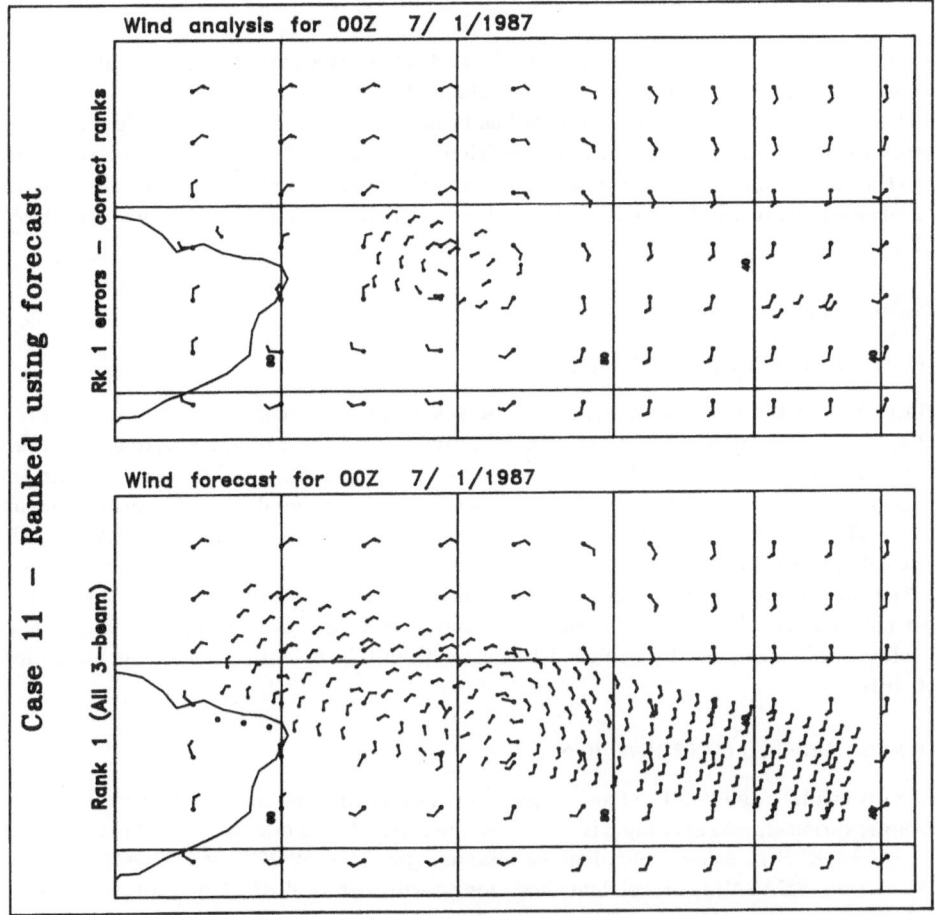

Figure 4: Example of simulated swath of ERS-1 winds and model forecast wind field (left) and its validation with associated analysis field (right). For clarity, only every third satellite and forecast model wind is plotted.

model analysis for the same verification time, and where the ERS-1 data are plotted are the occasions which were incorrectly ranked (the best rank is plotted).

In this example, the ambiguity removal has been done using the forecast field; the ranking errors are due to the incorrect forecast position of the low centre off the tip of Greenland. Cells to the south contained two very close ambiguities; errors in choosing wrongly here are not significant, as are winds at the centre of low and high pressure circulations, where the wind direction is indeterminate anyway.

5 Calibration

5.1 ENGINEERING CALIBRATION

ESA intends that the first 3-6 months after launch will be reserved as a calibration phase. Initially, the correct operation of the space hardware and the ground systems, including processing software, will be checked, followed by a full engineering calibration to obtain the most accurate σ^os. This will will involve the use of active transponders on the ground which allow a small number of absolute point measurements across the swath. For SEASAT, the Brazilian tropical rain forest was used as a natural, homogeneous target to assist in the calibration of the SASS. Recent aircraft experiments [11] have confirmed that these areas are also suitable for relative calibration (intra- and inter-beam) at C-band. It is hoped that an absolute calibration can be achieved to better than 0.5dB, with inter-beam biases less than 0.4dB.

5.2 GEOPHYSICAL CALIBRATION

In order to validate the wind vector output from the AMI wind scatterometer processing algorithms, various forms of comparison data are needed. These may be special observations in support of ERS-1, other coincident campaigns (possibly WOCE and TOGA, such as JASIN was for SEASAT), or conventional, long-term meteorological measurements from ships, buoys and other platforms. Grid point data, as derived from meteorological analyses could also be used.

5.2.1 *Conventional meteorological observations*

The UK Meteorological Office receives most global meteorological data rapidly over the Global Telecommunications System (GTS), and these observations are immediately placed in a comprehensive data-base for general use. Currently, data are available on-line for 5 days before being archived to tape. Typically, there are around 4000 surface wind observations from dedicated weather ships, buoys, merchant ships of the Voluntary Observing Fleet (VOF) and fixed platforms and oil rigs daily. However, it should be noted that:

- over 90% are in the Northern Hemisphere
- the majority of these are in the North Atlantic and North Sea
- many are close to coasts

- the frequency of observation is dependent on the particular platform, but most report several times a day (every hour for weather ships, every 3 or 6 hours for most of the others).

- very few of the VOF are instrumented, the wind force being estimated from the sea condition.

- nearly all instrumental averaging periods are short (2-10mins), therefore making satellite comparisons difficult to interpret.

- as a general rule, winds from the majority of platforms are probably accurate to 2-5ms^{-1} and 20-30°. Dedicated platforms such as ocean weather ships, oceanographic vessels and well maintained buoys will have the better accuracy. Comparisons of good quality wind measurements made between ships and buoys during JASIN [7] suggest accuracies of 1-2ms^{-1} and 10-20° are possible.

The number of occasions that surface-based observations will collocate with the scatterometer swath within 2-3 hours is very small, perhaps 130-150 per day [12]. There will be very few collocations in the Southern hemisphere. The number of collocations with weather ships and other good-quality oceanographic vessels and (relatively) short-lived buoys may well be under 10 per day. The rest will be from the VOF and oil rigs, etc., whose data are often of doubtful quality. There is also no guarantee that any of the Atlantic weather ships will still be operating in 1990-2. To offset this small number, several scatterometer measurements (especially on a 25km grid) may be compared with a single ship observation, and although these are not totally independent, they will provide more stable statistics. Data from special campaigns will give additional good quality surface data.

5.2.2 Grid point data

Another source of comparison data is the gridded winds derived from surface observations. These analyses (either from operational models run by ECMWF or the UK Meteorological Office for example, or from custom-made models) have the advantage of averaging out individual measurement noise and are more consistent spatially and temporally with the scatterometer measurement. They also are constrained to be meteorologically consistent with other parameters (e.g. surface pressure) and other atmospheric levels.

This technique (though only analysing the wind vectors) was successfully used for SEASAT to increase the number of comparisons for GOASEX and JASIN. Numerical analyses will certainly be used in the verification of ERS-1 data [13], but only after the they have been fully calibrated and then validated (using all available sources of comparison data) will they be assimilated into numerical models.

5.2.3 Aircraft data

Data from low-flying aircraft with suitable instrumentation may be used to supplement surface observations during the calibration phase. Consideration is being given to the use of meteorological data gathered from aircraft in support of special validation campaigns

Max. range: 7500 km depending on payload and altitude etc.

Duration: up to 14 hours

Min. height: down to 100ft (exceptionally 50ft in calm conditions and for short periods), by radar altimeter.

Wind vector: U,V components from INS, verified by Doppler radar and Navaids. Also fast response U,V,W turbulence vanes

SST: Barnes IR radiometer (8-14 microns)

Ambient air temperature and humidity from fast response, compensated Platinum resistance and hygristor probes.

Measurement	Accuracy	Sampling
U,V component	$\pm0.4\mathrm{ms}^{-1}$	40Hz
Wind speed	$\pm0.6\mathrm{ms}^{-1}$	40Hz
Wind direction	$\pm3°$	40Hz
Sea surface temperature	±0.3K	4Hz
Air temperature	±0.3K	20 or 4Hz
Altitude	$\pm1\%$	40Hz

Table 1: Measurement capabilities of the MRF Hercules.

because of the unique capabilities of such platforms. Their usefulness was amply demonstrated in TOSCANE-T campaign off Brittany in 1985 [14] and in the Mediterranean in 1986.

The Meteorological Research Flight (MRF) of the UK Meteorological Office operate a modified Hercules (Lockheed C-130) aircraft. This has extensive instrumentation for atmospheric observation of winds, temperature, humidity, clouds, aerosols and chemistry. In brief, the relevant features of such a platform for wind measurement include:

- accurate wind vector determination

- low level flight possible

- stable platform

- long range and endurance

- fully instrumented support data with sampling up to 40Hz [1]

- optimum sub-swath deployment

Detailed capabilities of the MRF Hercules relevant to wind measurement are given in Table 1, and [15] contains a discussion on the use of aircraft for meteorological observations, and gives details on all of the Hercules' sensors.

5.2.4 Problems in calibration

It is arguable that conventional sources of calibration data (principally from surface observations) are often of unknown or insufficient quality, and will be few and far between when paired with satellite data. Such surface data may be more suitable for the longer-term monitoring, tuning and validation of the wind extraction algorithms, both for the 'fast delivery' processing and off-line precision processing. For the calibration phase, it is likely that special campaigns, coordinated by ESA, will be necessary. A variety of platforms will provide a complementary dataset of near-surface observations for the initial calibration task.

One problem in interpreting the comparison of spacecraft and conventional measurements is their fundamental differences in observing technique. Ship or buoy wind measurements are made at a single point, and typically may not be averaged over more than 10 minutes. The satellite measurement, on the other hand, is spatially averaged over 50×50 km without time averaging. It has been estimated that a point measurement should be averaged for not less than an hour to be comparable with the scatterometer footprint [16]. Another problem is the low repetitivity of the spacecraft overpassing a ship (once every three days), even assuming the ship remains within the scatterometer swath. The VOF also try to avoid bad weather, thus making *in situ* data at high wind speeds and sea states very scarce.

The use of large aircraft like the Hercules can partially overcome some of these difficulties. Although they are not deployed for weeks on end in practice, their long range and flight endurance enable them to fly to a predicted overpass location, make fairly extended measurements across the whole swath, and could still be able to intercept another pass, if the region were at high latitudes. The region of coverage (within limits) can also be chosen

---

[1]40Hz is equivalent to 2.5m sampling at a flight speed of $100ms^{-1}$ (200 knots)

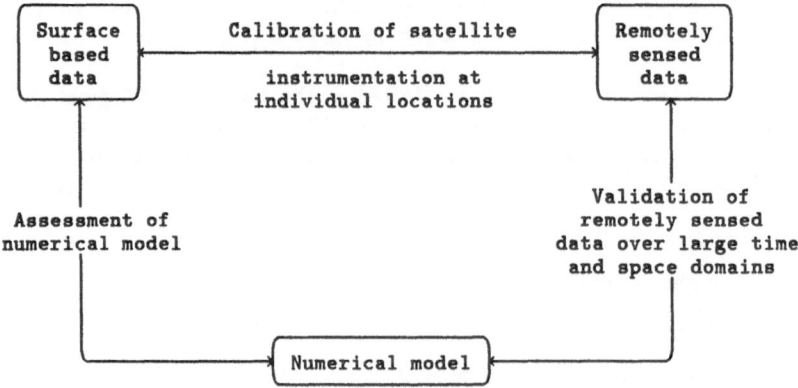

Figure 5: Three-way procedure to validate the ERS-1 data against conventional measurements and numerical models.

flexibly during the campaign. Cross-calibration with surface observations could be made in transit to and from the test area.

Although the flight level is not ideal (perhaps 30-50 metres, depending on wind strength), the bulk stability may be derived from the other instruments carried, and the equivalent 10-metre neutral stability wind (as "measured" by the scatterometer) estimated to an acceptable accuracy for calibration purposes.

6 Validation

After the calibration period, the model function coefficients in Equation 3 may need to be updated. Thereafter, validation, the process of keeping an objective eye on the quality of the satellite-derived data will be continuous, using a three-way comparison [13]. This is illustrated in Figure 5;

1. the numerical model is assessed against the surface data (and the surface data quality controlled against the model). This is, of course, a continuing procedure at operational forecasting centres, and independent of ERS-1.

2. the satellite data are directly compared with surface data where and when they coincide (not often, especially in the Pacific and Southern oceans). Such comparisons have been done for the SEASAT SASS with good results (e.g. against JASIN data, [7], [17] for instance).

3. the satellite data are compared with the model globally, particularly in areas where in-situ data are sparse. This will increase the number of comparisons, extend the range of their geographical coverage and be more likely to encounter extreme conditions.

Steps 2 and 3 can first of all be used to remove any gross biases in the scatterometer winds, and to tune a set of model coefficients for global use. Later, if there is sufficient evidence, different sets of coefficients may be determined for different conditions — such as for season, latitude and/or ocean basin.

Such an exercise is planned by the UK Meteorological Office, and after the satellite wind and wave data have been shown to be of sufficient quality, they will be incorporated into the atmospheric and wave numerical models, and could also be of use in storm warning and ship routing services.

7 Satellite data assimilation

Assimilation trials using SEASAT winds in numerical models have generally concluded that there is disappointingly little impact. This may be due to the relatively low-resolution models in use then, and the fact that they were restricted to the northern hemisphere, where other data are available. In one notable case, the so-called QE II storm, did show improvement in its forecast [18].

Simulated assimilation of SEASAT winds, currently being undertaken in the Forecasting Research Branch of the Meteorological Office, suggests that such satellite data could have most impact when fully utilised as a global dataset. The Operational Forecasting Branch are currently looking at ways to assimilate wave data into their wave model by using SEASAT altimeter data, and both of the above mentioned branches will shortly be collaborating on the detailed methods to be used to validate the ERS-1 wind and waves in near-real time over the whole lifetime of ERS-1. Also to be assessed is how to assimilate the satellite data into the forecasting models optimally, including whether ambiguity removal may best be done within this stage, rather than separately, as described in Section 4.2.

8 Flux determination

8.1 REQUIRED QUANTITIES

The variables that are required for air-sea interaction models are principally the turbulent fluxes of momentum, sensible heat (E_h), latent heat (E_e) and liquid water (precipitation). The momentum flux is often termed the *surface stress* (τ), and is a vector quantity.

For most purposes, direct measurement of these fluxes are too difficult to be done on a wide scale, so parameterisations must be used to estimate these quantities from data that can be obtained more easily. The following sections outline some ways of doing this, indicate their level of accuracy and applicability.

8.2 EDDY CORRELATION METHOD

The surface fluxes in a turbulently mixed boundary layer, sometimes called Reynolds fluxes, may be assumed to be independent of height, and from Monin-Obukov theory, may be expressed as:

$$\left.\begin{array}{rrcl}
\text{downwind surface stress} & \tau & = & -\rho\langle u'w'\rangle \\
\text{sensible heat} & E_h & = & \rho C_p\langle w't'\rangle \\
\text{latent heat} & E_e & = & L_e\langle w'q'\rangle \\
\text{friction velocity} & u_* & = & (\tau/\rho)^{1/2} = |\langle u'w'\rangle|^{1/2}
\end{array}\right\} \qquad (6)$$

where ρ is the air density, C_p is the specific heat of dry air at constant pressure and L_e is the latent heat of evaporation. u', w', t' and q' are the fluctuations of downstream and vertical velocity components, temperature and humidity, with $\langle\rangle$ denoting a time average.

Although the eddy correlation method is the most direct, and hence standard way of measuring these surface fluxes, it is also the most difficult to use because of the high frequencies required of the measurements, and, over the ocean, because of the platform motion. The eddy correlation method may be estimated from aircraft fitted with fast-response turbulence probes for the three wind components, temperature and humidity, as was done during JASIN [19]. The accuracy of this method has been estimated at around 10% for sensible heat flux and 15-20% for latent heat [20].

8.3   DISSIPATION METHOD

The dissipation method is most suited to oceanographic sites because vertical velocities are not required, and the sensors can be more robust since they do not need to respond to high frequencies. Additionally, the ship's velocity need not be removed from the relative wind speed. In steady horizontal flow, Large and Pond [20] show that:

$$\left.\begin{array}{lll} \text{friction velocity} & u_* & = & kZ\epsilon[\phi_m(Z/L) - Z/L]^{-1} \\ \text{sensible heat} & E_h & = & [kZu_*N_t/\phi_t(Z/L)]^{\frac{1}{2}} \\ \text{latent heat} & E_e & = & [kZu_*N_q/\phi_q(Z/L)]^{\frac{1}{2}} \end{array}\right\} \tag{7}$$

where $k = 0.4$ is von Kármán's constant, ϵ is the rate of molecular dissipation of turbulent kinetic energy and N_t and N_q are the temperature and moisture variance dissipation rates. Direct measurements of these latter quantities over the sea is difficult, but may be inferred from spectral density estimates of the downstream velocity, air temperature and absolute humidity at frequencies in the $f^{-\frac{5}{3}}$ region of the spectrum. The dimensionless stability parameter Z/L is determined through the bulk parameters. Since these estimates are not direct, they must be tested against the eddy correlation technique; Large and Pond [20] conclude that the dissipation method gives useful measurements, the fluxes of both methods agreeing to about the same order as the errors in the eddy correlation calculations.

8.4   BULK AERODYNAMIC METHOD

The bulk aerodynamic formulation is the most common and easy to apply method for estimating the surface fluxes since they can be parameterised in terms of mean surface quantities:

$$\left.\begin{array}{lll} \text{surface stress} & \tau_x & = & \rho C_D(u^2 + v^2)^{\frac{1}{2}}u \\ & \tau_y & = & \rho C_D(u^2 + v^2)^{\frac{1}{2}}v \\ \text{friction velocity} & u_*^2 & = & \left(\tau_x^2 + \tau_y^2\right)^{1/2}/\rho \\ \text{sensible heat} & E_h & = & \rho C_P C_T|\vec{v}|(T_s - T_a) \\ \text{latent heat} & E_e & = & \rho L C_Q|\vec{v}|(q_s - q_a) \end{array}\right\} \tag{8}$$

where τ_x, τ_y are the horizontal components of the momentum flux, u, v are the horizontal components of the wind velocity $\vec{v}$, T_a is the dry bulb temperature and q_a is the specific humidity of the air a few metres above the surface; T_s is the sea-surface temperature and q_s

the specific humidity of the air immediately above the surface (assumed to be saturated). ρ is the air density, C_P the specific heat at constant pressure and L is latent heat of evaporation.

Various determinations of the non-dimensional coefficients C_D, C_T and C_Q have been made; for equivalent neutral stability at a height of 10m, the formulations in [20] may be used:

$$\left.\begin{array}{ll} C_{DN} = \left\{ \begin{array}{ll} 1.14 \times 10^{-3}, & 3 < U_{10} < 10\text{ms}^{-1} \\ (0.49 + 0.065 U_{10}) \times 10^{-3}, & 10 \leq U_{10} < 25\text{ms}^{-1} \end{array} \right. \\[2mm] C_{TN} = \left\{ \begin{array}{ll} 1.13 \times 10^{-3}, & \text{unstable} \\ 0.66 \times 10^{-3}, & \text{stable} \end{array} \right. \\[2mm] C_{QN} = \quad 1.15 \times 10^{-3}, \quad \text{unstable} \end{array} \right\} \qquad (9)$$

where U_{10} is the wind speed at a height of 10m. The uncertainty in these coefficients is about 10% (15% for C_{QN}), about the same as the errors in measuring the mean quantities [21].

8.5   ADJUSTMENT FOR ANEMOMETER HEIGHT

Anemometers may be mounted at various heights on the different platforms — typically, drifting buoys may have them at no more than 2-3m above sea level (and in rough conditions, may be below the wave crests!). Moored buoy anemometers may be around 5m high, ships 5-25m, oil rigs 10-30m, and low-flying aircraft 30m upwards. Wind speed usually increases with height in the boundary layer — for neutral stability, and using bulk parameterization:

$$U(z) = \frac{u_*}{k} . \ln \left(\frac{z}{z_o} \right) \qquad (10)$$

where z_o is from these *in situ* sources, as well as with scatterometer winds, the wind speeds need to be adjusted to a common height. Empirically ([22], [23]):

$$z_o = \frac{0.3905}{u_*} + 1.604 \times 10^{-5} . u_*{}^2 - 0.017465 \qquad (11)$$

with u_* in cms^{-1} and z_o in cm. Equations 10 and 11 may be combined in an iterative process to yield estimates of u_* and z_o for a given wind speed, U, measured at height z.

For SEASAT, it was convenient to use 19.5m as the common height; for ERS-1, 10m has been chosen as a more widely used standard. The neutral stability wind speed at an equivalent height of 10m, U_{10}, may be found using Equation 10 with $z=10$m.

9   What are Fluxes used for?

Oceanographers are particularly interested in the surface wind stress, since this is the driving mechanism for the world's ocean circulations. In the shorter term, it affects the sea state. For air-sea interaction studies, the fluxes of sensible and latent heat are required, particularly in the longer term over months to years, when long time averages are taken. For oceanographers, the fluxes are used in heat budget calculations, and for meteorologists,

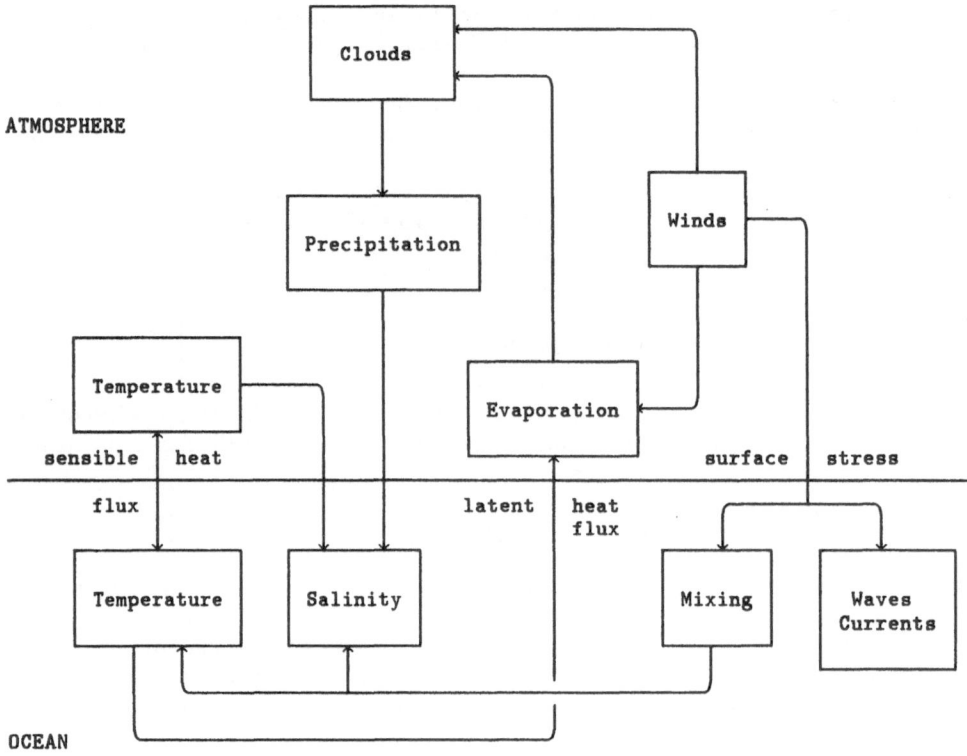

Figure 6: Major atmosphere-ocean coupling mechanisms

they determine the heat transfer into the lower atmosphere for general circulation models. Surface fluxes, together with an accurate knowledge of the sea-surface temperature, become even more important to climatologists attempting to model — and hence predict — phenomena such as El Niño events or droughts in the Sahel region of Africa.

The flux of water vapour from the ocean surface is a major source of the world's precipitation over both land and sea, and precipitation affects the sea surface salinity, and hence the stability of the upper oceans. Figure 6 shows these major coupling mechanisms between the atmosphere and ocean. Omitted from this diagram is any external influence (the sun's radiation) or modification of the fluxes by sea ice.

10 Accuracies

10.1 REQUIRED

For a variety of purposes, heat fluxes are required to an accuracy of the order of $10\mathrm{Wm}^{-2}$. In the tropics, this implies a knowledge of the SST to about 0.2-0.3°C, or about 0.7°C in mid-latitudes; air temperature (wet and dry bulb) to about 0.2°C (or specific humidity to 0.3gKg^{-1}), and surface wind speeds to about 10% [24].

Equivalent precipitation rates of about 1mm/day (about 20% of the annual rate) are required to meet the 10Wm^{-2} in terms of density changes, or 0.35mm/day in terms of latent heat flux.

10.2 SURFACE MEASUREMENTS

The accuracy of surface-based measurements is determined by the sampling frequency and length of time over which the fluxes are averaged. Since ship or buoy data are point measurements, it has been estimated [24] that at least 20 and preferably 100 or more samples per month are required to give adequate to reasonable heat flux accuracy over a 5° square (about 500km). Clearly, for much of the world's oceans, this is not achievable. Even in areas of adequate coverage (e.g. in shipping lanes) some of the raw data may be significantly in error.

For sea 'surface' temperature by bucket or engine intake, the relative accuracy is around 0.5-1°C. Errors for air temperature and humidity are difficult to assess due to heat island effects of the ship or salt contamination of the humidity sensors, and could be a few degrees and tens of percent respectively, particularly if the wind is light. Wind is rarely measured by anemometers on merchant ships, and is estimated using the Beaufort Scale, based on the sea state. Uncertainty in these winds could be around 2ms^{-1} for low wind speeds, increasing to possibly 5ms^{-1} at force 9. Due to the difficulties of siting anemometers on ships, winds measured in this way are estimated to have errors of 1-2ms^{-1}. Measurement of precipitation at sea is even more difficult due to the ships motion and the disturbed air flow around a gauge, and errors may well exceed 20%. In tropical areas, the point measurements may be highly unrepresentative of monthly totals over an area.

The above comments apply in particular to ships of the Voluntary Observing Fleet (VOF), where expensive instrumentation and its careful siting cannot be justified. Better quality observations are possible from dedicated research platforms — for instance, wind accuracies of about 0.7ms^{-1} have been estimated for JASIN ships and buoys.

10.3 SATELLITE MEASUREMENTS

While surface-based measurements are only representative of the local conditions, but can be sampled (relatively) frequently, satellite data (usually) has good global coverage at the expense of only sampling a given area about twice a day — depending on an instrument's swath width — for polar orbiters. In terms of samples per month, one satellite instrument gives roughly the same number of measurements as for surface data, but is at least globally consistent.

Satellite measurements of SST can be made with the AVHRR-type of instrument to better than 1°C, but only with clear skies; the SMMR on SEASAT had a similar accuracy in limited conditions. The ERS-1 ATSR instrument is expected to achieve accuracies of a few tenths of a degree when averaged over 50km areas, but again cannot give results in completely cloudy areas.

Although sounding instruments can retrieve atmospheric temperatures throughout most of the troposphere to 1-1.5°C, they cannot distinguish between surface temperature and air temperature a few metres above the surface. Equally, satellite instruments cannot measure near-surface humidity, so indirect approaches are necessary. The SMMR instrument on SEASAT has been shown to measure total integrated water vapour, W, reasonably accurately. Further, there appears to be reasonable correlation between monthly means of W and q_a [25] from which latent heat fluxes can be derived to about $35\,\mathrm{Wm}^{-2}$. The AMSU-B instrument, due to be flown on the NOAA series of polar orbiters in the mid-1990's is expected to retrieve profiles of humidity to 10% accuracy within the troposphere; this may allow better correlation with near-surface humidity and hence better monthly mean flux estimates.

The SEASAT SASS has ably demonstrated that scatterometer-type instruments are capable of achieving accuracies of $1\text{-}2\,\mathrm{ms}^{-1}$ and $10\text{-}20°$ for near-surface winds [7], though whether the empirical models used to derive the winds are globally applicable is open to question. The SEASAT altimeter and SMMR instruments also showed good agreement in deriving wind speed over a limited range of conditions. The ERS-1 scatterometer will probably have a similar accuracy, and by using different sets of model coefficients for off-line precision processing, maintain it under a variety of surface conditions.

Precipitation can be estimated from geostationary satellites by correlation with cloud top temperature, to perhaps 20-30% over the tropical oceans, but with less accuracy in mid-latitudes. Microwave instruments on polar orbiters can provide estimates of rain rate, but the large footprint of these instruments, and their infrequent sampling make them unsuitable to monitoring the tropics.

For all these types of measurement, satellite-derived quantities must be 'calibrated' against collocated *in situ* surface data. Although the latter are often called *surface truth*, as we have seen, the relative accuracies obtainable are often limited, and the absolute accuracies are mainly unknown. However, the satellite data can provide good spatial coverage of many required quantities to a reasonable accuracy, and estimates of other quantities, particularly where there is little or no *in situ* data. To improve this situation, there is still a need for more and better surface-based measurements, allowing a mixture of satellite and surface data each to be used to their best advantages.

11 Conclusions

The wind and wave data from ERS-1 will not be unique; by the mid 1990s, it is expected that there will be several other satellites, operated by the United States, Canada, Japan and China which could have similar instruments. Indeed, the decision is to be taken soon whether a follow-on ERS-2 can be funded within Europe. Further into the decade, the Polar Platform, part of the US Space Station project is likely to carry similar microwave instruments as part of its fully operational payload. Although ERS-1 strictly is not an operational satellite, ESA intends that it can demonstrate this capability during its 2-3 year life. The Meteorological Office, together with other agencies, has taken up that challenge, and intends to investigate the utility (or otherwise) of this new source of data, exploiting it fully, in numerical models and for storm warning, ship routing services and for climatology.

References

[1] TW Harrold. Ground clutter observed in the Dee weather radar project. *Meteorological Magazine*, 103:140–141, 1974.

[2] K Krishen. Correlation of radar backscattering cross sections with ocean wave height and wind velocity. *Journal of Geophysical Research*, 76:6528–6539, 1971.

[3] W Jones and LC Schroeder. Radar backscatter from the ocean: dependence on surface friction velocity. *Boundary Layer Meteorology*, 13:133–149, 1978.

[4] AE Long. Towards a C-band radar sea echo model for the ERS-1 scatterometer. In *Proceedings of First ISPRS Colloquium on Spectral Signatures in Remote Sensing, Les Arcs, France, December 1985*, pages 29–34, ESA SP-247, 1986.

[5] PM Woiceshyn et al. The necessity for a new parameterization of an empirical model for wind/ocean scatterometry. *Journal of Geophysical Research*, 91(C2):2273–2288, February 1986.

[6] RE Glazman. Wind-fetch dependence of Seasat scatterometer measurements. *International Journal of Remote Sensing*, 8(11):1641–1647, November 1987.

[7] D Offiler. A comparison of SEASAT scatterometer-derived winds with JASIN surface winds. *International Journal of Remote Sensing*, 5(2):365–378, 1984.

[8] D Offiler. ERS-1 surface wind ambiguity removal by means of objective processing and subjective human intervention. July 1985. Final report to ESA Contract No. 6154/85/NL/BI.

[9] F Delsol and E Hellsten. ERS-1 surface wind ambiguity removal by objective techniques. 1985. Final Report to ESA Contract No. 6156/84/NL/BI.

[10] KJ Schwenzfeger. Algorithm for wind scatterometer data analysis ambiguity suppression. In *The Use of Satellite Data in Climate Models, 10-12 June 1985, Alpbach, Austria*, pages 35–36, ESA SP-244, 1985.

[11] ESA. Radar backscattering in the tropical rain forest. *Earth Observation Quarterly*, (21):6–7, March 1988. ESA Publications, Noordwijk, The Netherlands.

[12] AE Long, D Offiler, and T Wolff. Report of the Scatterometer Algorithm Development (SAD) Group. April 1984. ESA report ref. SAD01.

[13] PE Francis. The use of numerical wind and wave models to provide areal and temporal extensions to instrumental calibration and validation of remotely sensed data. In *ERS-1 Wind and Wave Calibration Workshop, Schliersee, FRG, 2-6 June 1986*, pages 53–55, ESA SP-262, September 1986.

[14] E Attema. Airborne and tower-based scatterometry during the PROMESS and TOSCANE-T campaigns. In *Proceedings of First ISPRS Colloquium on Spectral Signatures in Remote Sensing, Les Arcs, France, December 1985*, ESA SP-247, 1986.

[15] CJ Readings. The use of aircraft to study the atmosphere: the Hercules of the Meteorological Research Flight. *Meteorological Magazine*, 114:66–77, 1985.

[16] RA Brown. On a satellite scatterometer as an anemometer. *Journal of Geophysical Research*, 88(C3):1663–1673, 1983.

[17] NASA. *JASIN workshop report: Vol 1, Findings and conclusions*. Technical Report no. 80-62, Jet Propulsion Laboratory, Pasadena, 1980.

[18] DG Duffy and R Atlas. The impact of Seasat-A scatterometer data on the numerical prediction of the *Queen Elizabeth II* storm. *Journal of Geophysical Research*, 91(C2):2241–2248, February 1986.

[19] S Nicholls. Measurements of turbulence by an instrumented aircraft in a convective boundary layer over the sea. *Quarterly Journal of the Royal Meteorological Society*, 104:653–676, 1978.

[20] WG Large and S Pond. Sensible and latent heat flux measurements over the ocean. *Journal of Physical Oceanography*, 12:464–482, 1982.

[21] TH Guymer et al. Transfer processes at the air-sea interface. *Phil. Trans. R. Soc.*, 308:253–373, 1983.

[22] WJ Pierson. *Verification procedures for the SEASAT measurements of the wind vector with the SASS*. Technical Report, Cuny Institute of Marine and Atmospheric Sciences, City College, New York, 1978. Report to JPL Contract 954411.

[23] I Halberstam. Some considerations in the evaluation of SEASAT-A satellite scatterometer (SASS) measurements. *Journal of Physical Oceanography*, 10:623–632, April 1980.

[24] PK Taylor. The determination of surface fluxes of heat and water by satellite microwave radiometry and *in situ* measurements. In C Gautier and M Fieux, editors, *Large Scale Oceanographic Experiments and Satellites*, pages 223–246, D Reidel Publishing Co., Dordrecht, 1984.

[25] WT Liu. Estimation of latent heat flux with Seasat-SMMR, a case study in the N. Atlantic. In C Gautier and M Fieux, editors, *Large Scale Oceanographic Experiments and Satellites*, pages 205–221, D Reidel Publishing Co., Dordrecht, 1984.

GLOSSARY OF ACRONYMS

AGC	Automatic Gain Control
ALT-2	Proposed altimeter for Polar Platform
AMI	Active Microwave Instrument
AMSR	Advanced Microwave Scanning Radiometer
AMSU	Advanced Microwave Sounding Unit
ATM	Airborne Thematic Mapper
ATS	Applications Technology Satellite
ATSR	Along Track Scanning Radiometer
ATSR/M	Along Track Scanning Radiometer and Microwave Sounder
AVHRR	Advanced Very High Resolution Radiometer
AVHRR-2	Advanced Very High Resolution Radiometer, 5-channel version
BOMEX	Barbados Oceanographic and Meteorological Experiment
CCCO	Committee on Climate Change in the Ocean
CISK	Conditional Instability of the Second Kind
CNES	Centre National d'Etudes Spatiales
COSPAR	(International) Committee for Space Research
CUS	Central User Service
CZCS	Coastal Zone Colour Scanner
DFVLR	Deutsche Forschungs- und Versuchsanstalt fur Luft- und Raumfahrt e.v. (German Aerospace Research Establishment)
DMSP	Defence Meteorological Satellite Programme
ECMWF	European Centre for Medium Range Weather Forecasts
EECF	Earthnet ERS-1 Central Facility
EOF	Empirical Orthogonal Function
EOP	Earth Observation Programme
EOPAG	ERS-1 Operation Plan Advisory Group
EOPP	Earth Observation Preparatory Programme
EOS	Earth Observing System
ERS-1	Earth Resources Satellite 1 (European)
ESA	European Space Agency
ESMR	Electronically Scanned Microwave Radiometer
ESOC	European Space Operations Centre
ESRIN	Headquarters of EARTHNET Office
EUMETSAT	European Meteorological Satellite Organisation

FAO	Food and Agricultural Organisation
FDPS	Fast Delivery Product Station
FGGE	First GARP Global Experiment
GAC	Global Area Coverage
GARP	Global Atmospheric Research Project
GATE	GARP Atlantic Tropical Experiment
GCM	General Circulation Model
GEMS	Global Environmental Monitoring System
GEOS-3	Geodetic Satellite
GEWEX	Global Energy and Water Cycle Experiment
GMS	Geosynchronous Meteorological Satellite
GOASEX	Gulf of Alaska SEASAT Experiment
GOES	Geostationary Operational Environmental Satellite
GOFS	Global Ocean Flux Study
GOS	Global Observing System
GOSSTCOMP	Global Operational Sea Surface Temperature Computation
GPS	Global Positioning System
GRID	Global Resources Information Data Base
GTS	Global Telecommunication System
HBR	High Bit Rate
HCMM	Heat Capacity Mapping Mission
HIRS	High-Resolution Infrared Radiation Sounder
HIRS2	Second generation HIRS
HRV	High Resolution Sensor on SPOT
IABO	International Association of Biological Oceanography
IAHS	International Association of Hydrological Sciences
IAMAP	International Association of Meteorological and Atmospheric Physicists
ICES	International Council for the Exploration of the Sea
ICSU	International Council for Scientific Unions
IDHT	Instrumentation, Data Handling and Transmission
IFOV	Instantaneous Field of View
IGAG	International Global Atmospheric Chemistry Programme
IGBP	International Geosphere-Biosphere Programme
IGU	International Geophysical Union
ILP	International Lithological Programme
IOC	Intergovernmental Oceanographic Commission
ISLSCP	International Satellite Land Surface Climatology Project
ITOS	Improved TOS series
IUBS	International Union of Biological Sciences
IUGG	International Union of Geodesy and Geophysics
JASIN	Joint Air-Sea Interaction Project
JERS-1	Japanese Earth Resources Satellite
JSC	Joint Scientific Committee
JGOFS	Joint Global Ocean Flux Study

LBR	Low Bit Rate
LFMR	Low Frequency Microwave Radiometer
MCSST	Multi-Channel Sea Surface Temperature
MESSR	Multispectral Electronic Self-Scanning Radiometer
MIMR	Multi-Band Imaging Microwave Radiometer
MOP	Meteosat Operational Programme
MOS-1	Marine Observation Satellite (Japanese)
MRF	Meteorological Research Flight
MSR	Microwave Scanning Radiometer
MSU	Microwave Sounding Unit
NASA	National Aeronautics and Space Administration
NASDA	Japanese Space Agency
NATO	North Atlantic Treaty Organisation
NCAR	National Centre for Atmospheric Research
NEΔT	Noise Equivalent Temperature Difference
NESDIS	National Environmental Satellite Data and Information Services
NOAA	National Oceanic and Atmospheric Administration
NOAA-1	First NOAA Series Satellite, etc
NORDA	Naval Ocean Research and Development Activity
N-ROSS	Navy Remote Ocean Sensing System
NSCAT	NASA Scatterometer
NWS	National Weather Service
OBRC	On Board Range Compression
PAF	Processing and Archiving Facilities
PBL	Planetary Boundary Layer
POEM	Physical Oceanography of the Eastern Mediterranian
PRARE	Precise Range and Range Rate Equipment
RAE	Royal Aerospace Eatablishment
RAL	Real Aperture Radar
RTE	Radiative Transfer Equation
SAR	Synthetic Aperture Radar
SAR-C	C-Band Synthetic Aperture Radar
SASS	SEASAT Scatterometer
SCAMS	Scanning Microwave Spectrometer
SCAR	Standing Committee on Antarctic Research
SCATT-2	Scatterometer derived from ERS1 instrument
SCOPE	Scientific Committee on Problems of the Environment
SCOR	Standing Committee on Oceanographic Research

SIR-B	Second Shuttle Imaging Radar Experiment
SLAR	Side-Looking Airborne Radar
SMLE	Sub-Optimal Maximum Likelihood Estimation
SMMR	Scanning Multichanel Microwave Radiometer
SPOT	Satellite Probatoire de l'Observation de la Terre
SSM/I	Special Sensor Microwave Imager
SST	Sea Surface Temperature
SSU	Stratospheric Sounding Unit
STREX	Storm Transfer and Response Experiment
TIROS	Television and Infra-Red Observation Satellite
TOGA	Tropical Oceans Global Atmosphere
TOPEX	NASA Ocean Topography Experiment
TOS	TIROS Operational System
TOVS	TIROS Operational Vertical Sounder
TRMM	Tropical Rainfall Measuring Mission
TEJ	Tropical Easterly Jet
UNEP	United Nations Environmental Programme
UNESCO	United Nations Educational, Scientific and Cultural Organisation
URSI	International Union on Radio Science
VAS	VISSR Atmospheric Sounder
VISSR	Visible Infrared Spin-Scan Radiometer
VOF	Voluntary Observing Fleet
VTIR	Visible and Thermal Infrared Radiometer
VTPR	Vertical Temperature Profile Radiometer
WAM	Wave Model
WAMEX-79	West African Meteorological Experiment
WCDP	World Climate Data Programme
WCRP	World Climate Research Programme
WMCE	Western Mediterranean Circulation Experiment
WMO	World Meteorological Organisation
WOCE	World Ocean Circulation Experiment
WWW	World Weather Watch

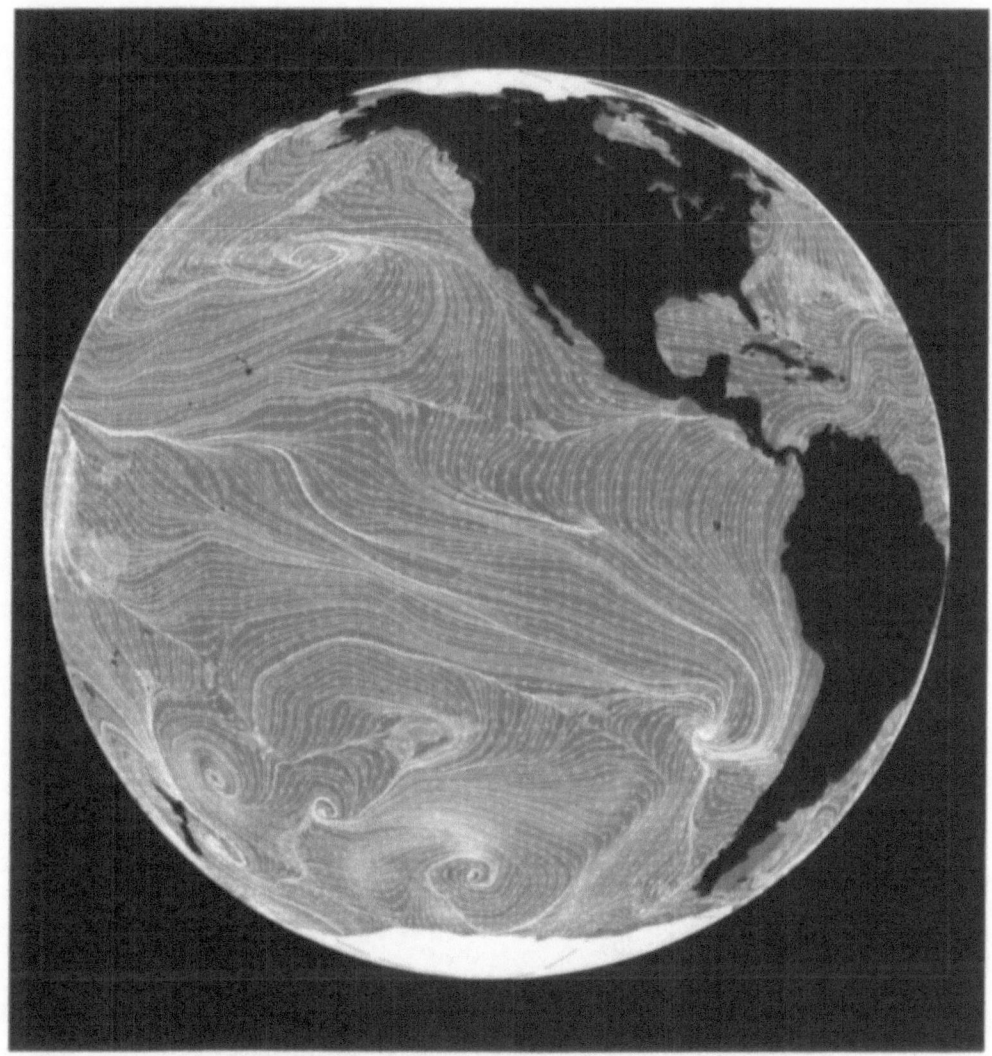

Plate 1.   Global wind map derived from 12 hours of SASS swaths (courtesy of S Peterherych, M Wurtele and P Woiceshyn). (R A Brown, chapter 5. See page 102 for reference.)

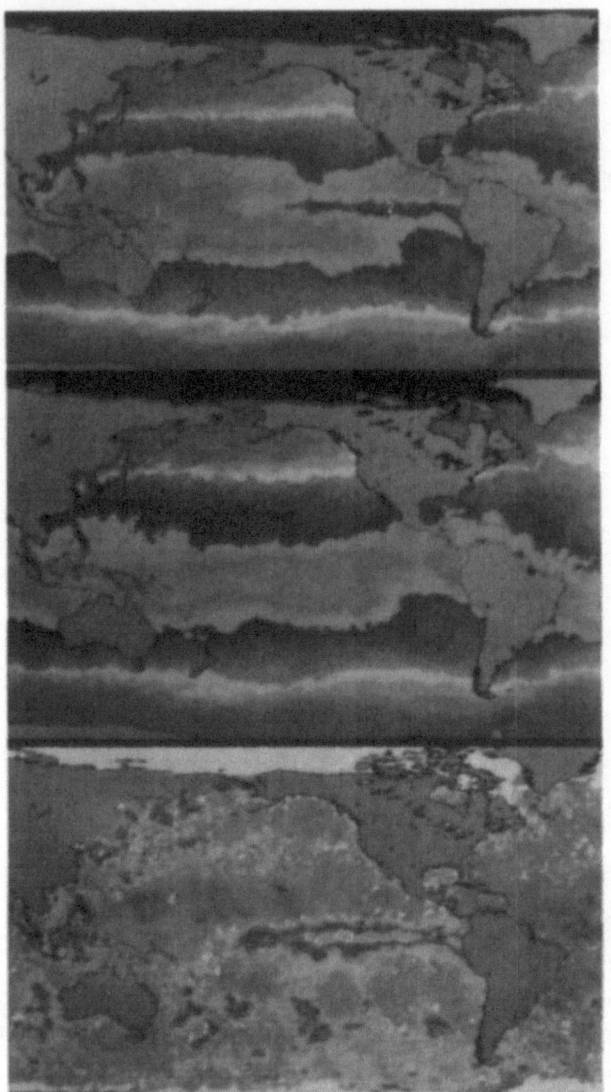

Plate 2. Sea Surface Temperature Signal of the 1982–1983 El Niño Event. Colour coded Pacific Ocean SST distributions, derived from the AVHRR/2 on NOAA-7, are shown for January 20 1984, during 'normal' conditions (top) and January 20 1983, during a strong El Niño event (centre); the difference between the two gives the El Niño SST anomaly (bottom). In the 'normal' situation, the warmest water is confined to the western Equatorial region (1) and a cold tongue stretches from the east (2), but during an El Niño event this zonal gradient is strongly reduced giving rise to warm anomalies to the east (3), in this case reaching 6K, and cold anomalies in the western tropics (4). [The images were produced by NOAA/NESDIS, and distributed by NASA]. (P J Minnett, Chapter 7. See page 159 for reference.)

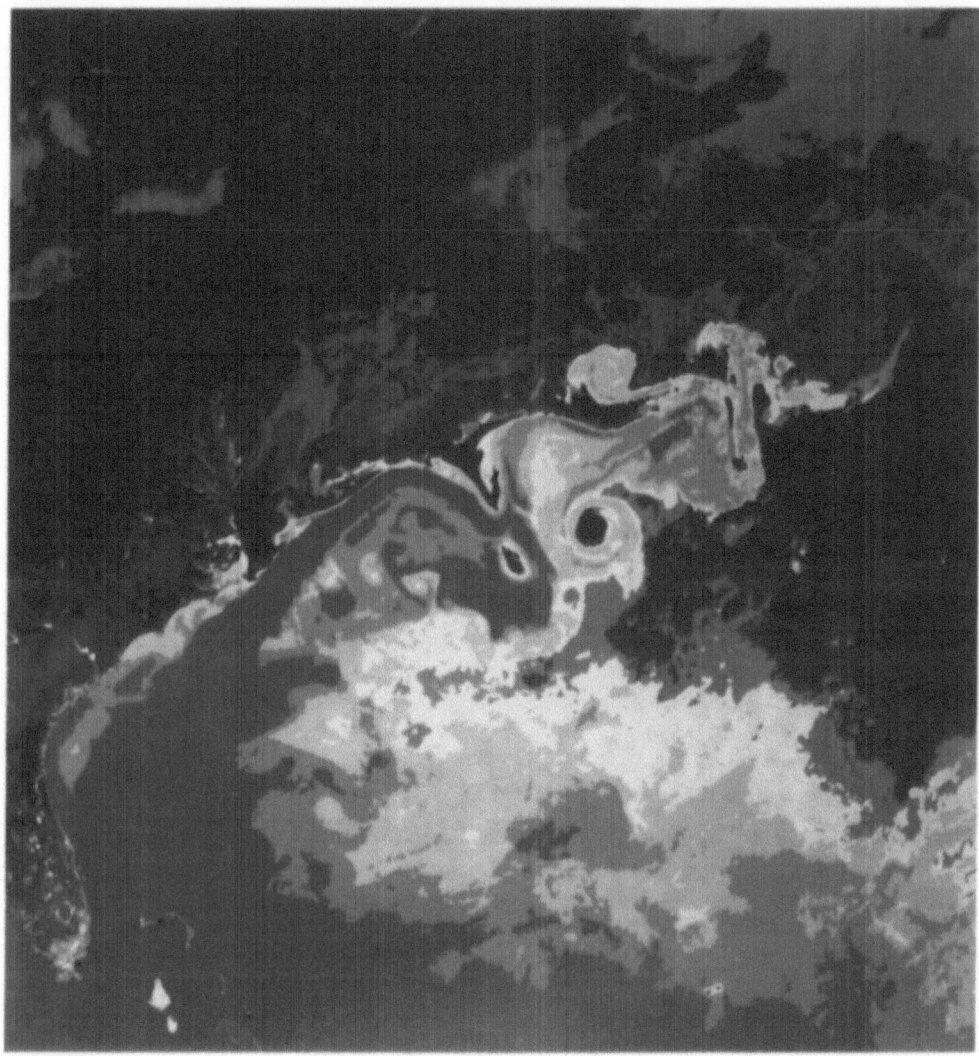

Plate 3.   Western North Atlantic Sea Surface Temperatures. The warm surface water of the Gulf Stream can clearly be seen extending to the northeast in this composite image derived from AVHRR data of the first week of April 1984. Instabilities in the Gulf Stream cause meanders which may 'pinch-off' to form rings with cold water cores. As the Gulf Stream progresses away from N. America it looses much heat to the atmosphere and mixes with cooler surrounding water. [Image generated at U. Miami]. (P J Minnett, Chapter 7. See page 159 for reference.)

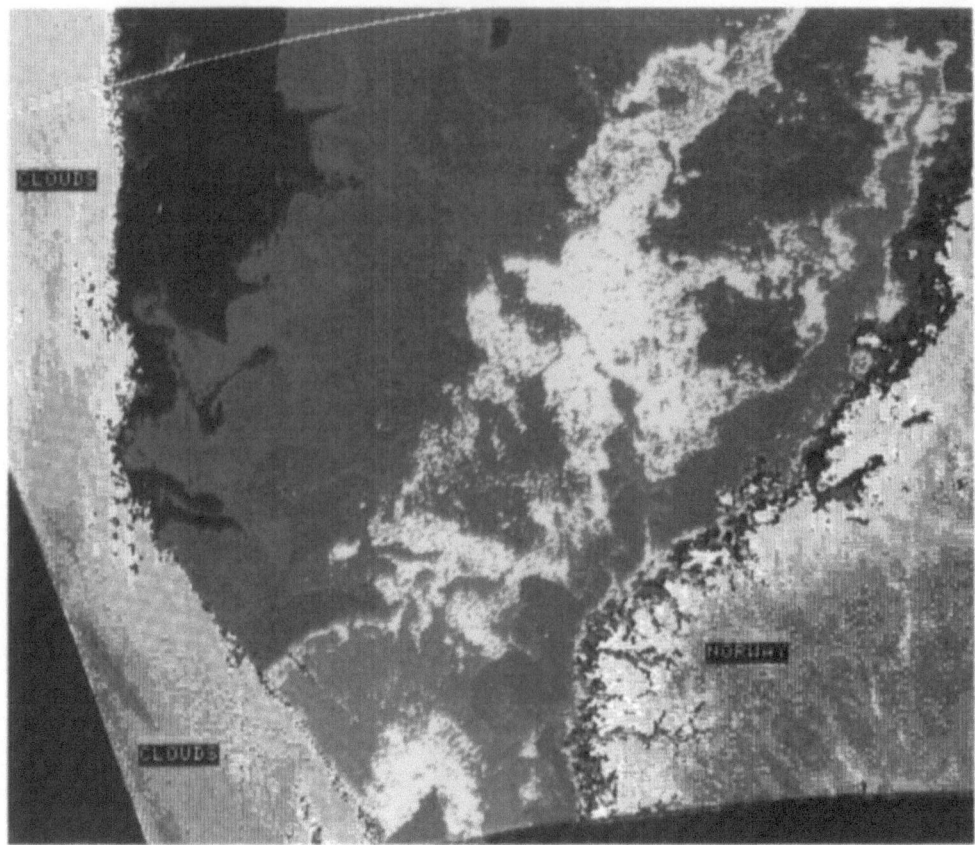

Plate 4. Surface Temperatures of the Norwegian Sea. Colour-coded SSTs derived from NOAA-7 data on 15 November 1984, 15:06 UTC, show the complex surface water mass structure in the Norwegian Sea. The northward flowing Norwegian Coastal Current carries warmer water of Baltic Sea origin along the coast, being joined by warm water from the N. Atlantic from the southwest. These two water masses appear to diverge again further north. To the west, partially obscured by cloud, is cold water of the Greenland Sea. The strong SST gradients show high spatial variability, clearly revealed only in high resolution imagery. (P J Minnett, Chapter 7. See page 159 for reference.)

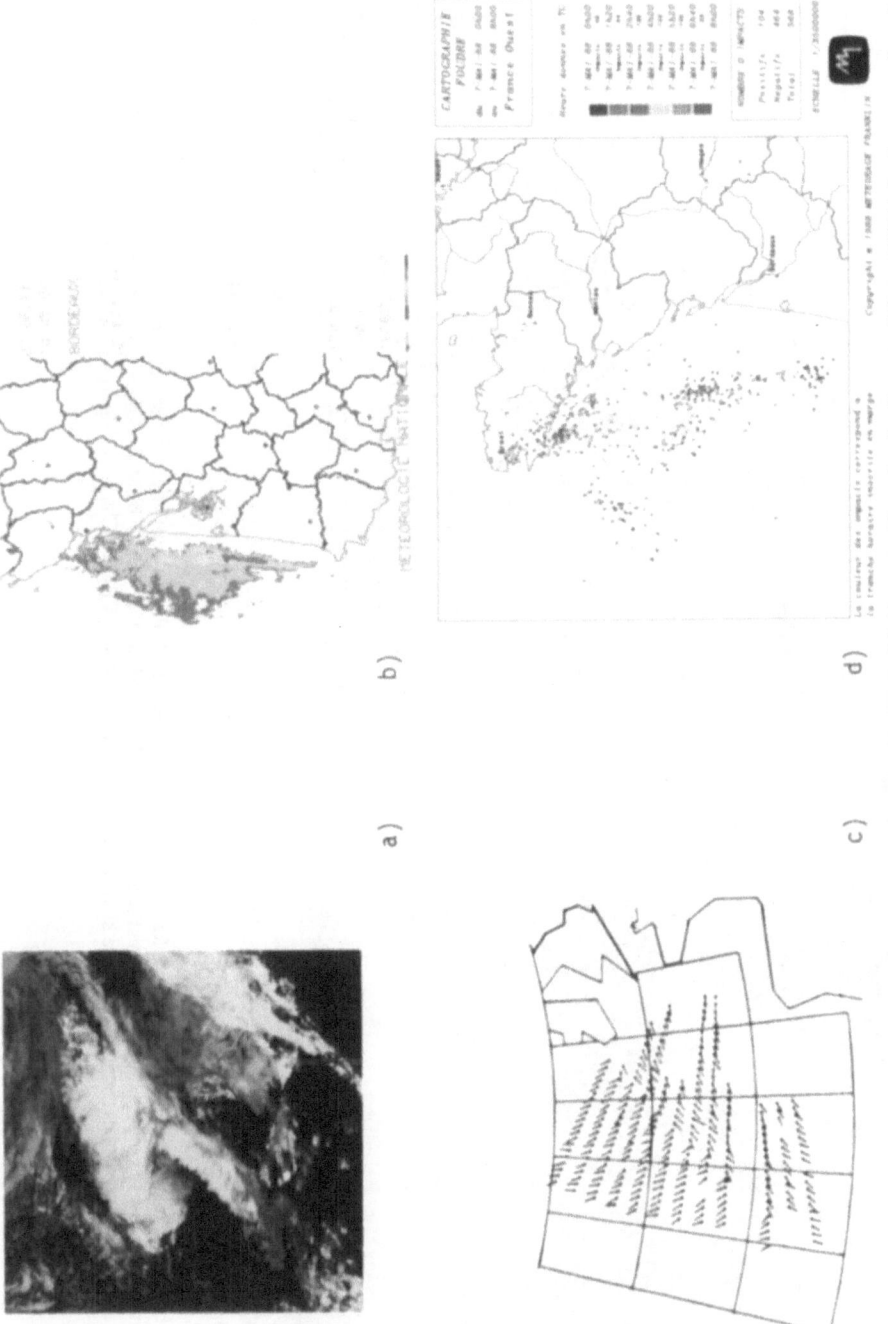

Plate 5. Marine observing systems for nowcasting. a): satellite (AVHRR C4) b): ground-based radar imagery (7 june 1987). c): skywave radar wind directions [ref 1] d): lightning detection system: times of detected impacts are colour-coded (courtesy of Météorage/France). (A Ratier, Chapter 13. See page 255 for reference.)

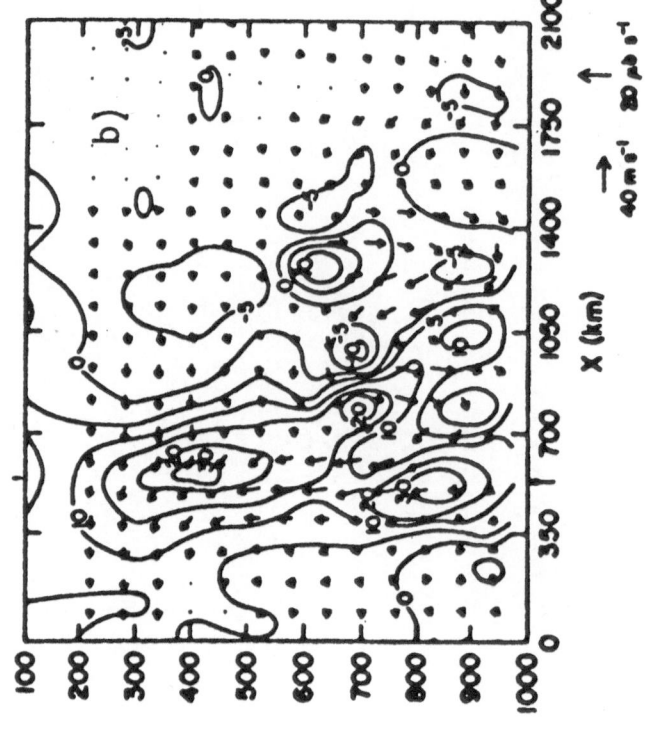

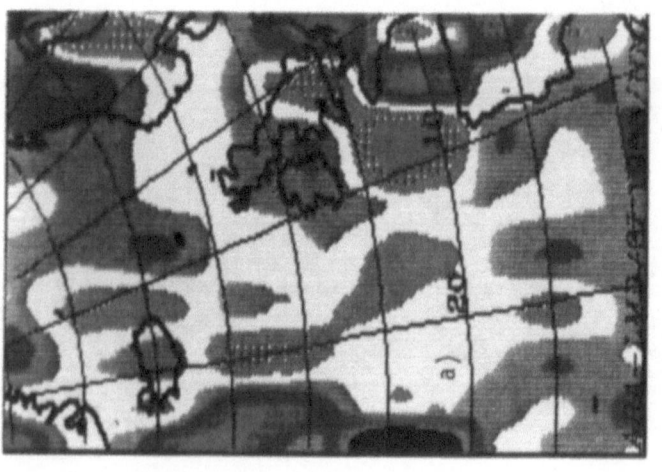

Plate 6. Strong banded vorticity patterns for two cases of severe weather. a) thermal vorticity map derived from NOAA 10 soundings Red and blue areas respectively correspond to positive and negative values. (from Chédin et al. 1988). b) Modelled cross frontal section of the QE2 storm case: relative vorticity isolines and circulation are plotted (from Anthes et al, 1984). (A Ratier, Chapter 13. See pages 257 and 264 for reference.)

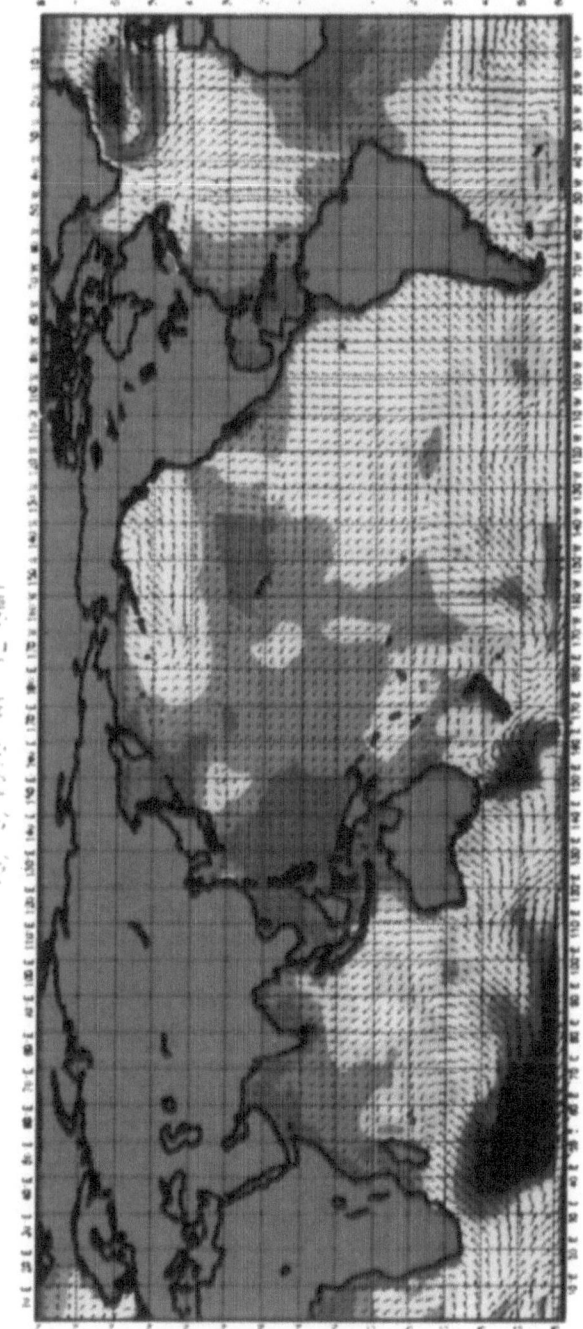

Plate 7. Swell significant wave height global field obtained by the WAM model using as input a wind field after assimilation of Seasat scatterometer data at ECMWF (Reading, UK). Courtesy of K Hasselmann (MPI, Hamburg, FRG). (E Oriol-Pibernat, Chapter 16. See page 326 for reference.)

LIST OF PARTICIPANTS

Professor D. O. Adefolalu, School of Science and Science Education, Federal University of Technology, Minna - Niger State, Nigeria.

Professor W. Alpers, Fachbereich 1 Physik, Universitat Bremen, Physik des Meeres, D-2800 Bremen 33, Germany.

Mr. Mudhaffar Abbas Al-Taee, Dept. of Applied Physics and Electronic & Manufacturing Engineering, University of Dundee, Dundee DD1 4HN, U.K.

Dr. J. M. Anderson, Dept. of Applied Physics and Electronic & Manufacturing Engineering, University of Dundee, Dundee DD1 4HN, U.K.

Mr. Anthony J. Andrews, McLean Building, Institute of Hydrology, Crowmarsh Gifford, Wallingford, Oxon OX10 8BB, U.K.

Professor J. Askne, Institutionen for Radio-och Rymdvetenskap, Chalmers Tekniska Hogskola, S-412 96 Goteborg, Sweden.

Dr. J. Bailey, Remote Sensing Unit, Dept. of Geography, University of Bristol, Bristol BS8 1SS, U.K.

Miss Katherine M. Baird, UJO7, Admiralty Research Establishment, Southwell, Portland, Dorset DT5 2JS, U.K.

Mr. Dariusz Balicki, University of Szczecin, Dept. of Remote Sensing and Marine Cartography, ul. Felcseke 3A, 71-412 Szczecin, Poland.

Dr. R. Bernard, CRPE, 38-40 Rue de General Leclerc, 92131 Issy-le-Moulineaux, France.

Mr. A. C. Bijlsma, Delft Hydraulics, P.O. Box 177, 2600 MH Delft, The Netherlands.

Mr. Per Erik Bjerke, Statoil, Forus, P.O. Box 300, N - 4001, Stavanger, Norway.

Dr. H. Bottger, European Centre for Medium Range Weather Forecasts, Sheffield Park, Reading RG2 9AX, U.K.

Commander William N. Bowman, Royal Naval School of Meteorology & Oceanography, Royal Naval Air Station, Culdrose, Helston, Cornwall, U.K.

Mr. Lars-Anders Breivik, Norwegian Meteorological Institute, P.B. 43, 0313 Oslo 3, Norway.

Dr. S. R. Brooks, GEC-Marconi Research Centre, West Hanningfield Road, Great Baddow, Chelmsford CM2 8HN, Essex, U.K.

Dr. Hans-Jurgen Brosin, Academy of Sciences of the GDR, Institute of Marine Research, Seestrasse 15, P.O. Box 38, Rostock-Warnemunde, DDR - 2530.

Professor R. A. Brown, Dept. of Atmospheric Sciences, AK-40, University of Washington, Seattle, Washington 98195, USA.

Mrs. J. Callison, Dept. of Applied Physics and Electronic & Manufacturing Engineering, University of Dundee, Dundee DD1 4HN, U.K.

Miss Marina Candouna, c/o E. Maier Reimer, Max Planck Institut fuer Meteorologie, Bundesstrasse 55, 2000 Hamburg 13, F. R. Germany.

Mr. R. G. Caves, 2Fr. 96 Spottiswoode Street, Edinburgh EH9 1BY, U.K.

Mr. Vipon Chopra, 1 Rusland Park Road, Harrow, Middlesex HA1 1UR, U.K.

Professor A. P. Cracknell, Dept. of Applied Physics and Electronic & Manufacturing Engineering, University of Dundee, Dundee DD1 4HN, U.K.

Dr. P. A. Davies, Dept. of Civil Engineering, University of Dundee, Dundee DD1 4HN, U.K.

Mr. Djavad Djavadi, Dept, of Applied Physics and Electronic & Manufacturing Engineering, University of Dundee, Dundee DD1 4HN, U.K.

Mr. Xavier Durrieu de Madron, Rue du Temple 2103, Noiraigue, Switzerland.

Mrs. Anke Eriksson, Institut fuer Geophysikalische Wissenschaften, Fachrichtung Meteorologie, Thielallee 50, 1000 Berlin 33, West Germany.

Professor R. Frassetto, Istituto per lo Studio della Dinamica delle Grande Masse, Palazo Papadopoli, 1364 San Polo, 30125 Venezia, Italy.

Dr. Joan Gosink, Geophysical Institute, University of Alaska, Fairbanks, Alaska 99775 - 0800, USA.

Dr. T. H. Guymer, IOS Deacon Laboratory, Wormley, Godalming, Surrey GU8 5UB, U.K.

Dr. A. Hollingsworth, European Centre for Medium Range Weather Forecasts, Sheffield Park, Reading RG2 9AX, U.K.

Dr. William Hsieh, Dept. of Oceanography, University of British Columbia, Vancouver, B.C., Canada V6T 1W5.

Mr. Weigen Huang, Dept. of Applied Physics and Electronic & Manufacturing Engineering, University of Dundee, Dundee DD1 4HN, U.K.

Mr. Ian T. Hunter, Maritime Weather Office, D F Malan Airport, Cape Town 7525, Republic of South Africa.

Mr. Leszek Hus, Dept. of Remote Sensing and Cartography of the Sea, University of Szczecin, Felczaka 3A, 71-412 Szczecin, Poland.

Mr. Sanatanam Kasturi, c/o Dr. I. S. Robinson, Dept. of Oceanography, University of Southampton, Southampton SO9 5NH, U.K.

Professor K. Katsaros, Dept. of Atmospheric Sciences, AK-40, University of Seattle, Washington 98195, USA.

Dr. Gunay Kocasoy, Bogaziçi University, Pollution Control Research Group, 80815 Bebek - Istanbul, Turkey.

Dr. G. J. Komen, KNMI, P.O. Box 201, Wilhelminalaan 10, 3730 AE de Bilt, The Netherlands.

Mr. Katsutoshi Kouzai, Dept. of Ocean Mechanical Engineering, Kobe University of Mercantile Marine, 5-1-1 Fukaeminami, Higashinada, Kobe 658, Japan.

Mr. Piotr Kowalczuk, ul. Kopernika 20, 09-402 Plock, Poland.

Mr. Romuald Kuzniar, 26 - 600 Radom, ul. Chrobrego 17A, Poland.

Mr. Roop Lalbeharry, Met Services Research Branch, Atmospheric Environment Service, 4905 Dufferin Street, Downsview, Ontario M3H 5T4, Canada.

Dr. John A. Leese, Institute for Naval Oceanography, Building 1103, Room 233, Stennis Space Center, Mississippi 39529 - 5005, USA.

Dr. Joao Antonio Lorenzzetti, Instituto de Pesquisas Espaciais, Av. dos Astronautas, 1758 Jrd. Granja, Sao Jose dos Campos, SP CEP 12201, Brazil.

Dr. Juan Jose Martinez-Benjamin, E.T.S. Ingenieros de Telecomunicacion, Jorge Girona Salgado s/n, Barcelona 08034, Spain.

Ms. Cecilie Mauritzen, Dept. of Earth, Atmospheric & Planetary Sciences, 54 - 1314, Massachusetts Institute of Technology, Cambridge, MA 02139, USA.

Dr. P. Minnett, Applied Oceanography Group, Saclant ASW Research Centre, Viale San Bartolomeo 400, I-19026 San Bartolomeo (SP), Italy.

Dr. E. Mollo-Christensen, Laboratory for Oceans, NASA/Goddard Space Flight Center, Greenbelt, Maryland 20771, USA.

Mr. D. Offiler, Satellite Meteorology Branch, Meteorological Office, London Road, Bracknell, Berks. RG12 2SZ, U.K.

Mr. Paulo B. Oliveira, Grupo de Oceanografia, Universidade de Lisboa, R. Escola Politecnica 58, 1200 Lisboa, Portugal.

Dr. E. Oriol-Pibernat, ESRIN Earthnet Programme Office, Via Galileo Galilei, 00044 Frascati, Italy.

Dr. L. Pedersen, Dept. of Electromagnetics, Technical University, Lyngby, Denmark.

Professor O. M. Phillips, Dept. of Earth and Planetary Sciences, The John Hopkins University, Baltimore, Maryland 21218, USA.

Mr. Robert L. Phillips, Dept. of Earth Sciences, University College Swansea, Singleton Park, Swansea SA2 8PP, U.K.

Mr. Jerzy Prajs, Dept. of Remote Sensing and Cartography of the Sea, University of Szczecin, Str. Felczaka 3A, 71-412 Szczecin, Poland.

Mr. A. Radford, European Centre for Medium Range Weather Forecasts, Sheffield Park, Reading RG2 9AX, U.K.

Mr. M. R. Ramesh Kumar, c/o Dr. I. S. Robinson, Dept. of Oceanography, University of Southampton, Southampton SO9 5NH, U.K.

Dr. C. G. Rapley, Mullard Space Science Laboratory, Holmburg St. Mary, Dorking RH5 6NT, Surrey, U.K.

Dr. A. Ratier, CNES, 18 Avenue Edouard Belin, 31055 Toulouse Cedex, France.

Dr. Clemens Simmer, Segeberger Landstrasse 169, D - 2300 Kiel 14, Federal Republic of Germany.

Ir. A. C. M. Stoffelen, K.N.M.I., Wilhelminalaan 10, 3732 GK de Bilt, The Netherlands.

Mr. Carlos Jose Direitinho Tavares, Instituto Nacional de Meteorologia & Geofisica, Rua C, Aeroporto de Lisboa, 1700 Lisboa, Portugal.

Mr. Jeremy P. Thomas, University College of North Wales, School of Ocean Sciences, Marine Science Laboratories, Menai Bridge, Gwynedd LL59 5ES, Wales, U.K.

Dr. A. O. Tooke, Dept. of Applied Physics and Electronic & Manufacturing Engineering, University of Dundee, Dundee DD1 4HN, U.K.

Dr. Brenda J. Topliss, Coastal Oceanography, Bedford Institute of Oceanography, P.O. Box1006, Dartmouth, Nova Scotia B2Y 4A2, Canada.

Mr. Piotr Trela, Dept. of Remote Sensing and Cartography of the Sea, University of Szczecin, Felczaka 3A, 71-412 Szczecin, Poland.

Mr. Tanay Sidki Uyar, Tubitak Scientific & Technical Research Council, Marmara Scientific & Industrial Research Institute, Mechanical & Energy Engineering Department, P.O. Box 21, Gebze 41401, Kocaeli, Turkey.

Dr. R. A. Vaughan, Dept. of Applied Physics and Electronic & Manufacturing Engineering, University of Dundee, Dundee DD1 4HN, U.K.

Dr. Paul A. Volz, 1805 Jackson Avenue, Ann Arbor, Michigan 48103, USA.

Mr. G. Whyte, Dept. of Applied Physics and Electronic & Manufacturing Engineering, University of Dundee, Dundee DD1 4HN, U.K.

Mr. Yu Yunyue, Physics Department, Qingdao University of Oceanography, 5 Yushanlu, Qingdao, People's Republic of China.